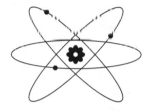 **HOW THE HIPPIES SAVED PHYSICS**

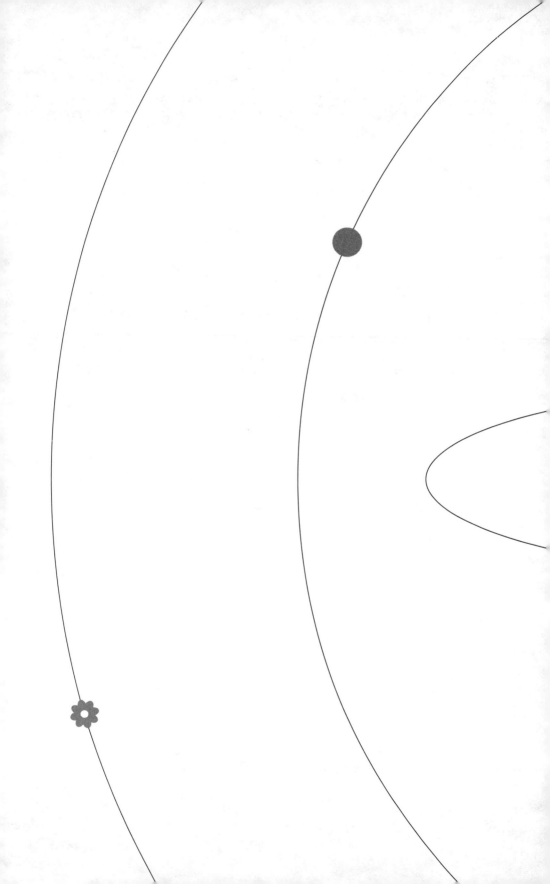

How the Hippies Saved Physics

SCIENCE, COUNTERCULTURE, AND THE QUANTUM REVIVAL

DAVID KAISER

W. W. Norton & Company

NEW YORK · LONDON

Grateful acknowledgment is made to the following for permission to reprint material: to Zane Kesey and the estate of Dr. Timothy Leary for permission to reprint the epigraph on p. vii, which originally appeared in Timothy Leary, "Preface," *Spit in the Ocean* 3 (Fall, 1977): 8–11; to Jack Sarfatti and Taylor & Francis, Ltd., for permission to reprint the epigraph to chapter 4, which originally appeared in Jack Sarfatti, "Implications of meta-physics for psychoenergetic systems," *Psychoenergetic Systems* 1 (1974): 3–8; and to the Melanie Jackson Agency, LLC, for permission to reprint quotations from the Richard P. Feynman papers.

For information about permission to reproduce selections from this book, write to Permissions, W. W. Norton & Company, Inc., 500 Fifth Avenue, New York, NY 10110

For information about special discounts for bulk purchases, please contact W. W. Norton Special Sales at specialsales@wwnorton.com or 800-233-4830

Manufacturing by RR Donnelley, Harrisonburg
Book design by Mark Melnik
Production manager: Devon Zahn

Library of Congress Cataloging-in-Publication Data
Kaiser, David
  How the hippies saved physics: science, counterculture, and the quantum revival / David Kaiser.
    p. cm.
  Includes bibliographical references and index.
  ISBN 978-0-393-07636-3 (hardcover)
1. Physicists—California—Berkeley—Biography.
2. Quantum theory. 3. Counterculture.
I. Fundamental Fysiks Group (Berkeley, Calif.) II. Title.
  QC15.K26 2011
  530.092′279467—dc22

W. W. Norton & Company, Inc.
500 Fifth Avenue, New York, N.Y. 10110
www.wwnorton.com

W. W. Norton & Company Ltd.
Castle House, 75/76 Wells Street, London W1T 3QT

1 2 3 4 5 6 7 8 9 0

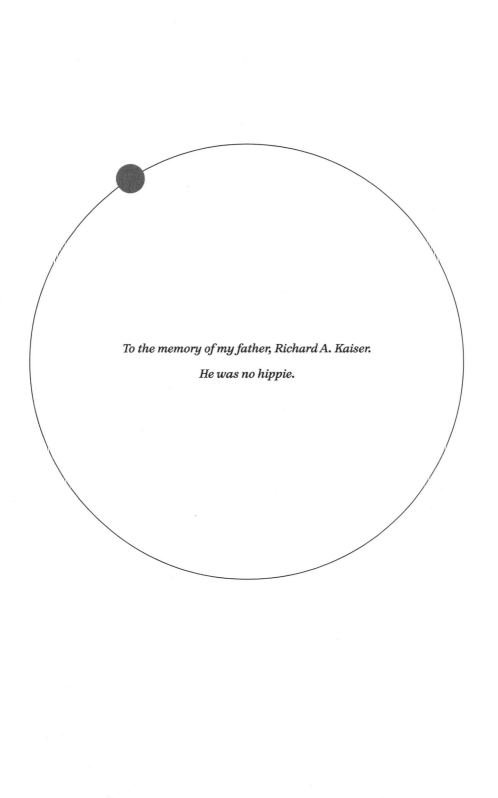

To the memory of my father, Richard A. Kaiser.

He was no hippie.

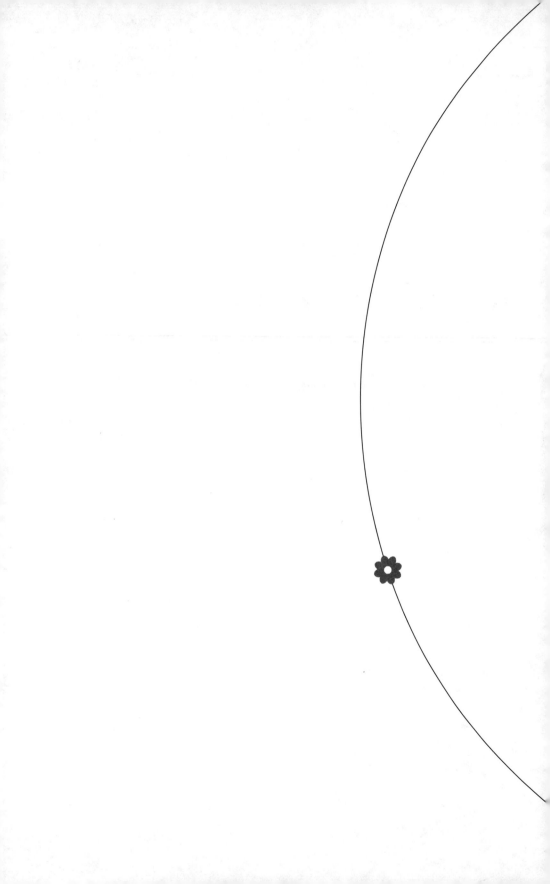

There must be thousands of young persons whose nervous systems were expanded and opened-up in the 1960's and who have now reached positions of competence in the sciences. . . . We expect the new wave of turned-on young mathematicians, physicists, and astronomers are more able to use their energized nervous systems as tools to provide new correlations between psychology and science.

—**Timothy Leary, 1977**

Table of √ Contents

Introduction

To most residents of Vienna, April 21, 2004, probably seemed like any other spring day in the Austrian capital. Students mulled over books in cafés; tourists delighted in the Hapsburg-era gardens, museums, and opera houses; and businesspeople scurried through their appointments. Amid the bustle, however, something magical happened. The city's mayor and the director of one of the city's largest banks collaborated on a breathtaking experiment. Working with physicists from the University of Vienna and a spin-off company, the mayor and banker performed the first electronic bank transfer using quantum cryptography. Specially prepared beams of light transmitted an unbreakable code—an encryption key—between the bank's branch office and city hall. If anyone else had tried to listen in on the signal, the eavesdropping would have been detected easily and unambiguously. More important, any attempt to breach security would have destroyed the sought-after signal, scrambling it into harmless, random noise. With these safeguards in place, the mayor's money wire went through without a hitch.[1]

Three years later, residents of Geneva, Switzerland, participated in a similar feat. Government officials, in cooperation with their own local physicists, employed quantum encryption to protect the transmission of electronic votes cast in the Swiss national election. As in Vienna, the communications remained perfectly secure. The laws of physics had made sure of it.[2]

Advances like these belong to the fascinating, flourishing field of quantum information science. An amalgam of topics with funny-sounding names—quantum computing, quantum encryption, quantum teleportation—the field sounds more like *Star Trek* with each passing year. These days quantum information science sports a multi-billion-dollar research program, ten thousand published research articles, and a variety of device prototypes. The field has leaped to the cutting edge of physics, catapulted by palpable enthusiasm among research scientists, industrial partners, and government agencies around the world.[3] Breathless coverage of the field can be found everywhere from the *New York Times* and the *Wall Street Journal* to *Wired Magazine* and *BusinessWeek*.[4]

The tremendous excitement marks the tail end of a long-simmering Cinderella story. Long before the huge budgets and dedicated teams, the field moldered on the scientific sidelines. To make the latest breakthroughs possible, researchers needed to grapple with quantum theory, physicists' famously successful (yet infamously strange) description of matter and energy at the atomic scale. The equations had been around since the 1920s. But figuring out how to *interpret* those equations, to parse the symbols in words and scrutinize just what they implied about the mysterious workings of the microworld—that interpretive task had long since fallen out of favor. During the middle decades of the twentieth century, most physicists recoiled from such philosophical labor. They treated the interpretation of quantum theory as a fringe topic, a fine leisure-time diversion for retired researchers in their dotage, perhaps, but not the sort of activity on which rising stars should spend their time. Thirty years ago, readers who were interested in the unsettled debates over the interpretation of quantum theory had to hunt in some out-of-the-way places. In 1979, some of the most extensive coverage appeared in an unpublished memorandum from the Central Intelligence Agency and a feature article in *Oui* magazine. The latter—no publication of the French embassy—was *Playboy*'s answer to *Penthouse*. Both items focused on work by physicists at the center of this story. The porn magazine's discussion was by far the better researched and more accurate of the two.[5]

Lost from view in today's hoopla is a story, equal parts inspiring and bizarre, of scientific striving in the face of long odds. The intellectual bedrock of quantum information science—the ideas that undergird today's quantum-encrypted bank transfers and electronic voting—took form in a setting that couldn't have been more different from the ivory tower of academe or the citadels of business and politics. In fact, the break-throughs in Vienna and Geneva ultimately owe their origins to the hazy, bong-filled excesses of the 1970s New Age movement. Many of the ideas that now occupy the core of quantum information science once found their home amid an anything-goes counterculture frenzy, a mishmash of spoon-bending psychics, Eastern mysticism, LSD trips, CIA spooks chasing mind-reading dreams, and comparable "Age of Aquarius" enthusiasms. For the better part of a decade, the concepts that would blossom into developments like quantum encryption were bandied about in late-night bull sessions and hawked by proponents of a burgeoning self-help movement—more snake oil than stock option.

•

The woolly pursuits of the 1970s hearkened back to an earlier way of doing physics and of being a physicist. The roots of quantum information science stretch all the way back to the golden age of theoretical physics of the 1920s and 1930s, when giants like Albert Einstein, Niels Bohr, Werner Heisenberg, and Erwin Schrödinger cobbled quantum mechanics together. From their earliest wranglings, they found themselves tangled up with all sorts of strange, counterintuitive notions. Many have become well-known catchphrases like "wave-particle duality," "Heisenberg's uncertainty principle," and "Schrödinger's cat." Each signaled that atom-sized objects could behave fantastically different from what our usual experience would suggest. To Einstein, Bohr, and the rest, it seemed axiomatic that progress could only be made by tackling these philosophical challenges head on. Manipulating equations for their own sake would never be enough.[6]

That style of doing physics did not last long. The clouds of fascism gathered quickly across Europe, scattering a once-tight community. The ensuing war engulfed physicists around the world. Torn from their prewar routines and thrust into projects of immediate, worldly significance—radar, the atomic bomb, and dozens of lesser-known gadgets—physicists' day-to-day activities in 1945 bore little resemblance to those of 1925. Over the next quarter century, Cold War imperatives shaped not just who received grants to pursue this or that problem; they left an indelible mark on the world of ideas, on what counted as "real" physics. Physicists in the United States adopted an aggressively pragmatic attitude. The equations of quantum mechanics had long since lost their novelty, even if their ultimate meaning still remained obscure. The pressing challenge became to put those equations to work. How much radiation would be emitted from a particular nuclear reaction? How would electric current flow through a transistor or a superconductor? As far as the postwar generation of physicists was concerned, their business was to calculate, not to daydream about philosophical chestnuts.[7]

Before the war, Einstein, Bohr, Heisenberg, and Schrödinger had held one model in mind for the aspiring physicist. A physicist should aim, above all, to be a *Kulturträger*—a bearer of culture—as comfortable reciting passages of Goethe's *Faust* from memory or admiring a Mozart sonata as jousting over the strange world of the quantum.[8] The physicists who came of age during and after World War II crafted a rather different identity for themselves. Watching their mentors stride through the corridors of power, advising generals, lecturing politicians, and consulting for major industries, few sought to mimic the otherworldly, detached demeanor of the prewar days. Philosophical engagement with quantum theory, which had once seemed inseparable from working on quantum theory itself, rapidly fell out of fashion. Those few physicists who continued to wrestle with the seemingly outlandish features of quantum mechanics found their activity shoved ever more sharply to the margins.[9]

Before there could be a field like quantum information science—and long before demonstrations like those in Vienna and Geneva could even be imagined—a critical mass of researchers needed to embrace a different mode of doing physics once more. They had to incorporate philosophy, interpretation, even bald speculation back into their daily routine. Quantum physicists needed to daydream again.

●

Rarely can we date with any precision the ebbs and flows of scientists' research styles or intellectual approaches. Yet these transitions—the how's and why's behind major shifts in a scientific field's reigning questions and methods—have long held a special fascination for me. We see laid bare in these moments a messy alchemy, intermixing the world of institutions with the world of ideas. Brilliant insights and dazzling discoveries take their place alongside political decisions, funding battles, personal rivalries, and cultural cues. These many ingredients combine to make one agenda seem worth pursuing in a particular time and place— and worth teaching to students—while quietly eclipsing other questions or approaches that had beckoned with equal urgency only a few years earlier.

In the case of the interpretation of quantum mechanics, which ultimately spawned quantum information science, we may detect just such a seismic shift in the 1970s. The physics profession in the United States suffered the lashings of a perfect storm between 1968 and 1972. Internal audits at the Department of Defense led to massive cutbacks on spending for basic research, which had financed, directly or indirectly, nearly all graduate training in physics for decades. Desperate for more soldiers to feed the escalation of fighting in the Vietnam War, meanwhile, military planners began to revoke draft deferments for students—first for undergraduates in 1967, then, two years later, for graduate students as well— reversing twenty years of draft policies that had kept physics students in their classrooms. Across the country, the Cold War coalition between the Pentagon and the universities crumbled under wave after wave of teach-ins and sit-ins, ultimately lost in a tear gas fog. Amid the turmoil, the nation's economy slid into "stagflation": rising inflation coupled with stagnant economic growth. All at once, physicists faced massive budget cuts, a plummeting job market, and vanishing student enrollments.[10]

As the Cold War nexus of institutions and ideas collapsed, other modes of being a physicist crept back in. The transition was neither smooth nor painless. Caught in the upheavals, a ragtag crew of young physicists banded together. Elizabeth Rauscher and George Weissmann, both graduate students in Berkeley, California, founded an informal discussion group in a fit of pique and frustration in May 1975. From their earliest years they had been captivated by books about the great revolutions of modern physics: relativity and quantum theory. They had entered the field with heads full of Einstein-styled paradoxes; they, too, dreamed of tackling the deepest questions of space, time, and matter. Yet their formal training had offered none of that. By the time they entered graduate school, the watershed of World War II and the hyperpragmatism of the Cold War had long since shorn off any philosophical veneer from physics students' curricula. In place of grand thoughts, their classes taught them narrow skills: how to calculate this or that physical effect, rather than what those fancy equations might portend about the nature of reality.

The two students had ties to the Theoretical Physics Division of the Lawrence Berkeley Laboratory, a sprawling national laboratory nestled in the Berkeley hills. They decided to do for themselves what their teachers and textbooks had not. Reserving a big seminar room at the lab, they established an open-door policy: anyone interested in the interpretation of quantum theory was welcome to attend their weekly meetings, joining the others around the large circular table for free-ranging discussions. They continued to meet, week in and week out, over the next three and a half years. They called themselves the "Fundamental Fysiks Group."

Their informal brainstorming sessions quickly filled up with like-minded seekers. Most members of the Fundamental Fysiks Group found themselves on the periphery of the discipline for reasons beyond their immediate control. Although they held PhDs from elite universities like Columbia, the University of California at Los Angeles, and Stanford, their prospects had dried up or their situations had become untenable with the bust of the early 1970s. Adrift in a sea of professional uncertainty, the young physicists made their way to Berkeley. Finding themselves with time on their hands and questions they still wanted to pursue, they gravitated toward Rauscher and Weissmann's group. They met on Friday afternoons at 4 P.M.—an informal cap to the week—and the spirited chatter often spilled late into the night at a favorite pizza parlor or Indian restaurant near campus.

The group's intense, unstructured brainstorming sessions planted seeds that would eventually flower into today's field of quantum information science; they helped make possible a world in which bankers and politicians shield their most critical missives with quantum encryption. Along the way, members of the Fundamental Fysiks Group, together with parallel efforts from a few other isolated physicists, contributed to a sea change in how we think about information, communication, computation, and the subtle workings of the microworld.

Despite the significance of quantum information science today, the Fundamental Fysiks Group's contributions lie buried still, overlooked or forgotten in physicists' collective consciousness. The group's elision from the annals of history is not entirely surprising. On the face of it, they seemed least likely to play any special role at all. Indeed, from today's vantage point it may seem shocking that anything of lasting value could have come from the hothouse of psychedelic drugs, transcendental meditation, consciousness expansion, psychic mind-reading, and spiritualist séances in which several members dabbled with such evident glee. History can be funny that way.

•

While the physics profession foundered, members of the Fundamental Fysiks Group emerged as the full-color public face of the "new physics" avant-garde. Hovering on the margins of mainstream physics, they managed to parlay their interest into a widespread cultural phenomenon. They cultivated a new set of generous patrons, ranging from the Central Intelligence Agency to self-made entrepreneurs like Werner Erhard, guru of the fast-expanding "human potential movement." With money pouring in from these untraditional sources, the Fundamental Fysiks Group carved out new institutional niches in which to pursue their big-picture discussions. Most important became the Esalen Institute in Big Sur, California, fabled incubator of all things New Age. For years on end, members of the group organized workshops and conferences, freely mixing the latest countercultural delights—everything from psychedelics like LSD to Eastern mysticism and psychic mind-reading—with a heavy dose of quantum physics.

To many journalists at the time, the Fundamental Fysiks Group seemed too good to be true. What better reflection of the times than to see physicists grappling with the problems of consciousness, mysticism, and the paranormal? The earliest coverage showed up in underground arenas dedicated to celebrating, not just reporting, the latest countercultural twists and turns. On the heels of his critically acclaimed films *The Godfather* and *American Graffiti*, for example, filmmaker Francis Ford Coppola bought the fledgling *City of San Francisco* magazine. One of its earliest issues after Coppola's renovation devoted a two-page spread to several core members of the Fundamental Fysiks Group, focusing on how the "new physicists" were busy "going into trances, working at telepathy, [and] dipping into their subconscious in experiments toward psychic mobility," all the better to understand subtle quantum effects.[11] A few months later some members of the group heard from Timothy Leary, the former Harvard psychology professor turned poster boy for New Age antics and all things psychedelic. At the time Leary was still in a California jail on drug charges, though he had hardly stopped working. Together with novelist and counterculture icon Ken Kesey (of *One Flew Over the Cuckoo's Nest* and "Merry Pranksters" fame, and the inventor of the "Electric Kool-Aid Acid Tests"), Leary was busy editing a special issue of the quirky Bay Area magazine *Spit in the Ocean*, and he was eager to publish some of the far-out essays that the hippie physicists had submitted.[12] Soon after that, one of the core members of the Fundamental Fysiks Group, Jack Sarfatti, showed up on the cover of *North Beach Magazine*, another San Francisco niche publication, in full guru mode: framed by a poster of Einstein and holding a copy of physicist George Gamow's autobiography, *My World Line*. When novelist and Beat generation hipster Herb Gold composed his memoirs of life among the likes of Allen Ginsberg and William S. Burroughs, the first off-scale personality to appear in the narrative was Sarfatti, holding forth on quantum physics in the Caffe Trieste, North Beach, San Francisco.[13] (Fig. I.1.)

The media coverage was by no means limited to these "tuned-in" venues. *Time* magazine ran a cover story about "The Psychics" with ample space devoted to Fundamental Fysiks Group participants. *Newsweek* covered the group a few years later. *California Living Magazine* ran a long story about the "*New* new physics," complete with head shots of

FIGURE I.1. The "new physicists" as counterculture darlings. *Left* (standing, left to right): Jack Sarfatti, Saul-Paul Sirag, Nick Herbert; (kneeling) Fred Alan Wolf, ca. 1975. *Right*: Jack Sarfatti as the eccentric genius of North Beach, 1979. (*Left*, courtesy Fred Alan Wolf; *right*, photograph by Robert L. Jones, courtesy Robert L. Jones and Jack Sarfatti.)

several group members. In May 1977, the group's Jack Sarfatti shared the podium with eccentric architect Buckminster Fuller and "five-stages-of-grief" psychiatrist Elisabeth Kübler-Ross as a keynote speaker at a "humanistic psychology" conference. Not long after that, the *San Francisco Chronicle* devoted a half-page article to Sarfatti, depicted as the latest in a long line of "eccentric geniuses" to set up shop in the city's bohemian North Beach area. Even newspapers as far away as the *New Hampshire Sunday News* covered the group's intellectual peregrinations. Virtually overnight, members of the informal discussion group had become counterculture darlings.[14]

•

One might be tempted to dismiss the Fundamental Fysiks Group and its antics as just one more fringe phenomenon: a colorful reminder of tie-dyed life in the 1970s, perhaps, but of little lasting significance. After all, as a sociologist observed as early as 1976, members of the group consistently posed questions and acknowledged experiences that would have "served to label the participants as mentally deranged" only a few years earlier.[15] Surely some *cordon sanitaire* separated the group from "real" physics.

When other sociologists turned attention to the Fundamental Fysiks Group—and related outcroppings of activity, such as studies of "plant empathy" or the international spoon-bending fad inspired by the apparently psychic feats of Israeli performer Uri Geller—they, too, framed the matter in terms of "demarcation."[16] The eminent philosopher Sir Karl Popper introduced the demarcation problem in the middle decades of the twentieth century: how do scientists draw boundaries between legitimate science and something else? The issue had little to do with truth or falsity. Popper readily acknowledged that many of today's scientific convictions will wind up as tomorrow's forgotten missteps. Popper was after something else, some set of criteria with which to distinguish proper scientific investigation from unscientific efforts. He had some searing examples in mind. As a young man he had experienced the convulsions that wracked daily life in his native Austria in the wake of World War I. The troubled times had inspired all manner of dogmatisms. He sought some means of separating Marxism, psychoanalysis, and astrology from the canons of scientific inquiry. What made the pursuit of those topics distinct from, say, Einstein's relativity?[17]

Since Popper's day, philosophers have spilled much ink in pursuit of those elusive demarcation criteria. Yet sociologists have countered with case after case, showing that scientists make judgments and draw boundaries in ways that rarely stack up with the philosophers' rarefied notions. Who is to say where the line should be drawn in any given instance? Popper's progeny never could establish any Maginot Line of legitimacy, some set of factors that might reliably separate real science from the imposter projects that had so exercised the great philosopher.[18]

The demarcation problem becomes acute in the case of the Fundamental Fysiks Group. Try as we might, we cannot cleave off the group or its activities from the "real" physics of the day. Many of the members' activities placed them on one end of a spectrum, to be sure. But no hard-and-fast dividing line separated them from legitimate—even illustrious—science. Members of the Fundamental Fysiks Group were entangled with mainstream physics on multiple levels, including people, patronage, and intellectual payoff. The group's marginal position and its multiple interactions with mainstream physics provide a unique view onto what it meant to do physics during the turbulent 1970s.

The hippie physicists of the Fundamental Fysiks Group help us map still larger transitions in American culture, beyond the shifting fortunes of physics. A few journalists in San Francisco and New York City coined the term "hippie" in the mid-1960s, searching for some way to describe the rising youth culture that was mutating beyond the "hipsters" of the 1950s Beat generation. With the media attention came the first waves of pushback. As California's then-governor Ronald Reagan put it in 1967, after the hippie scene in San Francisco's Haight-Ashbury district had become a national obsession, a hippie was someone "who dresses like Tarzan, has hair like Jane, and smells like Cheetah."[19] Reagan's quip lumped together groups whom scholars have recently labored to distinguish, often with Jesuitical precision. The left-leaning hippie movement, for example, had an uneasy relationship with the "New Left," the campus-based liberal and increasingly radical political movement associated with the Students for a Democratic Society and (ultimately) the Weather Underground. Members of the New Left aimed at organized political intervention, inspired by the civil rights movement and stoked by the escalation of the Vietnam War. The campus radicals often looked with dismay on their hippie counterculture cousins, for whom political organizations of any stripe seemed so very unhip. While the political types signed petitions and planned rallies, most hippies sought to "drop out."[20]

The hippie counterculture sported a playful worship of youth, spontaneity, and "authenticity," a personal striving often facilitated by heavy use of psychedelic drugs. LSD, synthesized in a Swiss lab in the late 1930s, was first outlawed in the United States in 1966; possession of the drug was bumped up to a felony offense in 1968. Until that time, the psychedelic had fascinated straight-laced chemists and psychologists as well as long-haired hippies. The Central Intelligence Agency and the U.S. Army sponsored research on effects of LSD at government laboratories and reputable research universities throughout the 1940s and 1950s. Along with psychedelics enthusiast Ken Kesey, for example, the physicist Nick Herbert, who would become a founding member of the Fundamental Fysiks Group, was introduced to LSD by psychologists at Stanford University.[21] Only later, over the course of the 1960s, did the drug seep into wider circulation among hordes of "tuned-in" youth. Long after the drug had been criminalized, LSD and other psychedelics, like psilocybin (from "magic mushrooms"), remained staple elements of the hippie counterculture.[22]

New Age enthusiasms had also been mixed up in the hippies' heady brew right from the start: everything from Eastern mysticism to extrasensory perception (ESP), unidentified flying objects (UFOs), Tarot card reading, and more. Research on LSD during the 1950s was often reported in parapsychology journals in between articles on mind-reading and reincarnation.[23] Americans' awareness of Eastern religions and healing practices, such as acupuncture, grew sharply following 1965 revisions to U.S. immigration law, after which immigration from Asia soared (having previously been capped by tight quotas).[24] Some of the earliest underground tabloids of the budding counterculture—newspapers like the *Oracle*, peddled in San Francisco's Haight-Ashbury neighborhood beginning in 1966—featured news about yoga, astrology, and the occult alongside information on where to score the most potent psychedelic drugs.[25] According to close observers, the hippie counterculture and New Age movements in the United States had fused by the early 1970s, achieving a critical mass, self-awareness, and no shortage of critics.[26] Even so, the boundaries of the counterculture remained porous. One analyst likened it to a medieval crusade, a "procession constantly in flux, acquiring and losing members all along the route of march."[27]

The inherent tensions that historians have begun to identify within the hippie counterculture—leftist but not "New Left," curious about the workings of the world but tempted by psychedelic escapism—help explain the wide range of followers whom the Fundamental Fysiks Group inspired. Their efforts attracted equally fervent support from stalwarts of the military-industrial complex as from storied cultivators of flower power, from the Central Intelligence Agency, the Pentagon, and defense-contractor laboratories like the Stanford Research Institute to the Esalen Institute. Members of the Fundamental Fysiks Group exemplified these tensions themselves. Many threw themselves headlong into the New Age alchemy, even as they pursued serious questions at the heart of quantum theory. They shifted easily from weapons laboratories to communes, universities to ashrams.[28]

All the while, members of the Fundamental Fysiks Group pioneered a flood of publications about the new physics and its broader implications. Many sold handsomely; some netted national awards. Best known today are such cultural icons as *The Tao of Physics* (1975) by physicist and group member Fritjof Capra and *The Dancing Wu Li Masters* (1979) by the writer Gary Zukav, at the time an avid participant in the Fundamental Fysiks Group's discussions and roommate of one of its founding members. The group also experimented with alternate ways to spread their message, inspired by and modeled on the counterculture's underground press.[29] The group's efforts helped to bring sustained attention to the interpretation of quantum mechanics back into the classroom. And in a few critical instances, their work instigated major breakthroughs that—with hindsight—we may now recognize as laying crucial groundwork for quantum information science.

•

The group of hippies who formed the Fundamental Fysiks Group saved physics in three ways. First concerned style or method. They self-consciously opened up space again for freewheeling speculation, for the kind of spirited philosophical engagement with fundamental physics that the Cold War decades had dampened. More than most of their generation, they sought to recapture the big-picture search for meaning that had driven their heroes—Einstein, Bohr, Heisenberg, and Schrödinger—and to smuggle that mode of doing physics back into their daily routine.

Second, members of the Fundamental Fysiks Group latched onto a topic, known as "Bell's theorem," and rescued it from a decade of unrelenting obscurity. The theorem, named for the fiery Irish physicist John S. Bell, stipulated that quantum objects that had once interacted would retain some strange link or connection, even after they had moved arbitrarily far apart from each other. Bell used words like "nonlocality" and "entanglement" to describe his result. To many group members, the phenomenon seemed equally evocative of Buddhist teachings. As one group member put it in 1976, "Bell's theorem gives precise physical content to the mystic motto, 'we are all one.'"[30] Working in various genres and media, the Fundamental Fysiks Group grappled with Bell's theorem and quantum entanglement. They struggled to make sense of it, test out its limits, and see what it might imply. In the process, they forced a few of their physicist peers to pay attention to the topic, jousting with them over its ultimate implications. From these battles, quantum information science was born.

Bell's theorem and quantum entanglement seemed to suggest that one could use quantum theory to act at a distance, instantly. Nudge a particle here and its partner would instantaneously dance over there, regardless of whether it was nanometers or light-years away. But Einstein's relativity forbade any force or influence to travel faster than the speed of light. The Fundamental Fysiks Group pushed relentlessly on that boundary, the seemingly weak joint in the architecture of all we know about how the universe hangs together. They had many motivations. One was dogged pursuit of the big metaphysical questions, the constant refrain of "how could the world *work* that way?" But there was more. If faster-than-light signaling were possible (perhaps even inevitable), then physicists would need to broaden the discipline to include even larger questions. Was action at a distance really so different from clairvoyance, psychokinesis, or the Eastern mystics' emphasis on holism? Those were the stakes, at least as the Fundamental Fysiks Group saw them. Sitting in the Bay Area, as the counterculture and New Age movements burst into technicolor bloom, the deep mysteries of quantum physics reflected all-new hues.[31]

The hippie physicists' concerted push on Bell's theorem and quantum entanglement instigated major breakthroughs—the third way they saved physics. The most important became known as the "no-cloning theorem," a new insight into quantum theory that emerged from spirited efforts to wrestle with hypothetical machines dreamed up by members of the Fundamental Fysiks Group. Akin to Heisenberg's famous uncertainty principle, the no-cloning theorem stipulates that it is impossible to pro-duce perfect copies (or "clones") of an unknown or arbitrary quantum state. Efforts to copy the fragile quantum state necessarily alter it. The fact that unknown quantum states—like the beams of light fired down long fiber-optics cables in the Vienna and Geneva demonstrations—cannot be copied is what stops eavesdroppers in their tracks. Unlike ordinary signals, to which one might surreptitiously listen, quantum-encrypted communications simply cannot be tapped without destroying the desired signal. The no-cloning theorem thus gives force to quantum encryption: it provides the mechanism by which bank transfers and elec-tion results can be transmitted with perfect security. That much is well known among today's physicists and aficionados of quantum informa-tion science; the latest textbooks often feature the result in their open-ing pages.[32] Less well known is that the no-cloning theorem emerged directly from the Fundamental Fysiks Group's tireless efforts—at once earnest and zany—to explore whether Bell's theorem and quantum entanglement might unlock the secrets of mental telepathy and extra-sensory perception, or even enable contact with spirits of the dead.

Hence the brashness of my title, *How the Hippies Saved Physics*. Readers may well note a tinge of the same bravado, equal parts ironic and defi-ant, that animated Thomas Cahill's well-known study, *How the Irish Saved Civilization*.[33] The similarity is by design. Both books examine moments of great instability and decay in reigning institutions: the Roman empire on one hand, Cold War institutions of physics research on the other. In both accounts, an unlikely group of underdogs and castaways kept the torch of learning aflame, nursing a body of scholarship and a set of questions until the mainstream had recovered sufficiently to appreciate their importance and build on them again. That which required saving was "Western civilization" in Cahill's case; it was a commitment to deep questioning of quantum reality in mine. Cahill casts the Irish monks of the Middle Ages in twin roles: both as cultivators of Europe's lost heritage and as effective missionaries, replanting the seeds of learning throughout the Continent. This book focuses on down-and-out hippie physicists, whose passion for physics and for the big questions at the heart of quantum theory was implacable. They demonstrated impressive tenacity in the face of professional hardships; their zeal to share their findings and spread the word was unflappable.

Several critics of Cahill's account have rightly pointed out that the role of the medieval Irish is easily exaggerated. Other groups at the time proved equally adept at squirreling away the intellectual riches of Greece and Rome, tending to them, building upon them, and helping to replenish the stocks of learning throughout the European continent at a propitious moment.[34] So too with the physicists at the center of this story. By no means were the individuals upon whom I focus unique. Other outcroppings of like-minded investigators existed, and at times the various groups found each other and interacted.

Yet the Fundamental Fysiks Group—an ensemble cast from the start—played what can only be considered an outsized role. The ratio of their ambitious participation to the humbleness of their professional situation was especially striking. They weren't just chasing new gadgets, though they certainly had these in mind and even marched a few steps down the patent-filing road. Their goal remained far more grand: changing an entire worldview. I find this mismatch between their soaring intellectual aspirations and their modest professional platform especially captivating. That they have left any mark at all—attenuated to be sure, and largely unrecognized amid today's breathtaking successes—should give current researchers, toiling in relative obscurity, some modicum of comfort. Members of the Fundamental Fysiks Group threw themselves into their investigations with gusto, keeping spirits high and enjoying every last minute of their quest. Surely there is a lesson in that.

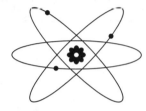 **HOW THE HIPPIES SAVED PHYSICS**

Chapter 1

"Shut Up and Calculate"

It was very different, when the masters of the science sought immortality and power; such views, although futile, were grand: but now the scene had changed. The ambition of the inquirer seemed to limit itself to the annihilation of those visions on which my interest in science was chiefly founded. I was required to exchange chimeras of boundless grandeur for realities of little worth.

—**Victor Frankenstein, character**
in Mary Shelley's *Frankenstein*

In the spring of 1974, a most unusual meeting took place. Two physicists—Fred Alan Wolf and Jack Sarfatti, who would soon become charter members of the Fundamental Fysiks Group—sat down with Werner Erhard in the lobby of the Ritz Hotel in Paris. Erhard, one of the leading exponents of the "human potential movement," was at the top of his game. His *est* workshops ("Erhard Seminars Training"), forerunner of today's self-help and personal-growth industry, had already grossed several million dollars and boosted Erhard to worldwide celebrity.[1] He had asked Wolf and Sarfatti to meet with him because he was fascinated by the way physicists attacked complicated and counterintuitive problems with rigor.[2]

The meeting did not get off to an auspicious start. Sarfatti felt restless, uninterested in the meeting; he had never heard of Erhard. Erhard's gaudy outfit, accessorized by a beautiful female admirer hanging on his sleeve, put Sarfatti off even more. Sarfatti asked what Erhard did. Erhard grinned and replied, "I make people happy." It was more than Sarfatti could take. Itching to leave, he said in a strong Brooklyn accent, "I think you're an asshole." As Sarfatti remembers it, Erhard rose from his chair—smile stretching from

1

ear to ear—embraced Sarfatti right there in the hotel lobby, and said, "I am going to give you money." Without knowing it, Sarfatti had used one of the catchphrases associated with Erhard's sprawling self-help venture. Soon the money began to flow: thousands of dollars, all from this most eager new patron of quantum physics.[3]

Erhard was not the first to seek enlightenment from the strange subject of quantum theory. Even more than relativity—with its talk of shrinking meter sticks, slowing clocks, and twins who age at different rates—quantum mechanics is a science of the bizarre. Particles tunnel through walls. Cats become trapped, half dead and half alive. Objects separated light-years apart retain telepathic links with one another. The seeming solidity of the world evaporates into a cloud of likelihoods. Long before Erhard, Wolf, or Sarfatti had arrived on the scene, the world's leading physicists had struggled to come to grips with quantum theory, to tease out just what it might mean. Many of their ideas sounded no less peculiar than the half-formed inklings that inspired Erhard on that fateful spring day.[4]

Quantum mechanics emerged over the first quarter of the twentieth century, honed primarily by Europeans working in the leading centers of theoretical physics: Göttingen, Munich, Copenhagen, Cambridge. Most of its creators—towering figures like Niels Bohr, Werner Heisenberg, and Erwin Schrödinger—famously argued that quantum mechanics was first and foremost a new way of thinking. Ideas that had guided scientists for centuries were to be cast aside. Bohr constantly spoke of the "general epistemological lesson" of the new quantum era. The disjuncture of cause from effect, Heisenberg's uncertainty principle, wave-particle duality—all required explicit, extended philosophical engagement, so these leaders proclaimed. They differed, often passionately, over which philosophical schools of thought might best clarify the new material. Some invoked the writings of eighteenth-century scholar Immanuel Kant; others quoted aphorisms from Hindu holy scriptures, or "Upanishads"; some even dabbled in Jungian depth-psychology. The subject's leading detractors, such as Albert Einstein, likewise agreed that quantum mechanics had to meet stringent philosophical tests. Mathematical self-consistency and agreement with experiments were important, but hardly sufficient.[5]

During this heady period, grown men argued into the night, trying to make sense of a series of puzzles and paradoxes. Names were called; tears were shed. At one point, an ailing Schrödinger sought refuge in bed while visiting Bohr's Institute for Theoretical Physics in Copenhagen. Unable to let a disputed matter of interpretation rest, Bohr hounded the poor Austrian at his bedside, repeating, "But surely Schrödinger, you must see . . ."[6]

That style of working on quantum mechanics faded fast after World War II. Especially in the United States, the war and its aftermath shaped how generations of new physicists were trained. Ultimately, the war changed what it meant to be a physicist. The Cold War completed the transformation, winnowing the range of acceptable topics and admissible approaches. Very quickly, philosophical inquiry or open-ended speculation of the kind that Bohr, Einstein, Heisenberg, and Schrödinger had considered a prerequisite for serious work on quantum theory got shunted aside. "Shut up and calculate" became the new rallying cry.[7]

Yet the Cold War consensus proved to be no more eternal than the prewar style had been. As the fortunes of physics plummeted in the late 1960s and early 1970s, sending academic physics departments into a tailspin, new intellectual possibilities opened up. Buoyed by cash from new patrons like Erhard, small clusters of physicists, including Wolf, Sarfatti, and their colleagues in the Fundamental Fysiks Group, labored to carve out a new identity for themselves and for the science they loved so much.

•

Back in the 1920s, sticking points seemed to abound in the new quantum theory. Every time physicists tried to make sense of their hard-won equations, new and bizarre challenges tumbled forth. One experiment captured the lion's share of peculiarities. It came to be known as the "double-slit experiment." Champions of quantum mechanics trotted it out time and again to sharpen their understanding of the issues involved.

Bohr and Heisenberg, for example, featured it in some of their earliest expositions of quantum mechanics.[8] Critics likewise saw much of value in the experiment, goading their colleagues to admit how preposterous their explanations sounded. Schrödinger—caught between the warring camps, with his own uneasy relationship to the equations he had produced—recognized the pedagogical value of the double-slit experiment for clarifying many of the core mysteries of quantum mechanics, and featured it prominently in lectures during the 1930s.[9] Since that time, generations of physicists have followed Schrödinger's lead. In fact, readers of the trade magazine *Physics World* recently voted the double-slit experiment the single most beautiful experiment of all time. In their view, it edged out heavyweight contenders from Galileo to Newton, and even a classic dating from ancient Alexandria, all of which also made the top ten.[10]

In an essay for Einstein's seventieth birthday, published in the late 1940s, Bohr used the double-slit experiment as the leitmotiv of his decades-long debate with Einstein.[11] Years earlier, Einstein had helped to launch the quantum revolution, introducing several crucial concepts. In fact, the Nobel Prize committee cited only his contributions to quantum theory when granting his award in 1921, remaining mum on relativity. Then, in one of the delicious ironies of the history of science, Einstein reversed course and turned his back on his own creation. (The irony was not lost on Einstein. "After all," he wrote to Schrödinger, "many a young whore turns into an old praying sister, and many a young revolutionary becomes an old reactionary.") He brandished the double-slit experiment in private correspondence to drive home his criticisms as early as April 1926, and in more public settings the following year.[12]

Fearing that their friendly squabbles over quantum theory had become too ethereal or detached from the real world over the years, Bohr worked with an artist to make his position more concrete when preparing his essay for Einstein's birthday. The resulting images had the look and feel of engineering diagrams, all bulky bolts and heavy planks. In Bohr's reconstruction, the double-slit experiment centered around an apparatus like the one in Figure 1.1, a thick wall with two slits hollowed out. A sliding latch was installed in front of one of the slits, so that physicists could choose whether to leave that slit open or shut. Behind the wall stood a recording screen—it could be photographic film or some other means of detection—bolted securely in place.

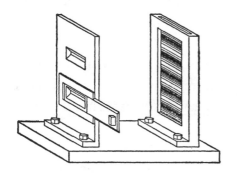

FIGURE 1.1. Niels Bohr's depiction of the double-slit apparatus. (Cropped from Bohr [1949], 219. Reproduced with permission of Open Court Publishing Company, a division of Carus Publishing Company.)

Einstein and Bohr each knew well what would happen if they shined a light on the wall when both slits were open. Bohr included a picture in his birthday essay. (Fig. 1.2.) If the light source were far enough away, the light waves would approach the wall-with-slits in a simple configuration that physicists call a "plane wave," with all the crests and troughs lined up neatly in rows. Most of the light from the source would be blocked by the wall. The light that passed through the narrow slits would fan out in a new pattern, arcing in semicircular waves toward the recording screen. The crests and troughs of the two curving light waves, emanating from the open slits, would no longer be lined up with each other. In some locations along the recording screen, the crest from one wave

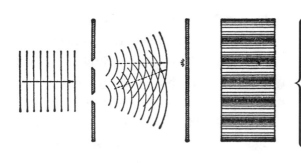

FIGURE 1.2. The double-slit apparatus and inter-ference pattern. (Cropped from Bohr [1949], 216. Reproduced with permis-sion of Open Court Publish-ing Company, a division of Carus Publishing Company.)

would arrive in step with the crest from the other, adding up to make a bright spot on the photographic film. In other locations, however, the crest from one wave would arrive with the trough of the other. At those spots, the light waves from each slit would cancel each other out, leaving no mark on the film. And so it would go as one moved down the recording screen: alternating light and dark bands known as an "interference pattern."

Bohr pressed on. One of the biggest surprises in quantum physics was that the same quintessential interference pattern arose when one fired tiny particles, such as electrons, at a wall with two slits. Each particle seemed to behave like a tiny billiard ball when released from the source on one side of the room and detected at the screen on the other side. Yet upon shooting tens, hundreds, or thousands of electrons at the twice-slitted wall, the locations at which each tiny electron was detected matched the wavelike interference pattern. That would never happen with ordinary billiard balls. When thrown at a wall with two slits, the balls would cluster in two clumps, one behind each of the open slits. The billiard balls would never arrange themselves in the alternating interference pattern. Even more strange, physicists could choose to shoot a thousand electrons at the wall one at a time, an hour apart. After all the electrons had made their way through the apparatus, the pattern of light and dark patches on the recording screen—marking where each individual electron had arrived, one at a time—would appear just as if physicists had sent light waves to interfere. (Fig. 1.3.)

Physicists had managed to conduct laboratory demonstrations of the effect as early as 1927.[13] Einstein pressed his colleagues at an informal conference that year to explain: what did the waving? Certainly not the electrons themselves, at least not without straining credulity. Each had been fired one at a time, so no two electrons could have interacted with each other (say, by repelling each other with their electric charge). Each had been detected as a tiny particle; none showed up at the recording screen as a washed-out wave. The distance between the slits was much larger than the electrons themselves, so it hardly made sense to think that an electron passed through both slits at the same time and inter-

FIGURE 1.3. Three snapshots of the detection of individual photons after they have passed through a barrier with slits. The photographs show results after 1/30 of a second (*left*), 1 second (*middle*), and 100 seconds (*right*). Each photon, or quantum of light, gets detected as an individual particle, and yet the pattern that builds up over time reveals wavelike interference. (Courtesy Robert Austin and Lyman Page, Princeton University.)

fered with itself on the other side. Einstein clearly enjoyed watching his colleagues squirm. Like two giddy schoolboys, Einstein and a close friend passed notes back and forth while one defender of quantum theory after another tried to fend off Einstein's challenges. "Don't laugh!" his friend scribbled. Einstein's prescient reply: "I laugh *only* at the naiveté [of the proponents of quantum theory]. Who knows who will be laughing in the coming years."[14]

Einstein's sparring partners were laughing soon enough. Bohr, Heisenberg, and their colleagues cobbled together an interpretation of what was happening in the double-slit experiment. Every quantum system, they reasoned, had an associated "wavefunction," which they labeled with the Greek letter, Ψ (pronounced "psi"). The values that the wavefunction assumed in different locations, and the way those values changed over time, were governed by a new equation first introduced by Schrödinger in 1926. Schrödinger's equation was similar in mathematical form to well-known equations that described wave behavior, such as water waves on the ocean. Max Born—Einstein's friend and Heisenberg's mentor—advanced an interpretation that same year that Ψ was related to probability. In particular, the probability for detecting a quantum object at a particular time and place was given, in Born's account, by the absolute square of the associated wavefunction: Probability = $|\Psi|^2$. In the double-slit experiment, according to this interpretation, the electron's wavefunction spread out like a wave and went through both slits, leading to the characteristic interference pattern.[15]

So were the electrons behaving like particles or waves? The answer—which brought a smile to Niels Bohr's face every time he walked a new audience through the experiment—was "all of the above." Einstein was less amused. "The Heisenberg-Bohr tranquilizing philosophy—or religion?—is so delicately contrived," he complained in a letter to Schrödinger in May 1928, that "for the time being, it provides a gentle pillow for the true believer from which he cannot very easily be aroused. So let him lie there. But"—he left no doubt—"this religion has so damned little effect on me."[16]

Heisenberg and Bohr had more tricks up their sleeves; they weren't finished with the double slit yet. They considered modifying the apparatus, to be able to measure through which slit an individual electron passed. Despite all the talk of wavefunctions, after all, each electron was emitted and detected like a tiny particle; surely each electron must have passed through one slit or the other, just like ordinary billiard balls would do. That notion could be tested, they explained, by placing some other tiny particles behind one of the slits. If an electron passed through that slit en route to the recording screen, then some of the test particles would get scattered, like pins tossed about by a bowling ball, signaling the electron's passage through the slit. If, on the other hand, none of the test particles were scattered, then the electron must have passed through the other slit.

It sounded simple enough. And it would have worked, too, but for one catch, known as Heisenberg's uncertainty principle. Soon after Schrödinger and Born worked out the basic rules for manipulating Ψ, Heisenberg demonstrated that the new equations behaved in some unexpected ways, totally unlike the usual physics of particles or waves. Certain pairs of quantities, such as position and momentum or energy and time, could never be specified with unlimited precision at a single instant. The more precisely a quantum object's position was specified, the less precisely its momentum could be, and vice versa. According to Heisenberg, in other words, we can never know exactly where an object is and where it is going at the same time.[17]

During lectures at the University of Chicago in 1929, in one of his earliest deployments of the uncertainty principle, Heisenberg demonstrated why the slit detector could not work as advertised. To yield a reliable measurement of whether an electron passed through a particular slit, the test particles would have to be clumped tightly behind that slit. The uncertainty in their position, in other words, would have to be much smaller than the distance between the two slits. That small uncertainty in position, in turn, would correspond to a large uncertainty in their momentum. The incoming electron thus would careen into a collection of test particles that already had some large uncertainty in their momentum; this would translate into a correspondingly large uncertainty in the electron's momentum following the collision. Heisenberg needed just a few lines of algebra to show that the collision would jostle the electron's path just enough to smear out the sharp peaks and valleys of the interference pattern. In fact, if every electron could be measured to pass through one slit or the other, the resulting detection pattern would revert to two broad peaks, one behind each slit; all wavelike interference would vanish. On the other hand, reducing the uncertainty in the electron's momentum after scattering, to retain the interference pattern, could only be done by increasing the uncertainty of the test particles' position—by such an amount that no one would know whether they had been clumped behind one slit, the other, or both.[18]

To Bohr, the paradox of the slit detector exemplified a more general feature of quantum mechanics. Ask a "particle-like" question—"through which slit did the particle pass?"—and you will always receive a particle-like answer ("slit A" or "slit B"). Ask a "wavelike" question—"how does Ψ behave in the region between the slits and the detectors?"—and you will always receive a wavelike response ("in a state interference, crests canceling troughs in some places and amplifying crests in others"). Bohr coined the term "complementarity" for his emerging philosophy. Explanation in the quantum realm, he maintained, required the constant juxtaposition of statements that were themselves mutually exclusive, the particle "yin" always paired with the wavelike "yang." (In 1947, when the king of Denmark anointed Bohr with the prestigious Order of the Elephant, Bohr needed to produce a family coat of arms for display in the Frederiksborg Castle near Copenhagen. He placed the classical Chinese yin-yang symbol at its center.) Einstein had little patience for this kind of talk. The goal of physics, he maintained his entire life, was to determine how the world works on its own, independent of the questions we happen to ask of it. Writing to Schrödinger, Einstein mocked Bohr's increasingly oracular outbursts as those of a "a ridiculous little Talmudic philosopher."[19]

Einstein had other bones to pick. Max Born had suggested—and nearly all quantum physicists came to agree—that the square of the wavefunction yielded a probability. But neither Born nor anyone else had succeeded in pressing beyond mere probabilities. For Einstein, this seemed an intolerable shortcoming. He made a few false starts of his own, at one point jotting a rushed note to Born to announce that he had found an interpretation of Ψ that did not resort to probabilities; but each of these efforts fell short of the mark. In the meantime, Einstein only accorded quantum mechanics what he called "transitory significance," despite his many contributions to the subject. "I still believe in the possibility of giving a model of reality," he explained in a lecture at Oxford in 1933, "a theory, that is to say, which shall represent events themselves and not merely the probability of their occurrence."[20] Writing to Born, he was even more direct. "Quantum mechanics is certainly imposing," he began. "But an inner voice tells me that it is not yet the real thing. The theory says a lot, but does not really bring us any closer to the secret of the 'old one.' I, at any rate, am convinced that *He* is not playing at dice." Einstein had no beef with the logical self-consistency or the empirical successes of quantum mechanics. In the right hands, he acknowledged,

Schrödinger's equation and Born's interpretation of Ψ could produce stunningly accurate descriptions of the overall outcomes of large collections of events, such as where, on average, thousands of electrons that had been fired at a barrier would be detected. But the quantum formalism could never reconstruct those aggregate results on a case-by-case basis; it could never explain why the electron in experimental run 867 happened to pass through one slit rather than the other and wind up at a particular location.[21]

Einstein's frustrations reached the boiling point in the summer of 1935. He exchanged a series of letters that summer with Erwin Schrödinger, each egging the other on with his discontent over the direction quantum physics had taken. Building on suggestions from Einstein, Schrödinger crystallized their position with a thought experiment that came to be known as "Schrödinger's cat." In what he called a "ludicrous example," Schrödinger pushed the problem of only being able to calculate probabilities to the extreme. Imagine a cat, Schrödinger instructed readers of his resulting article, "enclosed in a steel chamber, together with the following infernal machine": a small source of radioactive material next to a Geiger counter, which would be able to detect any radioactive decays. Rigged up to the Geiger counter would be a hammer. Should the Geiger counter detect even a single radioactive decay, it would release the hammer, which would strike a bottle of poison, killing the cat. Suppose, Schrödinger continued, that the radioactive material had a probability of one-half to decay within an hour. The best that quantum mechanics could say was that after one hour had elapsed, the cat locked inside the box would be in the strangest of conditions: "in equal measure, the living and the dead cat are (*sit venia verbo* [pardon the expression]) blended or smeared out." Neither dead nor alive, the cat would be in some weird quantum mixture of half-dead-and-half-alive, a condition with no analogue in ordinary experience. But, Schrödinger and Einstein emphasized, no one had ever seen a cat in such a horrid state. Surely, they were convinced, there must be more to physics than mere probabilities.[22]

Bohr, in contrast, delighted in the new probabilistic framework, reaching back to his undergraduate studies of Kant and Kierkegaard to craft a new quantum worldview. Heisenberg, too, found ample fodder for philosophizing in the turn to probabilities. The son of a classicist, Heisenberg enlisted ancient concepts of being and becoming, or "potentia," from the likes of Plato and Aristotle. Puzzling through the uncertainty principle, he liked to recall later in life, had sent him scrambling for his copy of Plato's *Timaeus*. (To Heisenberg's close friend and collaborator Wolfgang Pauli, such claims smacked of mere posturing. Pauli declared in a letter to Bohr that Heisenberg was in fact "very unphilosophical.")[23] Indeed, Bohr, Heisenberg, Pauli, and their colleagues like Max Born became convinced that their new quantum theory ushered in an entirely new philosophical age. Bohr announced at every opportunity that his "either-or" interpretation of the quantum realm, complementarity, was a "general epistemological lesson," to be applied liberally across the entire gamut of human learning, from biology and psychology to anthropology. Typical example: according to Bohr, we can either experience the free flow of our own thoughts, or observe ourselves in the process of thinking, but not both at the same time. Soon after the onset of the Cold War, Max Born was moved to liken capitalism and communism to particle and wave, destined for a quantumlike complementarity.[24]

Einstein would have none of it. "This epistemology-soaked orgy ought to come to an end," he wrote to a colleague at one point. Setting aside the wider speculations in which the quantum theorists indulged so freely— traipsing from natural sciences to social sciences, religion, politics, and beyond—Einstein still harbored deep reservations about their interpretation of the physics. Their embrace of probabilities was especially troubling. Such a probabilistic description might well be useful, Einstein granted, but it was hardly fundamental. "My own opinion," he confided to a correspondent late in 1939—nearly fifteen years after the breakthroughs by Heisenberg, Schrödinger, Bohr, and Born—was that "we will return to the task to describe real phänomena in space and time (not only probabilities for possible experiment)." By that time, most of the younger generation had stopped worrying about Einstein's quibbles. Yet others, closer in age to Einstein (such as Schrödinger), came to share Einstein's dissatisfaction with quantum mechanics. All agreed that mysteries like the double-slit experiment demanded serious philosophical attention. The fate of physics depended on it.[25]

•

The creators of quantum mechanics formed a tight-knit community. At its center, roughly a dozen physicists occupied what sociologists would call a "core set." Surrounding the core, only a few dozen more published on the topic anywhere in the world during the critical period of the mid-1920s. The main players knew each other well. They continually crossed paths at Bohr's institute in Copenhagen, Born's center in Göttingen, or the informal conferences sponsored by the industrialist-turned-philanthropist Ernest Solvay. Quantum physicists criss-crossed Europe by rail, dropping by for visits that lasted days, weeks, or months. "Kramers was here for eight days," Born wrote to Einstein in typical fashion in July 1925, "and Ehrenfest. . . . Last week Kaptiza from Cambridge was here, and Joffé from Leningrad." "If it is agreeable to you," Schrödinger wrote to Einstein a few years later, "I would be glad to come over sometime to talk" more in person about Bohr's latest ideas.[26] When not in the same town, they kept up their conversations by letter, tens of thousands of which have survived. Over the years, scholars have dutifully inventoried, archived, microfilmed, and translated these letters, subjecting them to the kind of line-by-line scrutiny once reserved for Scripture. The letters reveal just how earnestly the early quantum physicists worked to interpret their new formalism, day in and day out. Clustered in small, informal groups, they struggled to put flesh on the new equations, to wrap their heads around how the world could possibly *work* that way.[27] (Fig. 1.4.)

The same philosophical impulse shaped their earliest pedagogical writings. Some textbooks included entire chapters with titles like "Quantum mechanics and philosophy." Other textbook authors paused within their expositions to pronounce the death of the Kantian "thing-in-itself," or to weigh the consequences of Heisenberg's uncertainty principle for scientists' age-old quest for objectivity.[28] The young American physicists who learned quantum mechanics at the feet of the European masters

FIGURE 1.4. Niels Bohr and Albert Einstein deep in conversation about the mysteries of quantum mechanics while visiting the house of a mutual friend in 1930. (Photographs by Paul Ehrenfest, courtesy Emilio Segrè Visual Archives, American Institute of Physics.)

likewise agreed that the material demanded philosophical attention. They often broke with their teachers' preferred philosophies—American instructors turned most often to the homegrown philosophy of Harvard physicist Percy Bridgman, rather than the rarefied heights of Plato, Kant, or Kierkegaard. But they, too, demanded that their students sit with the quantum weirdness during the 1920s and 1930s and hone their own philosophical response. General examinations from across the country, required for graduate students to advance to candidacy for a PhD, routinely pressed students to compose essays about wave-particle duality, the double-slit experiment, and related matters. Throughout the 1930s, reviewers held the latest American textbooks on quantum mechanics accountable for their philosophical orientation and exposition.[29]

The landscape changed sharply after World War II. In the early 1950s, Einstein—having moved to the United States twenty years earlier, fleeing fascism in Europe—surveyed the scene with despair. The problem was no longer his colleagues' "tranquilizing philosophy"; it was their ardent lack of interest in philosophy altogether.[30] Graduate students at Caltech were caught equally off guard. Having dutifully pored over reports from their predecessors about what to expect on the general examination, the new generation felt cheated. One complained that all the effort he had "invested in analysis of paradoxes and queer logical points was of no use in the exam." Others recorded how their questions had avoided matters of interpretation altogether, focusing instead upon a narrow set of stock problems. (Forget about philosophy and just give the "usual spiel," came one student's advice to those who would take the examination after him.) Essay questions disappeared from graduate students' written exams across the country, replaced by a coterie of standard problems to calculate. Textbook reviewers in the United States began to praise books on quantum mechanics that "avoided philosophical discussion" or omitted "philosophically tainted questions." Enough with the "musty atavistic to-do about position and momentum," stormed MIT's Herman Feshbach in 1962.[31]

Much had changed. The hateful policies of Mussolini and Hitler had chased scores of intellectuals out of Europe. Nearly a hundred physicists and mathematicians followed Einstein's lead and resettled in the United States during the 1930s. Born and Schrödinger rode out the war in Edinburgh and Dublin, respectively, while a few—including, most famously, Heisenberg—remained behind in the Nazi Reich. By the close of the 1930s, quantum physicists had been scattered across the globe, their days of riding the rails in pursuit of further banter gone forever.[32]

The new world that these émigrés found, meanwhile, was changing fast under their feet. With memories of fascism still fresh, dozens of them joined the Allied war effort, alongside their new American and British colleagues. During the war, physicists all over the world—but especially in the United States—received a crash course in "gadgetry," their new shorthand for the special flavor of research and development conducted side by side with engineers and military planners. Radar, the proximity fuse, solid-fuel rockets, and especially the atomic bomb project ripped academic physicists from their ivory towers and thrust them into a grubby world of grease and pumps, gauges and lathes. The round-the-clock pressure to produce working gadgets in time to impact the course of the war left little leisure for philosophizing. Physicists learned to put their heads down, ignore philosophical tangents, and wring numbers from their equations as quickly as possible. When Edward Teller lectured on quantum mechanics at Los Alamos—the central scientific laboratory of the atomic bomb project—for the gaggle of students and lab hands whose education had been interrupted by the war, he raced through the interpretive material so quickly that he replaced the fabled double slit with a *single* slit on the blackboard, from which the crucial interference pattern would never arise! Here, in stark relief, was the new face of war-forged pragmatism.[33]

The wartime relationships continued unabated after the war, especially as the Cold War with the Soviet Union hardened into a fact of life in the late 1940s. Defense agencies swamped the previous sources of funding for physics, keeping physicists' attention tethered close to the demands of national security. Only a small minority spent the bulk of their time working on weapons after the war. Yet across the United States, from bustling research universities to tiny liberal-arts colleges, nearly all academic physicists became enrolled in a massive Cold War project: to produce more physicists, at an ever-increasing rate, to ensure that the nation's supply of technical workers was trained and ready should the Cold War ever turn hot. Leading policymakers freely equated the country's population of physicists with a "standing army." In the course of a single speech in 1951, for example, a top member of the Atomic Energy Commission managed to describe physicists as a "war commodity," a "tool of war," and a "major war asset," to be "stockpiled" and "rationed." Analysts at the Bureau of Labor Statistics agreed. "If the research in physics which is vital to the nation's survival is to continue and grow," they asserted in a 1952 report, "national policy must be concerned not only with keeping the young men already in the field at work but also with insuring a continuing supply of new graduates."[34] Adding fuel to the fire, a series of reports published in the mid-1950s, which had been bankrolled secretly by the Central Intelligence Agency, seemed to suggest that the Soviet Union was training new scientists and engineers even more quickly than the United States. Coming at a propitious moment politically—one was published just two weeks after the Soviets' surprise launch of the first Sputnik satellite, in October 1957—these reports helped shake loose another billion dollars from Congress (more than $7 billion in 2010 dollars) to support graduate training in "defense" fields like science and engineering.[35]

The Cold War imperative for scientific "manpower" had immediate effects on enrollments. Backed by expansive fellowship programs and special draft deferments, classrooms in American physics departments bulged faster than any other field. Nearly all fields were growing exponentially after World War II, thanks to a backlog of veterans returning to the nation's campuses, supported by programs like the GI Bill. Yet physics outpaced them all, its graduate-level enrollments doubling nearly twice as quickly as all other fields combined. By the outbreak of fighting in the Korean War, American physics departments were producing three times as many PhDs per year as the prewar highs—a number that would only climb higher, by another factor of three, after Sputnik.[36]

The astronomical growth had an immediate effect on teaching. Enrollments in stock courses for graduate students, such as introductory quantum mechanics, swelled to more than 100 students in physics departments from MIT to Berkeley. Such classroom numbers, Berkeley's department chair exclaimed to his dean, were "a disgrace and should not be tolerated at any respectable university."[37] Despite a frenzy of faculty hiring, student-to-faculty ratios ballooned in physics departments across the country. Professors routinely complained that the bloated enrollments trampled out any sense of the prewar "intimacy" between faculty and students. Students agreed. "The classes are so large that there is little or no individual contact between student and teacher," complained one graduate student in Harvard's department after the war.[38]

Faced with such runaway growth, physics professors across the country revamped their teaching style. They began to accentuate those elements that could lend themselves to high-throughput pedagogy, pumping record numbers of students through their courses. First to go was the discussion-based, qualitative, philosophical inquiry into what quantum mechanics *meant*. Staring out at the sea of faces in their stadium-seating classrooms, many instructors felt they had little choice. (Fig. 1.5.) "With these subjects," explained one frustrated professor in 1956, "lecturing is of little avail." He had in mind once-central topics like the meaning of the uncertainty principle, Bohr's complementarity, and the consequences for causality of the probabilistic turn. "The baffled student hardly knows what to write down, and what notes he does take are almost certain to horrify the instructor, who perspicaciously usually resolutely refuses to question his students on these topics." And so, this commentator concluded with regret, when it came to "the philosophical issues raised by quantum mechanics . . . the student never has a chance to gauge their depth." A few years later, another critic weighed in. A lion of the interwar era who had emigrated from Europe to the United States, he accused his American colleagues of confusing what was "easy to teach"—the "technical mathematical aspects" of quantum mechanics, which could be chopped up and parceled out on problem sets and exams—with the conceptual, interpretive material that students needed most.[39]

FIGURE 1.5. Enrico Fermi lecturing to physics graduate students in the early 1950s. (Photograph by Samuel Goudsmit, courtesy Emilio Segrè Visual Archives, Goudsmit Collection, American Institute of Physics.)

The few traces that remain from the nation's physics classrooms bear these observations out. Comparing lecture notes from graduate-level courses on quantum mechanics from across the country, each dating from the 1950s, reveals a stark pattern. An increase by a factor of three in enrollments correlated with a decrease by a factor of five in the proportion of time spent on interpretive or philosophical material. In short, the larger the class, the less time spent talking through the big issues at the heart of quantum mechanics. Textbooks followed a similar trend. As physics enrollments continued to climb well into the 1960s, the proportion of essay questions plummeted to around 10 percent of all problems embedded in new textbooks. Faced with skyrocketing enrollments, no one had time to grade such verbiage. What students and faculty needed, opined a Berkeley physics professor in 1965, were more textbooks like Leonard Schiff's successful *Quantum Mechanics*. The Berkeley physicist had used the first edition, from 1949, as a student, and he looked back on it fondly. "The book kept me sufficiently busy to prevent pseudo-philosophical speculations about the True Meaning of quantum mechanics"—just the ticket for the new classroom realities. He urged the publisher to bring out a new edition of Schiff's book. By trimming what had already been paltry discussion of interpretive matters, the new edition could be larded even more fully with tough calculations, to keep the new generation busy. (The publisher brought out the new edition in 1968 to widespread acclaim from reviewers; it sold well.)[40] Countries that had similar physics enrollment patterns—major Cold War players like the United Kingdom and the Soviet Union—produced remarkably similar textbooks. Other European countries, like France, West Germany, and Austria, spent much more time rebuilding after the war and did not experience the same bulge in physics classrooms. Physicists in those countries continued to write textbooks in the prewar fashion, featuring long excursions into philosophy and stuffed with juicy essay questions.[41]

The enrollment-driven pragmatism, so stark in American physics departments after World War II, was anything but a "dumbing down." The second and third editions of Schiff's acclaimed textbook, for example, contained homework problems—aimed at entry-level graduate students—that would have stumped leading physicists only a decade or two earlier. The quarter century during which this Cold War style reigned witnessed an extraordinary buildup of calculating skill. All the same, an intellectual trade-off slipped by unnoticed, with wide-ranging implications. For every additional calculation of baroque complexity that physics students tackled during the 1950s and 1960s, they spent correspondingly less time puzzling through what all those fancy equations meant—what they implied about the world of electrons and atoms. The fundamental strangeness of quantum reality had been leeched out.

•

Not everyone in the United States adopted the mantra of "shut up and calculate" after the war. But the few groups that tried to retain the prewar style rapidly became exceptions that proved the rule. Throughout the hot summer months of 1954, for example, about a dozen physicists gathered in New York City to discuss the foundations of quantum mechanics. Even as most of their colleagues were too busy rewriting their lecture notes, editing their textbooks, and revising their examinations to drop nearly any mention of such interpretive material, this group pressed on, unconvinced that all was well with the central pillar of modern physics.[42]

More than a fascination with quantum mysteries brought these physicists together. Most shared the same politics as well. The group had been convened by Hans Freistadt, a native of Vienna who had fought in the U.S. Army during World War II. By the early 1950s he was an instructor at the sleepy Newark College of Engineering in New Jersey, the latest stop in his wanderings following his dramatic testimony before the Joint Congressional Committee on Atomic Energy back in 1949. Yes, he had confirmed, he was a member of the Communist Party, and yes, he continued, he had indeed received one of the first fellowships from the Atomic Energy Commission (AEC) to pursue graduate studies in physics. His studies had concerned strictly unclassified material. All the same, the headline-grabbing revelation, and the political firestorm that ensued, nearly ended the AEC fellowship program.[43]

Joining Freistadt to ponder quantum mysteries in the 1954 discussion group was Byron Darling. Until recently a tenured professor of physics at Ohio State University, Darling found himself out of work after testifying before the House Un-American Activities Committee (HUAC) during its March 1953 investigation of "Communist methods of infiltration" of the educational system. The committee had accused Darling of past membership in the Party; he pleaded the Fifth Amendment. Although he had signed his university's anti-Communist loyalty oath, had answered every question put to him by the university's investigating committee, and had stated categorically that he was not nor had ever been a member of the Communist Party, Ohio State dismissed him for failing to answer all of HUAC's questions. He left Columbus for New York City, where he passed the time during the summer of 1954 talking about possible alternatives to quantum theory with Freistadt and company, before taking up a new post at the University of Laval in Quebec.[44]

Nearly all the other members of Freistadt's discussion group shared a clear leftist orientation. Some had even left tenure-track jobs in the United States to work overseas for a few years, returning just in time to join the discussions in New York that summer. Freistadt's seminar produced two publications, both written by him. The first, published in the Marxist cultural magazine *Science and Society* before the sessions began, was filled with predictable talk of the "doctrinaire" thinking shown by "modern scientists in capitalist countries," whose "positivist obscurantism" had landed quantum theory in its current state of "crisis." The other, a technical review article on a variant of quantum mechanics, was published as a supplement to an Italian physics journal and promptly forgotten for the next twenty years. Fired from jobs or castigated in the media for their alleged political activities, the group members' politics and their unpopular research interests each marked them as clearly outside the discipline's mainstream. In that climate, they could find little traction for their work.[45]

Twenty years later, another informal discussion group convened, likewise bent on exploring the big metaphysical questions raised by quantum mechanics. Like Freistadt's group, the Fundamental Fysiks Group, established in Berkeley in the spring of 1975, was peopled with physicists on the margins. Yet for all the similarities, the two groups left rather different footprints. Where Freistadt's group toiled in obscurity, members of the Fundamental Fysiks Group became media darlings, publishing a series of best-selling books and leaving a genuine imprint on physics research and curricula throughout the country.

The divergence in outcomes for the two discussion groups, otherwise so similar in makeup and structure, illuminates how quickly conditions had changed for physicists by the early 1970s. Politics had thwarted the career trajectories of most members of the Fundamental Fysiks Group, but not the personal politics of red-baiting as in Freistadt's day. Rather, they were caught at the wrong place at the wrong time, bystanders of a systematic political upheaval that rocked the physics profession from top to bottom. Freistadt's circle had labored on the fringes of boom times for American physicists. By the time members of the Fundamental Fysiks Group found each other, the boom times had turned to bust.

When trouble came for physicists, it came fast. All too quickly, the assumptions that had driven the enrollment boom broke down. As tensions with the Soviets cooled and resources dried up, military patrons and congressional leaders revisited long-standing priorities. No longer did calls ring out to produce scientific "manpower" at all costs. The Pentagon's return on decades of investment in open-ended basic research—which had justified, and paid for, nearly all graduate training in physics—struck a new generation of analysts as rather lackluster. Years into the slog of the Vietnam War, meanwhile, antiwar protesters grew more brazen, taking over campus buildings and planting pipe bombs, all part of a campaign to force the Pentagon out of the higher-education business. (Physics laboratories provided some of their favorite targets, potent symbols of the "mutual embrace" between academic scientists and military paymasters.) Caught between hardnosed Pentagon accountants on the one hand and raised-fist radicals on the other, physics had nowhere to go but down.[46]

Nearly every field suffered cutbacks in the realigned political and budgetary landscape, but none more than physics. Since World War II, the discipline had become more reliant than any other on federal funding. When trouble hit, physicists' enrollments plummeted faster and deeper than any other field: down fully one-third from their peak in just five years, falling to one-half by decade's end. (Fig. 1.6.) Demand disappeared even more quickly. Records from the Placement Service of the American Institute of Physics tell the grim tale. The service had arranged job interviews between prospective employers and physics students since the early 1950s. As late as the mid-1960s, the service had registered more employers than students looking for jobs. By 1968, the balance had tipped: 989 applicants registered, with only 253 jobs on offer. And then the bottom fell out. In 1971, the Placement Service registered 1053 applicants competing for just 53 jobs.[47]

Into that state of wreckage trod the young physicists who would form the Fundamental Fysiks Group. Like it or not, they would not follow physics careers like the ones their teachers had enjoyed. The ways and means of being a physicist came unmoored in a way they hadn't been for two generations. No longer would the attitude of "shut up and calculate" hold sway unchecked. Sitting around the large conference table at the Lawrence Berkeley Laboratory, with few other demands on their time, they sought to recapture the sense of excitement, wonder, and mystery that had attracted them to physics in the first place, just as it had animated the founders of quantum mechanics. They might not have enjoyed secure employment, but they fervently believed one thing: physics could be fun again.

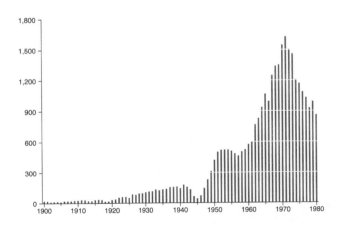

FIGURE 1.6. Number of physics PhDs granted in the United States, 1900–1980. (Illustration by Alex Wellerstein, based on data from the American Institute of Physics and the National Science Foundation.)

Chapter 2

"Spooky Actions at a Distance"

In a theory in which parameters are added to quantum mechanics to determine the results of individual measurements, without changing the statistical predictions, there must be a mechanism whereby the setting of one measuring device can influence the reading of another instrument, however remote. Moreover, the signal involved must propagate instantaneously.

—John S. Bell, 1964

One recent development dominated the Fundamental Fysiks Group's deliberations: a striking theorem published in the mid-1960s by the Irish physicist John S. Bell. The iconoclastic Bell had long nursed a private disquietude with quantum mechanics. His physics teachers—first at Queen's University in his native Belfast during the late 1940s, and later at Birmingham University, where he pursued doctoral work in the mid-1950s—had shunned matters of interpretation just as vehemently as their American colleagues did at the time. The "ask no questions" attitude frustrated Bell, who remained unconvinced that Bohr had really vanquished the last of Einstein's critiques long ago and that there was nothing left to worry about. At one point in his undergraduate studies, his red shock of hair blazing, he even engaged in a shouting match with a beleaguered professor, calling him "dishonest" for trying to paper over genuine mysteries in the foundations, such as how to interpret the uncertainty principle. Certainly, Bell would grant, quantum mechanics worked impeccably "for all practical purposes," a phrase he found himself using so often that he coined the acronym "FAPP." But wasn't there more to physics than FAPP? At the end of the day, after all the wavefunctions had been calculated and probabilities plotted, shouldn't quantum mechanics have something coherent to say about nature?[1]

In the years following his impetuous shouting matches, Bell tried to keep these doubts to himself. At the tender age of twenty-one he realized that if he continued to indulge these philosophical speculations, they might well scuttle his physics career before it could even begin. He dove into mainstream topics, working on nuclear and particle physics at Harwell, Britain's civilian atomic energy research center. Still, his mind continued to wander. He wondered whether there were some way to push beyond the probabilities offered by quantum theory, to account for motion in the atomic realm more like the way Newton's physics treated the motion of everyday objects. In Newton's physics, the behavior of an apple or a planet was completely determined by its initial state—variables like position (where it was) and momentum (where it was going)—and the forces acting upon it; no probabilities in sight. Bell wondered whether there might exist some set of variables that could be added to the quantum-mechanical description to make it more like Newton's system, even if some of those new variables remained hidden from view in any given experiment. Bell avidly read a popular account of quantum theory by one of its chief architects, Max Born's *Natural Philosophy of Cause and Chance* (1949), in which he learned that some of Born's contemporaries had likewise tried to invent such "hidden variables" schemes back in the late 1920s. But Bell also read in Born's book that another great of the interwar generation, the Hungarian mathematician and physicist John von Neumann, had published a proof as early as 1932 demonstrating that hidden variables could not be made compatible with quantum mechanics. Bell, who could not read German, did not dig up von Neumann's recondite proof. The say-so of a leader (and soon-to-be Nobel laureate) like Born seemed like reason enough to drop the idea.[2]

Imagine Bell's surprise, therefore, when a year or two later he read a pair of articles in the *Physical Review* by the American physicist David Bohm. Bohm had submitted the papers from his teaching post at Princeton University in July 1951; by the time they appeared in print six months later, he had landed in São Paulo, Brazil, following his hounding by the House Un-American Activities Committee (HUAC). Bohm had been a graduate student under J. Robert Oppenheimer at Berkeley in the late 1930s and early 1940s. Along with several like-minded friends, he had participated in freewheeling discussion groups about politics, worldly affairs, and local issues like whether workers at the university's laboratory should be unionized. He even joined the local branch of the Communist Party out of curiosity, but he found the discussions so boring and ineffectual that he quit a short time later. Such discussions might have seemed innocuous during ordinary times, but investigators from the Military Intelligence Division thought otherwise once the United States entered World War II, and Bohm and his discussion buddies started working on the earliest phases of the Manhattan Project to build an atomic bomb. Military intelligence officers kept the discussion groups under top-secret surveillance, and in the investigators' eyes the line between curious discussion group and Communist cell tended to blur. When later called to testify before HUAC, Bohm pleaded the Fifth Amendment rather than name names. Over the physics department's objections, Princeton's administration let his tenure-track contract lapse rather than reappoint him. At the center of a whirling media spectacle, Bohm found all other domestic options closed off. Reluctantly, he decamped for Brazil.[3]

In the midst of the Sturm und Drang, Bohm crafted his own hidden variables interpretation of quantum mechanics. As Bell later reminisced, he had "seen the impossible done" in these papers by Bohm. Starting from the usual Schrödinger equation, but rewriting it in a novel way, Bohm demonstrated that the formalism need not be interpreted only in terms of probabilities. An electron, for example, might behave much like a bullet or billiard ball, following a path through space and time with well-defined values of position and momentum every step of the way. Given the electron's initial position and momentum and the forces acting on it, its future behavior would be fully determined, just like the case of the trusty billiard ball—although Bohm did have to introduce a new "quantum potential" or force field that had no analogue in classical physics. In Bohm's model, the quantum weirdness that had so captivated

Bohr, Heisenberg, and the rest—and that had so upset young Bell, when parroted by his teachers—arose because certain variables, such as the electron's initial position, could never be specified precisely: efforts to measure the initial position would inevitably disturb the system. Thus physicists could not glean sufficient knowledge of all the relevant variables required to calculate a quantum object's path. The troubling probabilities of quantum mechanics, Bohm posited, sprang from averaging over the real-but-hidden variables. Where Bohr and his acolytes had claimed that electrons simply did not possess complete sets of definite properties, Bohm argued that they did—but, as a practical matter, some remained hidden from view.[4]

Bohm's work had captivated members of Hans Freistadt's 1954 discussion group, that bunch of bedraggled leftist physicists who dove into quantum physics and philosophy as a welcome break from their run-ins with HUAC and related red-baiters. In fact, Freistadt devoted his long review article to Bohm's approach to hidden variables. Quite independently, Bohm's papers fired Bell's imagination as well. Soon after discovering them, Bell gave a talk on Bohm's papers to the Theory Division at Harwell. Most of his listeners sat in stunned (or perhaps just bored) silence: Why was this young physicist wasting their time on such philosophical drivel? Didn't he have any real work to do? One member of the audience, however, grew animated: Austrian émigré Franz Mandl. Mandl, who knew both German and von Neumann's classic study, interrupted several times; the two continued their intense arguments well after the seminar had ended. Together they began to reexamine von Neumann's no-hidden-variables proof, on and off when time allowed, until they each went their separate ways. Mandl left Harwell in 1958; Bell, dissatisfied with the direction in which the laboratory seemed to be heading, left two years later.[5]

Bell and his wife Mary, also a physicist, moved to CERN, Europe's multinational high-energy physics laboratory that had recently been established in Geneva. Once again he pursued cutting-edge research in particle physics. And once again, despite his best efforts, he found himself pulled to his hobby: thinking hard about the foundations of quantum mechanics. Once settled in Geneva, he acquired a new sparring partner in Josef Jauch. Like Mandl, Jauch had grown up in the Continental tradition and was well versed in the finer points of Einstein's, Bohr's, and von Neumann's work. In fact, when Bell arrived in town Jauch was busy trying to strengthen von Neumann's proof that hidden-variables theories were irreconcilable with the successful predictions of quantum mechanics. To Bell, Jauch's intervention was like waving a red flag in front of a bull: it only intensified his resolve to demonstrate that hidden variables had not yet been ruled out. Spurred by these discussions, Bell wrote a review article on the topic of hidden variables, in which he isolated a logical flaw in von Neumann's famous proof. At the close of the paper, he noted that "the first ideas of this paper were conceived in 1952"—fourteen years before the paper was published—and thanked Mandl and Jauch for all of the "intensive discussion" they had shared over that long period.[6]

Still Bell kept pushing, wondering whether a certain type of hidden-variables theory, distinct from Bohm's version, might be compatible with ordinary quantum mechanics. His thoughts returned to the famous thought experiment introduced by Einstein and his junior colleagues Boris Podolsky and Nathan Rosen in 1935, known from the start by the authors' initials, "EPR." Einstein and company had argued that quantum mechanics must be incomplete: at least in some situations, definite values for pairs of variables could be determined at the same time, even though quantum mechanics had no way to account for or represent such values. The EPR authors described a source, such as a radioactive nucleus, that shot out pairs of particles with the same speed but in opposite directions. Call the left-moving particle "A," and the right-moving particle "B." A physicist could measure A's position at a given moment, and thereby deduce the value of B's position. Meanwhile, the physicist could measure B's momentum at that same moment, thus capturing knowledge of B's momentum and simultaneous position to any desired accuracy. Yet Heisenberg's uncertainty principle dictated that precise values for certain pairs of variables, such as position and momentum, could never be known simultaneously.[7]

Fundamental to Einstein and company's reasoning was that quantum objects carried with them—on their backs, as it were—complete sets of definite properties at all times. Think again of that trusty billiard ball: it has a definite value of position and a definite value of momentum at any given moment, even if we choose to measure only one of those properties at a time. Einstein assumed the same must be true of electrons, photons, and the rest of the furniture of the microworld. Bohr, in a hurried response to the EPR paper, argued that it was wrong to assume that particle B had a real value for position all along, prior to any effort to measure it. Quantum objects, in his view, simply did not possess sharp values for all properties at all times. Such values emerged during the act of measurement, and even Einstein had agreed that no device could *directly* measure a particle's position and momentum at the same time. Most physicists seemed content with Bohr's riposte—or, more likely, they were simply relieved that someone else had responded to Einstein's deep challenge.[8]

Bohr's response never satisfied Einstein, however; nor did it satisfy John Bell. Bell realized that the intuition behind Einstein's famous thought experiment—the reason Einstein considered it so damning for quantum mechanics—concerned "locality." To Einstein, it was axiomatic that something that happens in one region of space and time should not be able to affect something happening in a distant region—more distant, say, than light could have traveled in the intervening time. As the EPR authors put it, "since at the time of measurement the two systems [particles A and B] no longer interact, no real change can take place in the second system in consequence of anything that may be done to the first system." Yet Bohr's response suggested something else entirely: the decision to conduct a measurement on particle A (either position or momentum) would *instantaneously* change the properties ascribed to the faraway particle B. Measure particle A's position, for example, and—*bam!*—particle B would be in a state of well-defined position. Or measure particle A's momentum, and—*zap!*—particle B would be in a state of well-defined momentum. Late in life, Bohr's line still rankled Einstein. "My instinct for physics bristles at this," Einstein wrote to a friend in March 1948. "Spooky actions at a distance," he huffed.[9]

•

Fresh from his wrangles with Jauch, Bell returned to EPR's thought experiment. He wondered whether such "spooky actions at a distance" were endemic to quantum mechanics, or just one possible interpretation among many. Might some kind of hidden variable approach reproduce all the quantitative predictions of quantum theory, while still satisfying Einstein's (and Bell's) intuition about locality? He focused on a variation of EPR's setup, introduced by David Bohm in his 1951 textbook on quantum mechanics. Bohm had suggested swapping the values of the particles' spins along the x- and y-axes for position and momentum.[10]

"Spin" is a curious property that many quantum particles possess; its discovery in the mid-1920s added a cornerstone to the emerging edifice of quantum mechanics. Quantum spin is a discrete amount of angular momentum—that is, the tendency to rotate around a given direction in space. Of course many large-scale objects possess angular momentum, too: think of the planet Earth spinning around its axis to change night into day. Spin in the microworld, however, has a few quirks. For one thing, whereas large objects like the Earth can spin, in principle, at any rate whatsoever, quantum particles possess fixed amounts of it: either no spin at all, or one-half unit, or one whole unit, or three-halves units, and so on. The units are determined by a universal constant of nature known as Planck's constant, ubiquitous throughout the quantum realm. The particles that make up ordinary matter, such as electrons, protons, and neutrons, each possess one-half unit of spin; photons, or quanta of light, possess one whole unit of spin.[11]

In a further break from ordinary angular momentum, quantum spin can only be oriented in certain ways. A spin one-half particle, for example, can exist in only one of two states: either spin "up" or spin "down" with respect to a given direction in space. The two states become manifest when a stream of particles passes through a magnetic field: spin-up particles will be deflected upward, away from their previous direction of flight, while spin-down particles will be deflected downward. Choose some direction along which to align the magnets—say, the z-axis—and

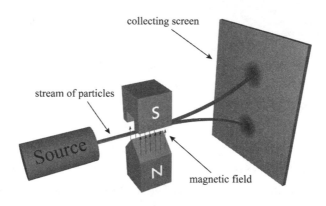

FIGURE 2.1. Device for measuring quantum particles' spin. Spin one-half particles, such as electrons, emerge from the source on the left and travel through the magnetic field, which points up from the north pole, *N*, of the magnet toward the south pole, *S*. Particles with spin up will be deflected upward from the original direction of flight and collect in one region of the collecting screen (or photographic plate); particles with spin down will be deflected downward. (Illustration by Alex Wellerstein.)

the spin of any electron will only ever be found to be up or down; no electron will ever be measured as three-quarters "up" along that direction. Now rotate the magnets, so that the magnetic field is pointing along some different direction. Send a new batch of electrons through; once again you will only find spin up or spin down along that new direction. For spin one-half particles like electrons, the spin along a given direction is always either +1 (up) or –1 (down), nothing in between.[12] (Fig. 2.1.)

No matter which way the magnets are aligned, moreover, one-half of the incoming electrons will be deflected upward and one-half downward. In fact, you could replace the collecting screen (such as a photographic plate) downstream of the magnets with two Geiger counters, positioned where the spin-up and spin-down particles get deflected. Then tune down the intensity of the source so that only one particle gets shot out at a time. For any given run, only one Geiger counter will click: either the upper one (indicating passage of a spin-up particle) or the lower one (indicating spin down). Each particle has a fifty-fifty chance of being measured as spin up or spin down; the sequence of clicks would be a random series of +1s (upper counter) and –1s (lower counter), averaging out over many runs to an equal number of clicks from each detector. Neither quantum theory nor any other scheme has yet produced a successful means of predicting in advance whether a given particle will be measured as spin up or spin down; only the probabilities for a large number of runs can be computed.

Bell realized that Bohm's variation of the EPR thought experiment, involving particles' spins, offered two main advantages over EPR's original version. First, the measurements always boiled down to either a +1 or a –1; no fuzzy continuum of values to worry about, as there would be when measuring position or momentum. Second, physicists had accumulated decades of experience building real machines that could manipulate and measure particles' spin; as far as thought experiments went, this one could be grounded on some well-earned confidence. And so Bell began to analyze the spin-based EPR arrangement. Because the particles emerged in a special way—spat out from a source that had zero spin before and after they were disgorged—the total spin of the two particles together likewise had to be zero. When measured along the same direction, therefore, their spins should always show perfect correlation: if A's spin were up then B's must be down, and vice versa. Back in the early days of quantum mechanics, Erwin Schrödinger had termed such perfect correlations "entanglement."[13]

Bell demonstrated that a hidden-variables model that satisfied locality—in which the properties of A remained unaffected by what measurements were conducted on B—could easily reproduce the perfect correlation when A's and B's spins were measured along the same direction. At root, this meant imagining that each particle carried with it a definite value of spin along any given direction, even if most of those values remained hidden from view. The spin values were considered to be properties of the particles themselves; they existed independent of and prior to any effort to measure them, just as Einstein would have wished.

Next Bell considered other possible arrangements. One could

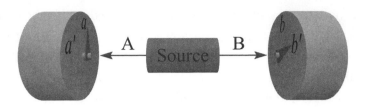

FIGURE 2.2. Bell's updated thought experiment, based on Bohm's version of the EPR setup. A source shoots out pairs of particles, A and B. Each detector has two directions along which it can measure a particle's spin, corresponding to the orientation of the magnets used to separate particles with spin up from those with spin down. As shown here, the apparatus is set to measure the spin of particle A along one direction (setting a) and the spin of particle B along a different direction (setting b'). (Illustration by Alex Wellerstein.)

choose to measure a particle's spin along any direction: the z-axis, the y-axis, or any angle in between. All one had to do was rotate the magnets between which the particle passed. What if one measured A's spin along the z-axis and B's spin along some other direction? (Fig. 2.2.) Bell homed in on the expected correlations of spin measurements when shooting pairs of particles through the device, while the detectors on either side were oriented at various angles. He considered detectors that had two settings, or directions along which spin could be measured. To keep track of all the possible combinations, he labeled the settings on the left-hand detector—which would measure the spin of particle A—as a and a': a for when the left-hand detector was oriented along the z-axis, and a' for when that detector was oriented along its other direction. Same for the right-hand detector, toward which particle B careened: b when the right-hand detector was oriented along the z-axis, and b' when it was oriented along its other direction. (Bell took the settings a' and b' to lie in the same direction: when the detectors were set to a' and b', every pair of particles would be measured as having opposite spin; same for when both detectors were set to a and b.)

Bell labeled the outcomes of each of these measurements. He denoted the measured outcome of the spin of particle A when the left-hand detector was in setting a as A, and the outcome when the left-hand detector was set to a' as A'; similarly for B and B' for the measurements on particle B. All of these measurement outcomes— A, A', B, and B'—were just plain numbers. In fact, they were particularly simple ones: because every spin measurement, along any direction, could only ever result in spin up or spin down, A, A', B, and B' could only ever equal +1 or –1. Bell could then consider various combinations of measurements, such as AB, the product of outcomes when the left-hand detector was set to a and the right-hand detector to b; or AB', which arose when the left-hand detector was set to a and the right-hand detector to b'. Since each measurement outcome (A, A', B, B') could only equal +1 or –1, the pairs—AB or AB', and so on—would likewise just equal +1 or –1. One could then consider a particular combination, S, built from all the various correlations that could arise:

$$S = AB - A'B + AB' + A'B' = (A - A')\,B + (A + A')\,B'$$

One of the terms in parentheses would always vanish, and the other would always equal +2 or –2. Perhaps in one instance $A = +1$ and $A' = +1$; then $(A - A') = 0$, and $(A + A') = 2$. Or it could be that $A = -1$ and $A' = +1$, so that $(A - A') = -2$ and $(A + A') = 0$. Since B and B' always equal +1 or –1, the combination, S, must always equal +2 or –2; no other value could ever arise. Bell imagined emitting a large number of particle pairs from the source, one pair at a time, and recording the measured outcomes at each detector (noting carefully the settings at each detector for each particular run). After many pairs of particles had been measured, one would expect to find the average value for S, S_{average}, to fall within the range $-2 \leq S_{\text{average}} \leq +2$: sometimes S would equal +2 and other times –2, so that the average of large numbers of runs should give some value in between.[14]

So far, so good. But Bell wasn't finished yet. As he demonstrated next, quantum mechanics made unambiguous predictions for the probabilities of various correlations between the spins of particles A and B as one varied the direction along which they were measured. For various choices of the angle between detector settings a and b' (or, equivalently,

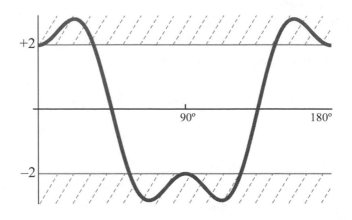

FIGURE 2.3. Predicted values for the quantity S, made up of combinations of spin measurements on particles A and B along various directions. The horizontal axis shows the angle between detector settings a and b' (or, equivalently, between a' and b). As Bell demonstrated, the assumption that particles A and B carried definite values for spin along each direction prior to measurement—as Einstein and his collaborators had urged—limited S to lie between +2 and –2. Yet the quantum-mechanical prediction for the correlation violated that bound by more than 40 percent for certain choices of angle. (Illustration by Alex Wellerstein, based on Aspect [2002], 130.)

between settings a' and b), quantum mechanics predicted clear violations of the innocuous-looking inequality, $-2 \leq S \leq +2$. In fact, for judicious choices of angle, the quantum predictions exceeded this bound by a sizable amount—more than 40 percent. In effect, quantum mechanics predicted that particles A and B should be *more strongly correlated* than the bound on S would allow. (Fig. 2.3.)

Using only a few lines of algebra, Bell thus proved that *no* local hidden-variables theory could ever reproduce the same degree of correlations as one varied the angles between detectors. The result has come to be known as "Bell's theorem." Simply *assuming* that each particle carried a full set of definite values on its own, prior to measurement—even if most of those values remained hidden from view—necessarily clashed with quantum theory. Nonlocality was indeed endemic to quantum mechanics, Bell had shown: somehow, the outcome of the measurement on particle B depended on the measured outcome on particle A, even if the two particles were separated by huge distances at the time those measurements were made. Any effort to treat the particles (or measurements made upon them) as independent, subject only to local influences, necessarily led to predictions different from those of quantum mechanics. Here was what Bell had been groping for, on and off since his student days: some quantitative means of distinguishing Bohr's interpretation of quantum mechanics from other coherent, self-consistent possibilities. The problem—entanglement versus locality—was amenable to experimental test. In his bones he hoped locality would win.[15]

In the years since Bell formulated his theorem, many physicists (Bell included) have tried to articulate what the violation of his inequality would mean, at a deep level, about the structure of the microworld. Most prosaically, entanglement suggests that on the smallest scales of matter, the whole is more than the sum of its parts. Put another way: one could know *everything* there is to know about a quantum system (particles A + B), and yet know *nothing* definite about either piece separately. As one expert in the field has written, entangled quantum systems are not even "divisible by thought": our natural inclination to analyze systems into subsystems, and to build up knowledge of the whole from careful study of its parts, grinds to a halt in the quantum domain.[16]

Physicists have gone to heroic lengths to translate quantum nonlocality into everyday terms. The literature is now full of stories about boxes that flash with red and green lights; disheveled physicists who stroll down the street with mismatched socks; clever Sherlock Holmes–inspired scenarios involving quantum robbers; even an elaborate tale of a baker, two long conveyor belts, and pairs of soufflés that may or may not rise.[17] My favorite comes from a "quantum-mechanical engineer" at MIT, Seth Lloyd. Imagine twins, Lloyd instructs us, separated a great distance apart. One steps into a bar in Cambridge, Massachusetts, just as her brother steps into a bar in Cambridge, England. Imagine further (and this may be the most difficult part) that neither twin has a cell phone or any other device with which to communicate back and forth. No matter what each bartender asks them, they will give opposite answers. "Beer or whiskey?"

The Massachusetts twin might respond either way, with equal likelihood; but no matter which choice she makes, her twin brother an ocean away will respond with the opposite choice. (It's not that either twin has a decided preference; after many trips to their respective bars, they each wind up ordering beer and whiskey equally often.) The bartenders could equally well have asked, "Bottled beer or draft?" or "Red wine or white?" Ask *any* question—even a question that no one had decided to ask until long after the twins had traveled far, far away from each other—and you will *always* receive polar opposite responses. Somehow one twin always "knows" how to answer, even though no information could have traveled between them, in just such a way as to ensure the long-distance correlation.[18]

•

From today's vantage point, Bell's theorem is of unparalleled significance. His proof that quantum mechanics necessarily implied nonlocality—that a measurement of particle A would instantaneously affect particle B, even if they were a galaxy apart—dramatized the philosophical stakes involved when trying to make sense of quantum reality. Bell's short article has accumulated more than 3200 citations in the professional scientific literature, an astonishing level of interest rivaled by roughly 1 out of every 10,000 physics papers ever published. Today Bell's theorem, and the entangled states at its core, is the centerpiece of everything from quantum computing, to quantum encryption, to quantum teleportation. (The special beams of light at the heart of the 2004 money transfer in Vienna consisted of entangled pairs of photons.) Without question, physicists, philosophers, and historians now see Bell's theorem, entanglement, and nonlocality as among the most important developments in quantum theory. As authors of a recent textbook put it, Bell's theorem and entanglement have become "a fundamentally new resource in the world that goes essentially *beyond* classical resources; iron to the classical world's bronze age."[19] (Fig. 2.4.)

All that lay far in the future when Bell was puzzling through his short paper back in the early 1960s. Bell worked out his theorem not at CERN,

{ FIGURE 2.4. John S. Bell in his office at CERN, 1982. (Courtesy CERN.)

but while on sabbatical in the United States. Indeed, he later recalled that it was only in the United States—where so few physicists showed any signs of interest in such topics—where he could achieve the isolation required to push through his thoughts and write up his papers. Bell left CERN in November 1963—arriving in the United States one day after John F. Kennedy had been assassinated, as it happened—and spent the year visiting the Stanford Linear Accelerator Center, the University of Wisconsin at Madison, and Brandeis University near Boston. He completed his review article on hidden variables first, and mailed it off to the *Reviews of Modern Physics*, in whose editorial office the manuscript mysteriously vanished, leading to an unheard-of two-year delay in its publication.[20]

At Brandeis he completed his second paper, "On the Einstein Podolsky Rosen paradox," containing his proof that quantum mechanics cannot be squared with locality. At the time, authors had to pay steep fees to cover the cost of publishing their articles in the venerable *Physical Review*, long the standard-bearer among the world's physics research journals. Bell was too shy to ask his American hosts to pay for an article so far removed from their research interests. So he submitted it to a brand-new journal with the curious title *Physics Physique Fizika*. Not only did the new journal waive page fees, but it actually *paid* authors to publish there—although the honoraria turned out to be nearly equal to the cost of ordering reprints. The journal's editors had high hopes that their new venture would help alleviate the information overload and hyperspecialization then afflicting the field, comparing it to a general-interest magazine like *Harper's*. In their opening editorial, the editors pledged to "try their very best to present a selection of papers which are worth the attention of all physicists." Bell's article appeared in the third issue of the fledgling journal, in November 1964.[21]

And then ... nothing. No activity or acknowledgment whatsoever. Bell's paper, deemed worthy of "the attention of all physicists" by the journal's editors, did not receive so much as a single citation in the literature for four long years—and then it was passing mention in a one-page article. Slowly, slowly, citations to Bell's paper began to appear, like the irregular clicks of a Geiger counter: six in 1971, seven in 1972, three in 1973. A burst of sustained activity began only in 1976, when twenty to thirty new articles on the topic began to appear each year. By 1980, a quite respectable 160 articles had been published in the physics literature on Bell's theorem.[22]

During the mid- and late 1970s, pockets of interest coalesced, usually led by physicists who held a longtime interest in hidden variables and the interpretation of quantum mechanics. An active group emerged around hidden-variables theorist David Bohm, whose long journey following his McCarthy-era dismissal from Princeton had ended with him settled at Birkbeck College in London, following hops and skips to São Paulo, Brazil; the Technion Institute in Haifa, Israel; and Britain's Bristol University. A separate group clustered around Louis de Broglie and Jean-Paul Vigier in Paris; and a third group, spearheaded by Franco Selleri, shuttled among Bari, Catania, and Florence in Italy. Most of these physicists had been working on hidden variables and the interpretation of quantum mechanics for decades; Bell's theorem appeared an obvious extension of their long-standing interests. Acknowledgments in these many articles show a tight fabric of social interactions: members of each of these groups knew each other, frequently traded tips and critiques, and saw each other's latest papers as preprints long before they appeared in the journals. By 1980, in other words, an "invisible college" devoted to Bell's theorem had emerged, with centers of activity dotted throughout Western Europe.[23]

Surprisingly, the largest share of articles on Bell's theorem during this period came from physicists working in the United States—27 percent of all the articles, in fact, compared with 7 percent, 14 percent, and 19 percent from authors based in Britain, France, and Italy, respectively. All this despite the absence of any deep interest in foundational topics on American soil, hidden variables or otherwise. Nearly three-quarters of these U.S.-based articles (72 percent), meanwhile, came from regular participants in the Fundamental Fysiks Group, the earliest sessions of which had been devoted to Bell's work and quantum nonlocality. (If one includes authors who acknowledged help from members of the Fundamental Fysiks Group, the proportion rises to 86 percent.) Members of the ragtag discussion group proved to be among the most prolific early authors on Bell's theorem in the world. Against all odds, the earliest champions of Bell's theorem congregated in that most unphilosophical of spaces: a large seminar room in the Lawrence Berkeley Laboratory.[24]

Chapter 3

Entanglements

> The Fundamental Fysiks Group grew fast. . . . We played a big role
> at the laboratory, but never an acknowledged role.
>   —**George Weissmann, 2008**

Members of the Fundamental Fysiks Group followed twisting paths to Bell's theorem. Most were members of the Sputnik generation, and their careers had begun to unfold in the usual way. They had been drawn to the big metaphysical questions at the heart of modern physics, only to find their coursework crushingly pragmatic. Many graduated just as the bottom fell out of the physics profession, thwarting their expected career trajectories. One way or another they made their way to Berkeley, and to each other.

Group members' experiences map larger features of the professional terrain as physicists' Cold War bubble burst. The sheer variety of their experiences and the fast pace of change illuminate what daily life was like for young physicists caught in the tumult of the late 1960s and early 1970s. Their research interests, job prospects, and even senses of themselves and of what a life in physics could mean were refracted through an unusually turbulent time. Though they arrived at the Fundamental Fysiks Group by different routes, they shared a faith that deep philosophical questions, such as the implications of Bell's theorem and quantum entanglement, were worth asking.[1]

•

John Clauser sat through his courses on quantum mechanics as a graduate student at Columbia University in the mid-1960s, wondering when they would tackle the big questions. Like John Bell, Clauser quickly learned to keep his mouth shut and pursue his interests on the side. He buried himself in the library, poring over the EPR paper, Bohm's articles on hidden variables, even Hans Freistadt's early review article. Then in 1967 he stumbled upon Bell's paper in *Physics Physique Fizika*. The journal's strange title had caught his eye, and while lazily leafing through the first bound volume he happened to notice Bell's article. Clauser, a budding experimentalist, realized that Bell's theorem could be amenable to real-world tests in a laboratory. Excited, he told his thesis advisor about his find, only to be rebuffed for wasting their time on such philosophical questions. Soon Clauser would be kicked out of some of the finest offices in physics, from Robert Serber's at Columbia to Richard Feynman's at Caltech. Bowing to these pressures, Clauser pursued a dissertation on a more acceptable topic—radio astronomy and astrophysics—but in the back of his mind he continued to puzzle through how Bell's inequality might be put to the test.[2]

Before launching into an experiment himself, Clauser wrote to John Bell and David Bohm to double-check that he had not overlooked any prior experiments on Bell's theorem and quantum nonlocality. Both respondents wrote back immediately, thrilled at the notion that an honest-to-goodness experimentalist harbored any interest in the topic at all. As Bell later recalled, Clauser's letter from February 1969 was the first direct response Bell had received from any physicist regarding Bell's theorem—more than four years after Bell's article had been published. Bell encouraged the young experimenter: if by chance Clauser did manage to measure a deviation from the predictions of quantum theory, that would "shake the world!"[3]

Encouraged by Bell's and Bohm's responses, Clauser realized that the first step would be to translate Bell's pristine algebra into expressions that might make contact with a real experiment. Bell had assumed for simplicity that detectors would have infinitesimally narrow windows or apertures through which particles could pass. But as Clauser knew well from his radio-astronomy work, apertures in the real world are always wider than a mathematical pinprick. Particles from a range of directions would be able to enter the detectors at either of their settings, a or a'. Same for detector efficiencies. Bell had assumed that the spins of every pair of particles would be measured, every time a new pair was shot out from the source. But no laboratory detectors were ever 100 percent efficient; sometimes one or both particles of a pair would simply escape detection altogether. All these complications and more had to be tackled on paper, long before one bothered building a machine to test Bell's work. Clauser dug in and submitted a brief abstract on this work to the *Bulletin of the American Physical Society*, in anticipation of the Society's upcoming conference. The abstract appeared in print right before the spring 1969 meeting.[4]

And then his telephone rang. Two hundred miles away, Abner Shimony had been chasing down the same series of thoughts. Shimony's unusual training—he held PhDs in both philosophy and physics, and taught in both departments at Boston University—primed him for a subject like Bell's theorem in a way that almost none of his American physics colleagues shared. He had already published several articles on other philosophical aspects of quantum theory, beginning in the early 1960s.[5] Shimony had been tipped off about Bell's theorem back in 1964, when a colleague at nearby Brandeis University, where Bell had written up his paper, sent Shimony a preprint of Bell's work. Shimony was hardly won over right away. His first reaction: "Here's another kooky paper that's come out of the blue," as he put it recently. "I'd never heard of Bell. And it was badly typed, and it was on the old multigraph paper, with the blue ink that smeared. There were some arithmetical errors. I said, 'What's going on here?'" Alternately bemused, puzzled, and intrigued, he read it over again and again. "The more I read it, the more brilliant it seemed. And I realized, 'This is no kooky paper. This is something very great.'" He began scouring the literature to see if some previous experiments, conducted for different purposes, might already have inadvertently put Bell's theorem to the test. After intensive digging—he came to call this work "quantum archaeology"—he realized that, despite a few near misses, no existing data would do the trick. No experimentalist himself, he "put the whole thing on ice" until he could find a suitable partner.[6]

A few years went by before a graduate student came knocking on Shimony's door. The student had just completed his qualifying exams and was scouting for a dissertation topic. Together they decided to mount a brand-new experiment to test Bell's theorem. Several months into their preparations, still far from a working experiment, Shimony spied Clauser's abstract in the *Bulletin*, and reached for the phone. They decided to meet at the upcoming American Physical Society meeting in Washington, DC, where Clauser was scheduled to talk about his proposed experiment. There they hashed out a plan to join forces. A joint paper, Shimony felt, would no doubt be stronger than either of their separate efforts alone would be—the whole would be greater than the sum of its parts—and, on top of that, "it was the civilized way to handle the priority question." And so began a fruitful collaboration and a set of enduring friendships.[7]

Clauser completed his dissertation not long after their meeting. He had some downtime between handing in his thesis and the formal thesis defense, so he went up to Boston to work with Shimony and the (now two) graduate students whom Shimony had corralled into the project. Together they derived a variation on Bell's theme: a new expression, more amenable to direct comparisons with laboratory data than Bell's had been. Even as his research began to hum, Clauser's employment prospects grew dim. He graduated just as the chasm between demand and supply for American physicists opened wide. He further hindered his chances by giving a few job talks on the subject of Bell's theorem. Clauser would later write with great passion that in those years, physicists who showed any interest in the foundations of quantum mechanics labored under a "stigma," as powerful and keenly felt as any wars of religion or McCarthy-like political purges.[8]

Finally Berkeley's Charles Townes offered Clauser a postdoctoral position in astrophysics at the Lawrence Berkeley Laboratory, on the strength of Clauser's dissertation on radio astronomy. Clauser, an avid sailor, planned to sail his boat from New York around the tip of Florida and into Galveston, Texas; then he would load the boat onto a truck and drive it to Los Angeles, before setting sail up the California coast to the San Francisco Bay Area. (A hurricane scuttled his plans; he and his boat got held up in Florida, and he wound up having to drive it clear across the country instead.) All the while, Clauser and Shimony hammered out their first joint article on Bell's theorem: each time Clauser sailed into a port along the East Coast, he would find a telephone and check in with Shimony, who had been working on a draft of their paper. Then Shimony would mail copies of the edited draft to every marina in the next city on Clauser's itinerary, "some of which I picked up," Clauser explained recently, "and some of which are probably still waiting there for all I know." Back and forth their edits flew, and by the time Clauser arrived in Berkeley in early August 1969, they had a draft ready to submit to the journal.[9]

Things were slow at the Lawrence Berkeley Laboratory compared to the boom years, and budgets had already begun to shrink. Clauser managed to convince his faculty sponsor, Townes, that Bell's theorem might merit serious experimental study. Perhaps Townes, an inventor of the laser, was more receptive to Clauser's pitch than the others because Townes, too, had been told by the heavyweights of his era that his own novel idea flew in the face of quantum mechanics.[10] Townes allowed Clauser to devote half his time to his pet project, not least because, as Clauser made clear, the experiments he envisioned would cost next to nothing. With the green light from Townes, Clauser began to scavenge spare parts from storage closets around the Berkeley lab—"I've gotten pretty good at dumpster diving," as he put it recently—and soon he had duct-taped together a contraption capable of measuring the correlated polarizations of pairs of photons. (Photons, like electrons, can exist in only one of two states; polarization, in this case, functions just like spin as far as Bell-type correlations are concerned.) In 1972, with the help of a graduate student loaned to him at Townes's urging, Clauser published the first experimental results on Bell's theorem.[11] (Fig. 3.1.)

Despite Clauser's private hope that quantum mechanics would be toppled, he and his student found the quantum-mechanical predictions to be spot-on. In the laboratory, much as on theorists' scratch pads, the microworld really did seem to be an entangled nest of nonlocality. He and his student had managed to conduct the world's first experimental test of Bell's theorem—today such a mainstay of frontier physics—and they demonstrated, with cold, hard data, that measurements of particle

FIGURE 3.1. John Clauser and his contraption to test Bell's theorem at Berkeley, mid-1970s. (Courtesy Lawrence Berkeley National Laboratory.)

A really were more strongly correlated with measurements of particle B than any local mechanisms could accommodate. They had produced exactly the "spooky action at a distance" that Einstein had found so upsetting. Still, Clauser could find few physicists who seemed to care. He and his student published their results in the prestigious *Physical Review Letters*, and yet the year following their paper, global citations to Bell's theorem—still just a trickle—dropped by more than half.[12] The world-class work did little to improve Clauser's job prospects, either. One department chair to whom Clauser had applied for a job doubted that Clauser's work on Bell's theorem counted as "real physics."[13]

●

Elizabeth Rauscher paid attention to Clauser's results. By the early 1970s, Rauscher, a physics graduate student at the Lawrence Berkeley Laboratory, had developed a reputation around the lab somewhat similar to Clauser's: both enjoyed asking unusual questions. A professor introduced them while Clauser was enmeshed in his first experiment on Bell's theorem. Soon Rauscher was bringing others over to see Clauser's makeshift laboratory.[14]

Rauscher grew up in the Berkeley area. For as long as she can remember, she has been passionate about science. As early as age four she began to study nature. She recalls sitting outside to watch how the grass grows, or gazing up at night, mesmerized by the flickering stars. She loved to get her hands dirty trying to figure things out. She designed, built, and reassembled her own telescopes. Sometimes when she asked her parents questions, they didn't know the answers. That taught her an important lesson: "people don't just automatically grow up and know stuff." Even as a young girl, she realized that she would have to study hard. By the time she reached high school, she began to hang out at the Lawrence Berkeley Laboratory, eager to soak in all the science she could. Back in those days people rarely checked for badges or identification on the lab's shuttle bus, so she used to board the bus looking as confident as she could, trying to convince the others that she belonged there and knew where she was going. More often than not it worked, and she enjoyed traipsing around the laboratory to see what there was to see.[15]

Early on, Rauscher decided to study the very small and the very large, or, as she soon learned to specify, atoms and galaxies. She assumed that human consciousness—yet another facet of existence that she found endlessly fascinating—would lie somewhere in between. When the time came for college Rauscher enrolled at Berkeley and dove into her studies of physics. She published her first scientific article, on nuclear fusion, as an undergraduate, and completed a master's degree in nuclear physics early in 1965. Then other realities set in. She got married, had a son, and became the sole provider for herself and her family. To pay the bills, she took a job as a staff scientist at the Lawrence Livermore Laboratory, a weapons laboratory near Berkeley dating from the early years of the Cold War. Soon after arriving, she witnessed firsthand some of the earliest layoffs that presaged the physicists' coming employment crisis. She managed to hang on to her job at Livermore and, once her son was old

FIGURE 3.2. Elizabeth Rauscher in the control room of the Bevatron particle accelerator at the Lawrence Berkeley Laboratory during her graduate studies, 1977. (Courtesy Elizabeth Rauscher.)

enough for full-day schooling, she returned to school herself, pursuing her doctorate in physics at Berkeley. She worked at the Lawrence Berkeley Laboratory—at last an official member of the lab, no longer faking her way in as a curious and pugnacious high school student. (Fig. 3.2.) Her advisor was the legendary nuclear chemist Glenn Seaborg: Manhattan Project veteran, discoverer of plutonium, Nobel laureate, and recently retired director of the Atomic Energy Commission.[16]

Rauscher found herself asking questions in class that none of her male colleagues would broach. Since her undergraduate days, she had gotten used to being the only woman in a large class of men. At that time, nationwide, women earned only 5 percent of the undergraduate degrees in physics, and just 2 percent of the physics PhDs.[17] Rauscher was a rarity indeed. She had learned to cope by wearing tweedy dresses and keeping her hair short, but still she stuck out. The dowdy clothes could not disguise the fact that she was something of a free spirit. At one point some of the men in the department complained that she sang too loudly in the halls. As a graduate student, she received anonymous threatening telephone calls; her laboratory work was sabotaged. But she realized that one couldn't do anything in physics without a PhD, so she buckled down and decided not to let the intimidation curtail her ambition or her curiosity.[18]

While still at Livermore, for example, she joined (and soon chaired) the Livermore Philosophy Group, also known as the "Tuesday Night Club." As a flyer from 1969 advertised, the informal group sought to "enjoy the closeness and warmth of friendship, to laugh a little (a solely human response) and find a small meaning or measure of joy in lighthearted humor"—all within a government laboratory founded solely to design and test nuclear weapons. Once back at Berkeley as a graduate student, Rauscher continued in the same vein. She began offering short summer courses at the laboratory under the rubric "philosophy of science," focused on the relations between science and society. The lectures and "rap sessions" revolved around many of the same freighted questions that had energized young physicists and radicalized students across the country: "Who determines the direction of scientific developments?" and "What is the motivation and purpose behind LBL [Lawrence Berkeley Laboratory] and other national laboratories?" Rauscher repeated the popular course several summers at Berkeley (eventually under the umbrella of the University Extension program), as well as at the Stanford Linear Accelerator Center.[19]

In keeping with her broad interests, she wandered into Arthur Young's Institute for the Study of Consciousness in Berkeley soon after it opened, still puzzling through questions that she had first posed as a child. What was the nature of human thought, and where was its place in the cosmos? Young, an inventor and aeronautical engineer, had become captivated by all things occult: Eastern mysticism, alchemy, extrasensory perception. He recruited another young physics student, Saul-Paul Sirag, as soon as he opened his institute in August 1973, so the two could talk about physics and what it might have to say about these farther-flung topics. After crossing paths at the institute, Rauscher and Sirag launched into their own series of conversations.[20]

Sirag's journey to the institute had been a long one. Born in Dutch Borneo to missionary parents, he and his family spent nearly three years in Japanese internment camps on Java during World War II. (Only after the war did they learn that Sirag's father had perished in one of the camps; the men had been separated from the women and children.) Sirag made his way to Berkeley in 1961 to study physics as an undergraduate, but dropped out during his senior year to pursue theater throughout the Bay Area and New York City. He grew his hair long, looking like a cross between Albert Einstein and Jimi Hendrix. (Fig. 3.3.) When a befuddled reporter for the *New York Daily News* covered a local-interest story early in the "summer of love"—June 1967—Sirag was right in the thick of it. "Police were nowhere to be found" at Tompkins Square Park in New York's lower East Side, the reporter noted, "as some 500 weirdly dressed patrons of love-ins clapped their hands in rhythm as they marched behind 'The Grateful Dead,' a guitar playing group of mangy hippies from the West Coast." The newspaper ran a huge photo of Sirag ("a bearded, wild-haired apostle of the hip group") as he took in "the scene," to accompany the story.[21] His efforts in avant-garde theater nabbed a little notoriety as well. At least twice the *Village Voice* spotlighted Sirag in some of his local theatrical productions.[22] His big chance came when he was cast for a part in the original Broadway production of *Hair*, which opened in 1968; he had already acted in smaller venues for the play's director. But Sirag failed to scrounge up the $600 required to join the Actors' Equity union, so he was unable to accept the part.[23]

On the heels of that disappointment, Sirag returned to Berkeley. He sat in on classes at the university without enrolling, and he made ends meet by picking up freelance writing jobs for small underground newspapers. He was fascinated by modern physics—often the subject of his newspaper columns—and he leapt at the chance to work with Arthur Young at his new Institute for the Study of Consciousness. The institute provided more than cerebral stimulation; it doubled as living quarters for the perennially cash-strapped Sirag.[24]

Around the same time, Sirag met another soul on the quest for foundations at Young's institute: Nick Herbert. Herbert had studied

FIGURE 3.3. Saul-Paul Sirag catching up on the news, late 1970s. (Courtesy Nick Herbert.)

engineering physics as an undergraduate at Ohio State University in the late 1950s. While he was working on an internship at the Redstone Arsenal in Huntsville, Alabama, the Soviets launched Sputnik. (In Huntsville, Herbert enjoyed a private discussion with Wernher von Braun, then leading Redstone's effort to launch the first American satellite, on the basics of ballistic rocketry.) By the time Herbert graduated, at the top of his class, fellowships for graduate study in physics had become plentiful, and he decamped for Stanford in the fall of 1959. At Stanford he learned "no-nonsense 'shut up and calculate' quantum mechanics"; his course used Leonard Schiff's famously pragmatic textbook. He toiled away on nuclear scattering experiments, working with a small basement-bound particle accelerator, before completing his PhD in 1967.[25]

By that time the job market for physicists had begun to constrict. Herbert landed a one-year replacement job, teaching physics at tiny Monmouth College in Illinois. After that it was back to California with no job in sight. Desperate for income, he faked his resumé—listing make-believe experience as an electronics technician—and got a job on an assembly line manufacturing telephone equipment. Taking sick leaves, he snuck out periodically to interview for other jobs, including an opening for an industrial physicist at Memorex, the Silicon Valley consumer electronics manufacturer. When Herbert showed up "looking like an insane hippie," the personnel manager insisted that Herbert get screened by a San Francisco psychologist before taking a chance on hiring him. Herbert checked out okay and got the job at Memorex, just before the company hit a rough financial patch and fired most of its research staff (Herbert included). With effort Herbert next landed a job with the Smith-Corona Marchant Corporation in Palo Alto, California, working on then-new techniques for photocopying and ink-jet printing.[26]

During his off hours, Herbert daydreamed about all those big questions he had never seen covered in his physics classes. He had worked hard to master some of the beautiful, arcane mathematics of quantum theory. But the math had not clarified one basic question: what the hell did it all mean? How could the world of atoms be nothing but a puff of probabilities, and yet conglomerations of those atoms could create something as strong and unbending as the chair on which he sat?[27]

Around 1970, while those questions swirled in his head, one of Herbert's former roommates from graduate school introduced him to Bell's 1964 paper. Herbert was immediately fascinated. His gut reaction told him that the paper must be wrong, and he set out to disprove it. In the process he became one of Bell's earliest converts. By 1972, he had begun delivering talks about Bell's theorem and nonlocality throughout the Bay Area. Within a few months, he had produced the shortest derivation of Bell's theorem, pruned to its essentials. Arthur Young invited Herbert to give a talk at the grand opening of his Institute for the Study of Consciousness, and Herbert and Sirag immediately hit it off. Delighted by their shared interests in the foundations of quantum mechanics, Rauscher in turn introduced Herbert and Sirag to John Clauser, the experimentalist then enmeshed in trying to test Bell's theorem. Soon they were traipsing off together to inspect Clauser's apparatus and talk about the meaning of the experimental results. As Sirag recalls, Clauser was always eager to talk about Bell's theorem; at last a few people had begun to show some interest.[28] (Fig. 3.4.)

FIGURE 3.4. Nick Herbert explaining his ideas about quantum physics and consciousness, late 1970s. (Courtesy Nick Herbert.)

Clauser, Rauscher, and company also had the ear of Henry Stapp, a senior theoretical physicist on staff at the Lawrence Berkeley Laboratory. In fact, Stapp was in all likelihood the first physicist in the United States to pay attention to Bell's theorem. He had joined the Theory Division at the lab right from his Berkeley PhD back in the mid-1950s, and proceeded to make a name for himself in the area of theoretical particle physics. Early in his career he visited Zurich on the invitation of Wolfgang Pauli, the Nobel laureate who had helped build quantum theory during the 1920s and 1930s and who explored the subject's mysteries with psychoanalyst Carl Jung. Later Werner Heisenberg, of uncertainty-principle fame, invited Stapp to spend a few months in Munich. These European forays stoked Stapp's interest in philosophical matters, well beyond what his Berkeley training had inspired, and by the late 1960s Stapp had begun to dabble in the foundations of quantum mechanics.[29]

Stapp's dissertation had focused on spin correlations in proton-proton scattering experiments—his roommate in graduate school, an experimentalist, worked on the Nobel Prize–winning discovery of the antiproton in 1955—so Stapp was unusually attuned to the types of correlations at the heart of Bell's theorem. Somehow he noticed Bell's paper, off in its obscure journal. Around the same time, he was invited to contribute a chapter to a book entitled *Quantum Physics and Beyond*, which another physicist was putting together. Stapp used the opportunity to write about Bell's theorem. He submitted his paper, entitled "Correlation experiments and the nonvalidity of ordinary ideas about the physical world," in July 1968; its sole reference was to Bell's still-unknown 1964 article. The book's editor asked Stapp to shorten his paper, and "in a huff," Stapp pulled the piece altogether. When the book appeared three years later, no mention of Bell, entanglement, or nonlocality graced its 300 pages. Years later, Stapp released his 1968 preprint as an official report of the Lawrence Berkeley Laboratory. By that time, Bell's theorem and its implications had moved front-and-center of Stapp's research interests. He began lecturing on the topic around the Bay Area as early as 1970, and kept up a steady stream of publications on the topic.[30]

Stapp's investigations captured the imagination of George Weissmann, like Elizabeth Rauscher a graduate student in the Theory Division at Lawrence Berkeley Laboratory. Like so many of the others, Weissmann's path to Berkeley and to Bell's theorem had been anything but straightforward. From his earliest childhood, Weissmann had wanted to become a scientist. "I had since early youth seen science as the source for ultimate truth—this is my confession," he later explained.[31] Weissmann studied physics and mathematics as an undergraduate in his native Zurich, then traveled to Imperial College in London for a year of postgraduate study. He had always been interested in Einstein and his ideas about space and time; he focused on Einstein's general relativity in London and planned to continue in that vein for his doctoral studies. He applied to the top programs in the United States, especially those few that specialized in relativity like Princeton and Caltech. He also applied to Berkeley, more as a hedge than out of any special interest in the department. He was told that he needed to take the standardized test known as the Graduate Record Examination (GRE) as part of his application process, which he dutifully did. Unbeknownst to him, however, a postal strike in Britain meant that most of the American schools to which he had applied never received his test scores. Berkeley's physics department alerted him about the missing scores; he express-mailed a new copy and was accepted. The other departments simply rejected his application for being incomplete. And so Weissmann arrived in Berkeley in 1971.[32]

Like the others, Weissmann was deeply frustrated by the lack of engagement with foundational questions in his coursework. It was all "turn-the-crank stuff." He once asked his professor in a quantum mechanics course, "What is an electron?" The response—"Well, you can think of it as a wavefunction..."—hardly satisfied. "Quantum theory was deeply puzzling and mysterious," as he put it recently. "What was it telling us about the nature of the world? That question was never even raised." He gravitated toward the Berkeley staff scientist Henry Stapp, already deep into his work on the interpretation of quantum theory, and they engaged in long and involved discussions. "Stapp really taught me quantum theory," Weissmann recalls. He was similarly attracted to the work of Geoffrey Chew, a leading particle physicist and at the time the director of the Lawrence Berkeley Laboratory's Theory Division. Chew became Weissmann's main thesis advisor.[33]

During "secret seminars" in the Theory Division—an informal discussion group that Chew had devised back in the 1950s to encourage his students to work together without fear of faculty intrusion—Weissmann found himself chatting more and more with fellow-student Elizabeth Rauscher. Their discussions spilled beyond Chew's seminar. They discovered their shared interest in whether physics might someday clarify the workings of human consciousness. In short order, they decided to hold those conversations outside—literally between the two main physics buildings at the laboratory, out of earshot of their colleagues and professors—because philosophically rooted topics like the nature of consciousness were deemed "verboten" by mainstream physicists. They knew they were onto something. But they could sense that their wide-ranging interests—"very deep issues of the meaning of life and death, and existence and non-existence, and everything in between, from the macrocosmos to the microcosmos"—would not find a welcome home in a traditional physics environment.[34]

Rauscher's and Weissmann's advisor, Geoffrey Chew, proved more open-minded than most. Although he did not always share his students' enthusiasms, he encouraged their pursuits. At one point he received some essays from a then-unknown postdoc working in Europe named Fritjof Capra. Capra's essays explored parallels between modern physics and Eastern thought. Chew was not particularly interested; he passed them along to Rauscher and Weissmann with the comment, "This looks like your kind of stuff." It was indeed. Rauscher got so excited that she helped convince Chew to invite Capra to visit the Theory Division at the Berkeley lab as an unpaid associate. Chew agreed, and Capra visited with Chew and his group during the fall of 1973. He returned for a long-term stay beginning in April 1975.[35]

●

The last two core members of what would become the Fundamental Fysiks Group made their way to Berkeley from southern California—by way of Europe. Their career paths and intellectual identities had come unmoored by the great dislocations that shook the physics profession in the late 1960s and early 1970s. In their roundabout way, together they found their way to Bell's theorem and to a re-engagement with the types of questions that had attracted them to physics in the first place.

Fred Alan Wolf completed his undergraduate studies in engineering physics at the University of Illinois in 1957 and went straight to graduate school at the University of California at Los Angeles. When his advisor left campus on sabbatical in 1962, Wolf moved to the Lawrence Livermore Laboratory, the defense laboratory near Berkeley, to complete his dissertation. (He narrowly missed overlapping with Elizabeth Rauscher at the lab.) From there he was hired by General Atomic in La Jolla, a private defense contracting firm that specialized in all things nuclear. The group was just gearing up for its work on Project Orion, an audacious plan to use exploding H-bombs to power a rocket ship. Wolf received Q clearance, the top-secret rating reserved for nuclear weapons work, and dove in.[36] By 1964, it looked like he was set: he was able to lead a comfortable, middle-class lifestyle. He had gotten married; soon he and his wife had two children and a nice house in La Jolla, their suburban dream come true. Yet after a year or two at General Atomic, both Wolf and his supervisor realized that his heart just wasn't in the project. He had been pleased to see that a computer program he had written for the project, which simulated the behavior of a superhot gas of electrons and ions in the core of the exploding nuclear engines, produced good fits with the empirical data coming out of the laboratory. But on the whole, the project struck Wolf as too much gadgeteering: one more grand (even hubristic) monument of Cold War technology, but not one that would engage the really fundamental, metaphysical questions at the heart of modern physics.[37]

Just as he was beginning to lose enthusiasm for life at General Atomic, a recruiter came around looking for physicists to join the faculty at nearby San Diego State College (now University). The school was making a push to increase the proportion of its faculty engaged in top-flight research. Wolf decided to give teaching a try.[38] Demand was still high for physicists in those years, and San Diego State made it clear that they were glad to have him. Wolf enjoyed his colleagues and he enjoyed his teaching. (Fig. 3.5.) Like so many physicists at the time, he picked up some extra consulting work on the side: first for a U.S. Navy electronics laboratory in San Diego; next for the Aerospace Research Laboratory at the Air Force's Wright Patterson Base in Ohio. But things began to sour soon after Ronald Reagan became governor of California in 1967. "Suddenly all of politics became centered around tax relief," Wolf recalled recently. "And one way to achieve tax relief was to start cutting services, such as funding to the schools." Indeed, Reagan's first order of business was to demand an immediate 30 percent budget cut from the University of California, with comparable spending cuts across the state's higher education system. Faculty lines at San Diego State went unfilled when senior colleagues retired; class sizes began to rise; funding for teaching assistants began to dry up. Within a few years, Wolf says, he realized that "the apple was rotten."[39]

Other things began to happen around that time, too. Starting in 1970, strangers began dropping by his campus office—some of them students, others from the surrounding community—eager to tell him all about their

FIGURE 3.5. Fred Alan Wolf teaching quantum theory as a professor at San Diego State College, late 1960s. (Courtesy Fred Alan Wolf.)

"psychic visions" and "past-life reminiscences." He had never telegraphed any interest in such topics. Perhaps, he mused, he was the only professor who did not immediately kick such visitors out of his office, so word got around. In any case, the visits stirred something inside him. He began to reflect on his own physics training. "Somewhere in all of that education," he later wrote, "I had lost the magic. I simply accepted the physics education as an indoctrination." The narrow focus on learning how to "manipulate the math" had steered him away from pondering bigger questions—after all, "most physicists at that time dismissed philosophical questions as not worthy of thought." Increasingly frustrated by the cutbacks at San Diego State, and freshly inspired to explore deeper meanings in the heart of modern physics, he embarked on a worldwide tour during a 1971 sabbatical. He had lined up visiting appointments in Jerusalem, Berlin, and Paris. But his first stops were to India and Katmandu, where he had his first transcendental experience in a Buddhist temple.[40]

Soon after his return to San Diego, Wolf received an invitation to spend six months at Birkbeck College in London. The funding and invitation had been arranged by a fellow specialist in atomic physics. Most important for Wolf, however, was the chance to meet David Bohm, the American physicist who had been forced out of the United States during the height of the 1950s red scare, just as he was piecing together his hidden-variables alternative to quantum mechanics. (Bohm's work had inspired the young John Bell to tackle the question of quantum nonlocality.) While visiting at Birkbeck during the summer and fall of 1973, Wolf learned the ins and outs of Bohm's hidden-variables program; he also gleaned his first inklings about Bell's theorem and nonlocality. A colleague at the University of Paris, meanwhile, asked Wolf if he might fill in and teach some courses there, starting in January 1974. Wolf, still frustrated by the quick reversal of fortunes at San Diego State, gladly accepted. He remained in Europe, shuttling between Paris and London, for the next year.[41]

A few months after Wolf returned to San Diego State from his European sabbatical, the colorful antiwar protester and head "Yippie," Jerry Rubin, visited campus. Wolf had already met Rubin the previous year during his travels. Rubin asked Wolf what he was doing there; Wolf reminded Rubin that he was a professor of physics on the faculty. "Man, you don't belong here," came Rubin's reply. Not long thereafter Wolf squabbled with his department chair over what courses Wolf would teach the following year. With Rubin's words still ringing in his ears, Wolf decided he had had enough. He told off his department chair, marched straight to the dean's office, and tendered his resignation. From there he made his way up to the San Francisco area and fell in with the other members of the Fundamental Fysiks Group.[42]

Wolf shared many steps on this journey with his friend and San Diego State colleague Jack Sarfatti. Like Wolf, Sarfatti had started out along a then-familiar path: undergraduate physics major at Cornell; a short stay with the U.K. Atomic Energy Research Establishment at Harwell; followed by a master's degree from the University of California at San Diego in 1967 and a PhD from the University of California at Riverside in 1969. As a graduate student he attended a NATO summer institute on nonlinear physics in Munich, where he brushed shoulders with leaders of the field from the United States and Europe. Just three years later, he began lec-

FIGURE 3.6. Jack Sarfatti teaching physics at San Diego State College, early 1970s. (Courtesy Jack Sarfatti.)

turing in a series of physics summer schools sponsored by the National Science Foundation. By the early 1970s, having published a few articles in prestigious journals on quantum theory, elementary particles, and even some idiosyncratic ideas about miniature black holes, Sarfatti could list half a dozen distinguished physicists scattered across the United States, Britain, and France as references to vouch for the quality of his work.[43]

Sarfatti was hired right out of graduate school to teach at San Diego State, and given the office next to Wolf's. (Fig. 3.6.) Within a few years, the two were sharing quarters at home, too: much like the television sitcom *The Odd Couple*, they each got divorced around the same time and moved in together to share the rent. Fun times ensued. At one point, Wolf and Sarfatti borrowed a home-movie camera to shoot a short film together, along with students from one of Wolf's classes. A frolicking piece, Wolf and Sarfatti joked that it had been filmed by a blind Argentinian director. Shot on the beach in San Diego in 1971, the film explores themes of forbidden knowledge and the intersections of science and religion. Wolf wanders the beach in rabbinical garb; Sarfatti, clad only in a loincloth, struts around as Jesus Christ.[44] (Think Federico Fellini meets Mel Brooks.)

Again, like Wolf, Sarfatti began to lose enthusiasm for his position at San Diego State during the early 1970s, and indeed for the sterile direction in which he saw theoretical physics heading. He announced his new plans in a letter to renowned Princeton physicist John Wheeler in the spring of 1973. (Sarfatti had met Wheeler a few years earlier at one of the NATO summer schools.) Sarfatti declared that he would leave his "uninspiring institution" and seek out "the best possible environment to create a great and historic piece of physics. I feel impelled by *history*—a certain sense of destiny," he explained. ("I recognize that I may be suffering under some sort of 'crackpot' delusion, but I cannot accept that as likely. In any case, I must try," he averred.) He longed to find one of the "*few* places left where physics has not been 'polluted' by the emphasis on applications, etc."; some place where bold ideas on fundamental questions could still find a home.[45]

As if responding to this cri de coeur, the physics gods smiled on Sarfatti. Right around the time that Wolf's invitation to Birkbeck College in London arrived, Sarfatti received a telegram from Abdus Salam. Salam was director of the International Centre for Theoretical Physics in Trieste, Italy, and would soon win the Nobel Prize for his contributions to theoretical particle physics. Sarfatti had met Salam at Harwell in the mid-1960s, and Salam had been following some of Sarfatti's publications since then. In his telegram, Salam invited Sarfatti to spend the autumn of 1973 at the Centre in Trieste. And so, as Sarfatti put it, "like Bob Hope and Bing Crosby in the movies, 'On the Road to . . . ,' both Fred and I were unexpectedly on our way to Europe." They visited each other frequently, Sarfatti often dropping by London or Paris and staying with Wolf.[46] (Fig. 3.7.)

Quickly the paths converged. Saul-Paul Sirag read a paper by Sarfatti, written while Sarfatti was still in Europe, and told Elizabeth Rauscher about it. Rauscher struck up a correspondence with Sarfatti while he was in Trieste, and Sarfatti dropped by Arthur Young's Institute for the Study of Consciousness as soon as he returned to California. Meanwhile, Sarfatti had already met Fritjof Capra in Europe; they overlapped in London and Trieste. By spring 1975, with Wolf and Sarfatti back from Europe and Capra installed in Berkeley, all the pieces were in place. As Rauscher put

FIGURE 3.7. Jack Sarfatti (*left*) and Fred Alan Wolf (*right*) thinking big thoughts in Paris, 1974. (Courtesy Fred Alan Wolf.)

it recently, she had "the idea that it would be easier to learn about all this material"—nonlocality and its broader implications—"if we got together for informal discussions and lectures."[47]

As Rauscher had done years earlier with the Tuesday Night Club at Livermore, she quickly developed a routine. Every week she would meet with the Lawrence Berkeley Laboratory director. Given the general state of affairs at the laboratory, with morale chasing the budget cuts in a downward spiral, Rauscher usually had to listen to the director grumble at the end of yet another bad day. During a pause in the grousing, she would ask if she could reserve a large seminar room at the laboratory—in fact, the room that had once been the office of the laboratory's founder, the great Ernest O. Lawrence himself. With the lab director's blessing, she would next race upstairs to the audiovisual department to borrow an overhead projector, in case any of the discussion participants felt the need to share images with the group. And then, late each Friday afternoon, the room would begin to fill up with stragglers and seekers. "I had figured, if I'm going to figure out reality, I'd better get some teamwork going here, and that's what I did," Rauscher explained. And so, in May 1975, the Fundamental Fysiks Group was born.[48]

From Ψ to Psi

> In my opinion, the quantum principle involves *mind* in an essential way [. . . such that] *the structure of matter may not be independent of consciousness!* . . . Some component of the quantum probability involves the turbulent creative sublayer of ideas in the mind of the "participator."
>
> —Jack Sarfatti, 1974

Members of the Fundamental Fysiks Group were certainly fascinated, even mesmerized, by Bell's theorem and nonlocality. Yet when it came to what Einstein had called "spooky actions at a distance," most members concluded that Einstein hadn't known the half of it. For the Fundamental Fysiks Group had been founded not, in the first instance, to explore the meaning of Bell's theorem, but to plumb the foundations of quantum mechanics in search of explanations for parapsychological, or "psi," phenomena: extrasensory perception, psychokinesis, the works. For most members of the group, Bell-style nonlocality seemed tailor-made to explain curious, occultlike actions at a distance. Their interests in Bell's theorem and in psi phenomena blossomed side by side.

The young physicists of the Fundamental Fysiks Group launched their quest at a propitious moment. The Central Intelligence Agency, the Pentagon, and several defense laboratories across the United States were each hard at work on psi, spurred by fears of Soviet-bloc advances in mind reading and mind control. Leading representatives of the military-industrial complex had been on the parapsychology trail even before long-haired hippies embraced the New Age occult scene.[1] Like psychedelic drugs in the 1960s, which had likewise spread from quintessential Cold War settings to the wide and inchoate youth movement, the Bay

Area witnessed a strange conjunction in the early 1970s: cloak-and-dagger spycraft entwined with the latest enthusiasms of the flower children. In the middle of it all sat the Fundamental Fysiks Group.

•

Early on, members of the Fundamental Fysiks Group drew up a roadmap for their discussions. "Quantum reality" led inexorably to all manner of New Age speculations.[2] (Fig. 4.1.) At first glance such a mishmash of interests must surely look bizarre: PhD physicists from elite programs dabbling in the occult? Yet on a longer view the combination appears neither shocking nor unprecedented. Both mesmerism in the 1770s and spiritualism in the 1870s had become international sensations. In both cases, leading scholars from Madras and St. Petersburg to Paris, London, Boston, and New York had formed committees and staged public demonstrations. Learned periodicals and the popular press published tens of thousands of articles debating the reality of the purported effects and evaluating possible explanations drawn from the scientific canons of the day. Indeed, spiritualism—the claim that certain special individuals, particularly sensitive "mediums," could establish contact with the dead and, by translating the spirits' mysterious knockings on tables or rappings on walls into specific alphabetic codes, deliver messages from beyond the grave—proved to be just the opening gambit of a broad fin-de-siècle revival of all things occult. Telepathy, psychokinesis, and alchemy all moved to center stage. In Britain, major scientific authorities, including Lord Rayleigh, J. J. Thomson, William Ramsay, and William Crookes—several of whom went on to become Nobel laureates and presidents of the Royal Society—devoted decades of effort to investigating the latest claims. They urged skepticism, not outright dismissal: Rayleigh, Thomson, and many more sat through hundreds of séances, each time wondering whether *this time* they might hit upon unimpeachable evidence of genuine effects. Others, such as Crookes and decorated physicist Oliver Lodge, issued bold and repeated pronouncements about the reality of such "psychical" phenomena.[3]

The occult revival lasted well into the early decades of the twentieth

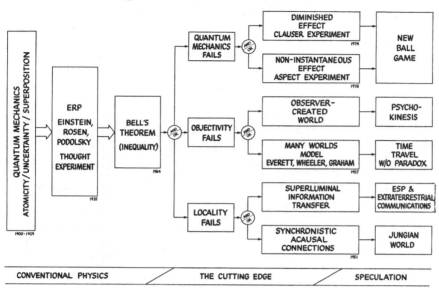

FIGURE 4.1. Saul-Paul Sirag's roadmap for the group's discussions, 1976.
(Courtesy Saul-Paul Sirag.)

century. New societies were formed, laboratories established, and journals launched. In fact, several founders of quantum mechanics wondered whether the strange behavior of the atomic realm might lead to still stranger phenomena. Erwin Schrödinger, for one, devoted extensive effort to understanding Eastern mysticism. In unpublished notes written just before his breakthrough with quantum mechanics, Schrödinger delved into Sanskrit etymology to clarify various Hindu beliefs. Several years later he lectured a Berlin journalist about "the Brahman doctrine that the all equals the unity of consciousness," admonishing that "it would be a vast error to believe that science knows any better or clearer answer [than the Brahman teachings] concerning these things."[4] Pascual Jordan, who helped develop quantum mechanics with Werner Heisenberg and Max Born during the 1920s, wrote a whole book about quantum physics, the Freudian unconscious, and parapsychology. The first edition appeared just as the ink was drying on the new quantum formalism; a second edition appeared two decades later. He published an English-language précis in 1955 in the *Newsletter of the Parapsychology Foundation*, describing how quantum theory could account for telepathy or clairvoyant visions of the future.[5]

Beginning in the early 1930s, Heisenberg's longtime friend and fellow quantum physicist Wolfgang Pauli worked closely with the psychoanalyst Carl Jung on a similar quest. For decades they sought to plumb hidden connections between quantum physics and what Jung famously called the "collective unconscious." Not only did Pauli join Jung in scholarly studies of the history of alchemy and mysticism but the fabled physicist and Nobel laureate also kept a diary of his dreams—eventually bulging with 400 entries—over which Pauli and Jung together pored in search of clues for how "even the most modern physics lends itself to the symbolic representation of psychic processes," as Pauli put it. Perhaps his log of symbol-laden dreams could point the way toward "deeper spiritual layers that cannot be adequately defined by the conventional concept of time."[6] Pauli wrote essays extolling the need to synthesize "rational understanding" with "the mystic experience of one-ness," achieving the same kind of complementarity that his generation had first formulated for particle and wave.[7]

Over time, however, the occult movement quietly faded from the mainstream, lumbering under the weight of so many decades of disappointments, spiked by occasional evidence of outright fraud. Whereas the Society for Psychical Research, founded in London in 1882, had quickly attracted renowned scientists and statesmen—several members of Parliament served as vice presidents of the Society alongside elite scientists, and even four-time British prime minister William Gladstone joined its ranks—by the 1950s the Society and its kin limped along on the sidelines. When the *Newsletter of the Parapsychology Foundation* announced in September 1955, for example, that "World-wide research moves ahead," few outside its dwindling membership seemed to notice or care. Just a few weeks earlier the journal *Science* had carried a devastating critique of "science and the supernatural," and commentators from the *New York Times* and *Time* magazine had gladly declared the field dead.[8]

When members of the Fundamental Fysiks Group rediscovered the occult twenty years later, and began to investigate psi phenomena from the vantage of cutting-edge physics, they were resurrecting a once-proud tradition. Like their routes to Bell's theorem, they bumped along crooked paths to psi. Or, as several of them would have it, perhaps it was all meant to be—just one more example of Jungian "synchronicity."[9]

Consider, for example, Jack Sarfatti's entrance into the psi world. In the summer of 1973, soon after receiving his invitation to the International Centre for Theoretical Physics in Italy, Sarfatti happened to read a story in the *San Francisco Examiner* about research under way at the Stanford Research Institute, or SRI.[10] SRI, much like defense-oriented laboratories at MIT and elsewhere, had been a flashpoint of student and faculty protest just a few years earlier. Much of the heat and light of the April 1969 marches and sit-ins at Stanford University focused on the vast array of classified military projects at SRI—everything from chemical weapons to Vietnam-era counterinsurgency techniques. Early in 1970, Stanford's trustees, eager to quell the protests, spun off SRI as a private research enterprise and divested the university's ties to it. SRI's researchers took their defense contracts with them, only to see contract revenues plummet as the Cold War bubble burst.[11]

The lean years brought new opportunities for some, including laser physicist and former Stanford lecturer Harold Puthoff. Puthoff had previously worked as a naval intelligence officer and a civilian researcher at the National Security Agency. He completed his PhD at Stanford in 1967 on a new type of tunable laser, and stayed on for several years to teach in Stanford's electrical engineering department, where he coauthored a textbook on quantum electronics. He joined SRI in 1969 and left the university the following year, when SRI was spun off; in short order his laser-research government contracts began to deflate. With time on his hands, he asked his SRI supervisor for permission to begin conducting a different set of experiments: tests of parapsychological effects. Puthoff was a devotee of Scientology at the time, a controversial set of beliefs that centers on mystical connections between mind and body. He had also dabbled in early rumblings of the California New Age scene during the 1960s, including workshops on gestalt therapy and consciousness expansion. Puthoff secured a grant from a private philanthropist who had made his money in the fried chicken business; with a check for $10,000 (more than $50,000 in 2010 dollars), he was up and running. He courted another laser physicist from Sylvania's research laboratory, Russell Targ, who had done some graduate-level work at Columbia but left before completing his PhD. Targ, too, had begun to sample some of the New Age offerings around the Bay Area. Together, Puthoff and Targ jumped into the psi business.[12]

Their big break came in September 1972, when the Israeli performer Uri Geller visited SRI to conduct laboratory tests of his psychic abilities. Geller claimed not only clairvoyance—the ability to read minds or receive signals from the future—but psychokinetic powers as well. His most famous feat: bending metal objects, such as spoons and keys, by focusing psychic energy in his fingers. Puthoff and Targ's psi lab at SRI, already up and running by the time promoters had brought Geller to the United States, seemed the perfect place to put Geller's powers to the test. Weeks of close observation ensued; hours of film were shot. Puthoff and Targ concluded that Geller had indeed demonstrated parapsychological abilities, such as reproducing drawings that had been sealed in an envelope, or guessing correctly—eight times in a row—the number on a die contained within a steel box.[13]

Even before the physicists at the Stanford Research Institute began to publish their results, their research started to make headlines around the Bay Area and beyond.[14] Thus it was that Sarfatti happened upon the newspaper article about their work in the summer of 1973, just a few weeks before he was to leave for Italy. Intrigued, he called SRI, was connected to the Electronics and Bioengineering Laboratory (as Puthoff and Targ's psi lab was called), and invited to come and see for himself. He spent a marathon day at the lab, seventeen hours in all, during which he met Puthoff and Targ as well as paranormal enthusiasts Brendan O'Regan and Edgar Mitchell, the latter a former astronaut who conducted telepathy experiments with friends on Earth while he orbited the moon during the Apollo 14 mission of February 1971. (Upon his return to Earth, Mitchell founded the Institute for Noetic Sciences in Palo Alto, California, to support parapsychological investigations; his institute had helped to bankroll Geller's visit to SRI in 1972.)[15]

Not long after Sarfatti's day-long visit at SRI, one of Uri Geller's close associates, the medical doctor and parapsychologist Andrija Puharich, published an admiring biography of Geller, entitled simply *Uri*. Puharich gave a copy to Sarfatti, who in turn loaned it to his mother. The book, combined with Sarfatti's recent introduction to Geller's feats at SRI, triggered a momentous shift in the young physicist. Puharich asserted in the book that Geller had received repeated telephone calls from a robotic-sounding voice that called itself "Spectra." The voice claimed to be an extraterrestrial computer orbiting the earth, contacting a small group of select individuals to help prepare for future contact. Upon encountering that passage, Sarfatti's mother told Jack that he, too, had received such telephone calls twenty years earlier, at the age of thirteen. The young Sarfatti had ignored, forgotten, or repressed all memory of the strange calls until his meeting with Puthoff and Targ, and his mother's reading of Puharich's book, brought it all screaming back to consciousness. From that point on, there was no going back: Sarfatti threw himself into the strange world of psi.[16]

During Sarfatti's first visit to the Stanford Research Institute psi lab, Brendan O'Regan had asked Sarfatti if he could introduce Geller to some of the European physicists whom Sarfatti was about to visit, so that the scientific tests could continue. Once Sarfatti joined his friend and San Diego State physics colleague Fred Alan Wolf in London a few months later, the two did just that. Their case was helped by Geller's own promoters, who had managed to book Geller on a live British Broadcasting Company television show in November 1973. A London-based mathematical physicist participated in the broadcast and declared Geller's feats to be genuine psychokinetic effects, in need of explanation from the world's physicists. By February 1974, with Wolf's and Sarfatti's help, renowned physicist and hidden-variables expert David Bohm and a colleague at

London's Birkbeck College had made contact with Geller and begun their own series of investigations, which would stretch over the course of the next year. (Bohm's colleague, an experimental physicist, had arranged Wolf's invitation to Birkbeck.) Sarfatti was in London during one of their sessions with Geller that June, and he dashed off a detailed press release. Not only had Geller again managed to bend metal objects (including, this time, one of Bohm's own keys), but he also produced a burst of radioactivity, from no known source, that sent a Geiger counter held in his hand clicking up to 150 times per second. The next day, Geller repeated the Geiger counter burst and bent the house key of skeptical observer and famous science-fiction author Arthur C. Clarke, while Clarke, Sarfatti, Bohm, and others looked on.[17] (Fig. 4.2.)

The results seemed clear. "My personal professional judgment as a PhD physicist," Sarfatti closed his press release, "is that Geller demonstrated genuine psycho-energetic ability at Birkbeck, which is beyond the doubt of any reasonable man, under relatively well controlled and repeatable experimental conditions." Bohm and his Birkbeck colleague agreed, publishing an account of their investigations in the top-flight scientific journal *Nature*. They urged caution against runaway theoretical

FIGURE 4.2. Physicists tested the psychic abilities of Israeli performer Uri Geller, first at the Stanford Research Institute in California (*left*) and later at Birkbeck College in London (*right*, with physicist David Bohm). (Photographs by Shipi Shtrang, courtesy Shipi Shtrang and Uri Geller.)

speculations, arguing that (as in the early stages of any scientific field) it was most important to establish a baseline of reliable empirical observations first. Sarfatti had a different idea. To him, the Geller tests forced physicists to return to the foundations of quantum mechanics. "The ambiguity in the interpretation of quantum mechanics," Sarfatti argued, "leaves ample room for the possibility of psychokinetic and telepathic effects." Most important, he elaborated, was the "intrinsically nonlocal" character of quantum theory. Drawing on a preprint of Bohm's own latest grapplings with Bell's theorem and nonlocality, as well as intriguing ideas from such giants of the discipline as Eugene Wigner and John Wheeler, Sarfatti argued that consciousness need not be separate from brute matter. Sarfatti maintained that quantum mechanics, properly understood, could provide a mechanism to account for psi effects like those exhibited by Uri Geller.[18]

•

A decade earlier, in an admittedly speculative move, Princeton's Nobel laureate Eugene Wigner had proposed that consciousness plays a central role in quantum mechanics. Left on its own, the quantum formalism seemed to imply an infinite regress of probabilities: an electron had a certain probability to be spin up or spin down; a detector had a certain probability to measure the particle's spin as being up or down; the detector's needle had a certain probability to point toward "up" or "down" on its display screen; and so on. This had become known as the "measurement problem" of quantum mechanics. What if, Wigner wondered, the consciousness of a human observer were the only thing that could break the regress and register a definite response: spin measured as up or down?[19]

Wigner introduced a simple thought experiment, often referred to as "Wigner's friend," to motivate his conclusion. Imagine that instead of conducting the spin measurement on the electron yourself, you ask a friend to do so. Until you interact with your friend by asking her what the measured outcome was, the best you can do is represent the total system—electron plus measuring device plus friend—by one quantum wavefunction. As far as you are concerned, when you ask your friend for the outcome, she will have a certain probability of responding "spin up" and a certain probability of responding "spin down." After the dust has settled, Wigner pressed on, you might go back and ask your friend, "What did you feel about the spin-measurement outcome before I asked you?" No doubt your friend would respond, "As I already told you, it was spin up (or spin down)." That is, as far as your friend is concerned, the outcome had already been settled before you bothered asking the question. Or, in quantum-mechanical parlance, the wavefunction for the system had already settled into one of its two possible states: electron spin up and friend in the state "I have measured the electron to be spin up"; or electron spin down and friend in the state "I have measured the electron to be spin down." That was *her* version of the wavefunction prior to your asking your question; yet your own version of the wavefunction was still stuck in a superposition of both possibilities. To Wigner, there could only be one proper wavefunction for the system—meaning that your friend's consciousness had already changed (or "collapsed") the wavefunction from a sum of possibilities to one definite outcome, even before you asked her about it. If you didn't believe that—if you clung to your own version of the wavefunction after her measurement was complete but before you asked her the outcome—then, feared Wigner, you would be forced to the "absurd" conclusion that your friend was "in a state of suspended animation" before she answered your question: caught, like Schrödinger's famous cat, between two irreconcilable states. Such suspensions were bad enough when attributed to cats; they simply would not do when applied to tenured professors at Ivy League institutions. "It follows," Wigner concluded, "that the being with a consciousness must have a different role in quantum mechanics than the inanimate measuring device." Or, more strongly: "Consciousness enters the [quantum] theory unavoidably and unalterably."[20]

Such talk stood out starkly from the pragmatic concerns with which most of Wigner's colleagues occupied themselves at the time. He came by it honestly. The Hungarian-born physicist had been trained on the Continent between the world wars; as a student, he had heard Einstein, Heisenberg, and others lecture on the still-new quantum mechanics.

Years later, his philosophical interests were rekindled when he took on Abner Shimony as a graduate student at Princeton. (Shimony, recall, later worked with John Clauser to rederive Bell's theorem in a form suitable for laboratory test.) Shimony came to Wigner directly from his own PhD in philosophy at Yale. From that time forward, Wigner devoted more and more of his attention to the foundations of quantum mechanics, corresponding frequently with pockets of physicists in Europe who chased these questions throughout the 1960s.[21]

Wigner soon acquired an interlocutor closer to home. His friend and Princeton colleague John Wheeler picked up on the theme of consciousness and quantum mechanics during the early 1970s. Wheeler, too, stood out from the pack. An American, he had come of age in the 1920s and 1930s, a time when Americans who wanted to become theoretical physicists still had to travel to Europe to "learn the music, and not just the libretto" of work in that field, as one of Wheeler's contemporaries famously put it.[22] Wheeler studied with Niels Bohr in Copenhagen in the 1930s and often hosted his mentor during Bohr's many extended visits to Princeton after the war. These contacts helped to stoke Wheeler's continuing philosophical engagement with quantum theory. Spurred further by Wigner's efforts, Wheeler emerged, by the mid-1970s, as one of the few leading physicists working in the United States who took the interpretation of quantum mechanics seriously.[23]

Wheeler argued for a view that he came to call the "participatory universe": observers participate in creating the reality they measure. As Wheeler argued, a physicist's decision to measure a particle's position rather than its momentum changes the objective properties of the real world. Wheeler emphasized the point at a conference at Oxford early in 1974. Quantum theory, he stipulated,

> demolishes the view we once had that the universe sits safely "out there," that we can observe what goes on in it from behind a foot-thick slab of plate glass without ourselves being involved in what goes on. We have learned that to observe even so minuscule an object as an electron we have to shatter that slab of glass. We

have to reach out and insert a measuring device. We can put in a device to measure position or we can insert a device to measure momentum. But the installation of the one prevents the insertion of the other. We ourselves have to decide which it is that we will do. Whichever it is, it has an unpredictable effect on the future of that electron. To that degree the future of the universe is changed. We changed it. We have to cross out that old word "observer" and replace it by the new word "participator." In some strange sense the quantum principle tells us that we are dealing with a participatory universe.[24]

To drive the point home to his physicist colleagues, Wheeler included a cartoon contrasting the old notion of an "observer" with his new idea of a "participator"—a cartoon he inserted into other conference talks over the next few months.[25] (Fig. 4.3.)

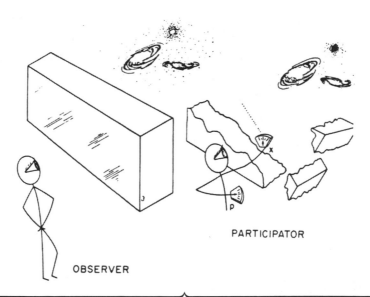

PARTICIPATOR

OBSERVER

FIGURE 4.3. Princeton physicist John Wheeler's cartoon version of the difference between the "old concept of the 'observer,'" and the "new concept of 'participator'" as required by quantum mechanics. (Patton and Wheeler [1975], 563. Reproduced with permission of Oxford University Press.)

Wheeler had grand ambitions for these "participators." Not only did they fix the reality of the here and now; they could even do so retro-actively. Wheeler returned to that old standby of quantum theory, the double-slit experiment, and gave it a new twist. Suppose, he argued, that the photographic plate behind the slits were mounted on a pivot. In one position, the plate would sit smack in the path of any particles that traversed the slits, thereby registering the familiar interference pat-tern. In another position, the plate could be swung clear of the particles' paths, so that they bypassed the plate altogether. In this second set-ting, the particles would continue on their way and encounter one of two sensitive detectors: one placed to detect only those particles that had traveled through the top slit, and the other placed to detect only those that had traveled through the bottom slit. Next the participator could tune down the intensity of the particle source so that only a single quantum was released at a time. The participator now had a choice. Insert the photographic plate into the particle's path and observe the famous quantum interference pattern—which could only arise if each particle effectively went through both slits at once. Or remove the pho-tographic plate and let the detectors determine whether a given particle had traveled through the top or the bottom slit. But here's the rub: the participator could decide to insert or remove the photographic plate *after* the particle had already passed through the slits! (Fig. 4.4.) Wheeler dubbed such scenarios "delayed-choice" experiments: a "last-instant free choice on our part," he explained, "gives at will a double-slit-interference record or a one-slit-beam count." The lesson? "The past has no existence except as it is recorded in the present. . . . The universe does not 'exist, out there,' independent of all acts of observation. Instead, it is in some strange sense a participatory universe." Not everyone was pleased with Wheeler's conclusion. Some anonymous reader highlighted this passage in the MIT library's copy of the conference proceedings, adding in the margin simply, "ugh."[26]

Wheeler still wasn't done. One could replace the benchtop apparatus of particle source, double-slit, and swiveling photographic plate with a cosmic substitute. Consider, he pressed on, streams of light imping-

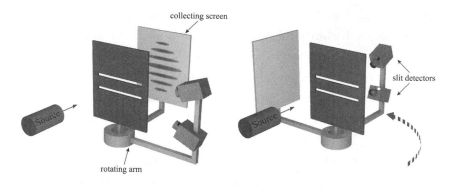

FIGURE 4.4. Physicist John Wheeler's "delayed-choice" thought experiment. After a particle has passed through the wall with double slits, an experimentalist may choose whether to leave the collecting screen behind the double slit, on which the particles will fill out the familiar interference pattern; or to swivel the slit detectors into place, which will determine through which slit each individual particle traveled. (Illustrations by Alex Wellerstein.)

ing on an earthbound participator from a faraway quasar, an intense astronomical light source billions of light-years away. In between the quasar and the Earth lies a galaxy, massive enough to bend the light's path and thus act as a "gravitational lens." (At the time Wheeler was writing, astronomers had recently identified just such a quasar-galaxy pair.) Some quanta of light, or photons, would travel directly from the quasar to Earth; others would travel a more circuitous route, starting off in a direction away from the Earth but getting bent back toward the Earth by the intervening galaxy. Now repeat the delayed-choice setup: by suitable arrangement of photographic plates and sensitive detectors, the participator could decide to measure by which route an individual photon traversed the cosmos (direct or via the path-bending galaxy); or she could decide to measure the quantum interference that comes from traversing both paths. "But the photon has already *passed* that galaxy billions of years before we made our decision." It was as if "we decide what the photon *shall have done* after it has *already* done it"—in this case, not microseconds before we make our choice, but billions of years before. Indeed, Wheeler emphasized, our decisions today can determine the past of a particle that was emitted long before there was even life on Earth.[27]

To Wheeler, the central feature of quantum theory—its participatory nature—thus explained not only the outcome of this or that experiment, but the emergence of the universe itself. He cited the pre-Socratic philosopher Parmenides and the Enlightenment philosopher George Berkeley, names that did not often appear, as they did in Wheeler's essay, nestled between citations to Einstein, Bohr, Richard Feynman, and Stephen Hawking. Building on all those authorities, Wheeler advanced his view: the participator "gives the world the power to come into being, through the very act of giving meaning to that world; in brief, 'No consciousness; no communicating community to establish meaning? Then no world!'" He continued, "On this view, the universe is to be compared to a circuit self-excited in this sense, that the universe gives birth to consciousness, and consciousness gives meaning to the universe." Or, as he returned to the theme a few years later, "Acts of observer-participancy—via the mechanism of the delayed-choice experiment—in turn give tangible 'reality' to the universe not only now but back to the beginning." In case his colleagues missed the point, Wheeler again turned to a whimsical cartoon. Like his "observer" and "participator" stick figures, Wheeler's self-actualizing universe continued to grace several of his talks and essays over the next few years.[28] (Fig. 4.5.)

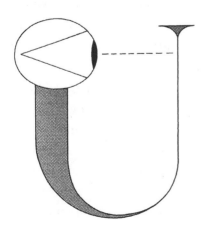

FIGURE 4.5. John Wheeler's vision of the entire universe as a "self-excited system brought into being by 'self-reference.'" (Patton and Wheeler [1975], 565. Reproduced with permission of Oxford University Press.)

Wheeler pushed this line vigorously throughout the 1980s, encouraging a number of physicists to conduct actual delayed-choice experiments. Yet his basic ideas on the matter had already jelled by the mid-1970s. They dribbled out in a series of little-noticed conference talks and preprints, attracting virtually no citations in the scientific literature for the remainder of the 1970s. They were little noticed, that is, except by a small number of people like Jack Sarfatti, who struck up an active correspondence with Wheeler and received Wheeler's latest musings by preprint and letter. In fact, Sarfatti and Wolf had tried to arrange unpaid visiting appointments with Wheeler at Princeton for their 1973 sabbaticals. "We understand that no financial support would be forthcoming during these hard times," they explained; desks and library cards would suffice. Only after Wheeler politely declined each of their repeated requests did they take off for Europe instead. Wheeler was relieved to learn of their European invitations. "I hated so much to seem unwelcome," he replied, which cut against his "natural eagerness to be hospitable." He wished them well and pledged to stay in touch, which indeed he did. Wheeler sent Sarfatti a preprint of his 1974 Oxford talk, for example, complete with its "participator" stick figure and self-actualizing universe cartoons, and it made a deep impression on Sarfatti. He began to cite it and build on its ideas even before Wheeler's essay had appeared in print.[29]

Sarfatti aimed to stitch these diverse ideas together. If every quantum object were interconnected with every other via quantum entanglement (as per Bell's theorem), and if consciousness played a central role in quantum mechanics (as Wigner and Wheeler had reasoned), then modern physics might provide a natural explanation for psi phenomena. From Wigner and Wheeler, Sarfatti took the point that everyone's consciousness participates in shaping quantum processes, both by deciding which observations to make and by collapsing the multiplying possibilities into definite outcomes. Sarfatti recast Wigner's main argument in terms of action and reaction. Surely matter can affect consciousness— LSD and other psychedelic drugs had made that lesson clear enough— so why not posit an equal and opposite reaction of consciousness on matter? To Sarfatti, such a move paid double dividends: it opened up a possible avenue for understanding psychokinesis, and it offered hope that Age of Aquarius students might come back to physics classrooms, finding new relevance in the subject.[30]

Most mental contributions to the behavior of quantum particles, Sarfatti continued, would be "uncoordinated and incoherent"—that is, they would each push in different directions and, on average, wash out. But, as Uri Geller seemed to demonstrate, certain talented individuals might possess "volitional control" such that they could impose some order on the usually random quantum motions. Some "participators" seemed to be more effective than others. Moreover, thanks to Bell's theorem, these individuals could exercise their control at some distance from the particles in question. In short: perhaps Geller could detect signals from far away or affect metal from across a room because the quanta in his head and the quanta far away were deeply, ineluctably entangled via quantum nonlocality. Bizarre? No doubt. But was it really any more outlandish than Wheeler's giddy flights?[31]

Sarfatti's first effort to bring Geller and psi into the rubric of quantum physics appeared as the lead article in the inaugural issue of a brand-new journal entitled *Psychoenergetic Systems*. Brendan O'Regan, whom Sarfatti first met at the Stanford Research Institute psi lab before departing for Europe, helped launch the journal to feature just this kind of reasoned— and, granted, speculative—investigation into effects beyond the usual boundaries of science. O'Regan cited historians and philosophers of science such as Thomas Kuhn and Gerald Holton, who had written about the spur to new scientific breakthroughs from previous accumulations of "anomalies," to bolster his claim that psi studies would usher in a whole new "paradigm" across the sciences. Pleased with Sarfatti's contribution, O'Regan appended a brief comment to the opening article, arguing that the exciting recent developments in quantum mechanics meant that "physics might have to invent psychic research, if it did not already exist." Over the next several years, the journal published many follow-up articles pursuing further relations between quantum mechanics and psi.[32]

Sarfatti's and O'Regan's enthusiasm was hardly dampened when, a few months after his first Birkbeck dispatch, Sarfatti began to distance himself from the Geller tests. His pro-Geller press release had been published in the weekly magazine *Science News*, and soon the magazine was inundated with letters. Most of the letter writers called for professional magicians to test Geller alongside of, or in place of, physicists. (Magicians had served as the most effective debunkers of spiritualist mediums back in the 1890s; and none other than the great magician Harry Houdini had devoted years of effort during the 1920s to debunking the claims of psychics, mediums, and other purveyors of the paranormal.)[33] Rising to the latest challenge, accomplished magicians such as James "The Amazing" Randi mobilized, proudly demonstrating that many of Geller's famous feats could be replicated by well-known sleight-of-hand tricks. After lunching with Randi and watching him bend spoons and affect the rate at which wristwatches' hands spun—all by the admitted power of conjuring, not psychokinesis—Sarfatti was moved to retract his earlier declaration in favor of Geller's powers. (Randi also explained how he could make Geiger counters burst with activity: by hiding a small source of beta-radioactivity up his proverbial sleeve.) "I do not think that Geller can be of any serious interest to scientists who are currently investigating paraphysical phenomena," Sarfatti explained in a new press release.[34]

In short order, the unease that Sarfatti articulated escalated into widespread controversy over Geller's psychic claims. "Super mystic or super fake?" asked *San Diego Magazine* a few years later, emphasizing in its feature article on Geller that the young Israeli performer seemed to be "eternally on trial." A kind of cold war of Geller publications had erupted: for every debunking effort by magicians like James Randi, there also appeared a new glowing endorsement. One of the latest had come from the Naval Surface Weapons Center in Silver Spring, Maryland, whose experts had proclaimed that "Geller has altered the lattice structure of a metal alloy that can *not* be duplicated. There is no present scientific explanation as to how he did this."[35] "The enormous spate of publications did little to quell the rising controversy," the San Diego journalist noted. "If anything, they simply added to the fire." Critics such as magician James Randi harrumphed that they had revealed Geller's psychic powers to be little more than skillful conjuring. Randi claimed (rather prematurely) that Geller's continued performances, even after magicians like Randi were on his trail, amounted to "digging his own grave." Yet all the while Geller's admirers, fanning out across the entire world, continued to keep faith in the authenticity of his proclaimed powers.[36]

As far as Sarfatti was concerned, his retraction applied to Geller, not to psi. Here again Sarfatti was following a well-trod path. Exactly a century earlier, William Crookes, J. J. Thomson, and the rest had followed the same procedure: whenever questions about a particular medium emerged, they dismissed that medium but not the notion of spiritualism in general. To Sarfatti, a whole universe of psi effects still beckoned, and their relationship with quantum mechanics remained to be explored. (More recently, Sarfatti has in effect retracted his retraction, maintaining that Geller did display at least some genuine psychic abilities.)[37]

Sarfatti worked out his many ideas in conversation with Fred Alan Wolf while they were in Europe. Like Sarfatti, Wolf enjoyed a front-row seat for an explosion of New Age activities. A high school friend and freelance writer got him interested in Uri Geller. When the friend learned that Wolf was heading off for Europe, he urged Wolf to seek out other notables of the New Age scene. Wolf obliged. Immediately upon arriving in Paris in January 1974, for example, Wolf looked up Carlo Suarès—painter, philosopher, and master of the ancient tradition of Kabbalah, or Jewish mystical numerology. Sarfatti joined Wolf for some of these chats with Suarès, and in no time Sarfatti began urging John Wheeler to contact Suarès himself because of the similarity of Wheeler's and Suarès's ideas about the structure of the universe.[38] A few months later, back in London, Wolf attended the "May Lectures," featuring presentations by New Age gurus like Andrija Puharich (of Uri Geller–studies fame) and Werner Erhard (the human-potential magnate). During his presentation at the May Lectures, Erhard announced that he wanted to meet physicists—he had been fascinated with the subject since boyhood, and he believed that physicists' rigorous training could offer insights beyond quantum theory. Wolf introduced himself to Erhard at intermission and was invited to a speakers-only workshop the next day.[39] Overnight, Wolf's circle of interlocutors widened considerably. When the time came to return to the States at the end of his sabbatical, Wolf reached California by way of Puharich's personal-residence-turned-psychic-laboratory in upstate New York. He spent several weeks there, attempting, on Puharich's request, to relate Puharich's psychic discoveries to Wolf's research in quantum physics. Wolf eventually admitted defeat—"I can't say that I discovered anything that would lend credence to their abilities," he later wrote— but only after undergoing his own out-of-body experience and enjoying stimulating discussion with Puharich and his followers. Along with Sarfatti, Wolf, too, was hooked.[40]

•

While Sarfatti and Wolf broadened their horizons in Europe, other physicists who would soon form the Fundamental Fysiks Group followed a complementary line of inquiry in Berkeley. Like Sarfatti, Nick Herbert and Saul-Paul Sirag were captivated by Wigner's suggestion about the central role of consciousness in quantum measurement. They turned Wigner's proposal on its head, asking what quantum theory implied about the nature of consciousness. Herbert had noticed a string of recent papers by a fellow physicist, Evan Harris Walker, in which Walker had begun to construct a theoretical model of consciousness. Walker, who had written a dissertation at the University of Maryland in 1964 on plasma physics and the behavior of charged bodies in motion, had made his career at the Ballistic Research Laboratories of the Army's Aberdeen Proving Ground in Maryland. Like Harold Puthoff (founder of the Stanford Research Institute psi lab), however, Walker began to find more time on his hands as defense spending on research waned. He began toying with Bohm's hidden variables during his off hours.[41]

Walker postulated that consciousness might be an infinite set of hidden variables, real but beyond direct physical observation. Like the hidden variables in Bohm's original model, these "c-variables," as Walker dubbed them, would determine the measured outcomes of quantum processes. And then he began to calculate. Quantum processes in the brain, such as electron tunneling across synaptic gaps between nerve endings, seemed to establish three distinct rates of data processing: subconscious (at a trillion bits per second); conscious (100 million bits per second); and "will" or volition (10,000 bits per second). This last, Walker suggested, could serve as a "data channel" for psi effects. According to quantum mechanics, all kinds of events could transpire, some with high probability and others with vanishingly low probability. Walker hypothesized that an individual might be able to consciously select an otherwise low-probability outcome, and use his or her will to arrange the c-variables so as to produce that outcome. After all, Bohm had introduced hidden variables into quantum theory precisely to replace probabilistic descriptions with definite, causal mechanisms. Walker's hypothetical process would involve no transfer of energy, he clarified, only information. Thus a psychic could in principle violate the second law of thermodynamics—creating a more-organized state out of a less-organized one—but not the conservation of energy. Moreover, thanks to Bell's theorem and long-distance entanglement, the low-probability event could take place miles away from the volitional brain that had willed it into being.[42]

Several of Walker's papers appeared in the *Journal for the Study of Consciousness*, which served as the house organ of Arthur Young's Institute for the Study of Consciousness in Berkeley, the meeting place and watering hole where several members of the Fundamental Fysiks Group frequently crossed paths. Herbert and Sirag were thus among Walker's earliest readers and devoted fans. Herbert was so impressed that he made dozens of photocopies of Walker's early articles to hand out to friends. (He was still working his day job at the copy machine company Smith-Corona Marchant, so he could make photocopies rather easily.) Sirag featured Walker's work in some of his freelance articles, and defended it from skeptical critics, including hidden-variables maven David Bohm. Undeterred, Sirag shot back: one of the most attractive features of Walker's model was that "just those things that are peculiar about quantum mechanics remind us of those things that are peculiar about consciousness, especially as exemplified in psi phenomena." Psi and ψ were made for each other, and, as far as Sirag was concerned, Walker's unified theory illuminated both aspects equally.[43]

Herbert and Sirag had the opportunity to tell Walker in person how much they admired his work when he came to give a talk to the Fundamental Fysiks Group in the mid-1970s. By that time they had explored Walker's work from a number of angles, well beyond mere pencil and paper. Most playful, no doubt, was the bizarre contraption that Herbert dreamed up with a colleague who worked at the Xerox PARC research laboratory, which he dubbed the "metaphase typewriter." Sirag, in a thinly veiled fictional account of what happened next—Nick Herbert became "Manny Hilbert"—explained that the metaphase typewriter had begun "as a joke, a tongue-in-cheek way of challenging the farfetched but intriguing theory of Harris Walker that consciousness functions as a set of Hidden Variables in a quantum mechanical system." Herbert reasoned that if Walker were correct, then the mind might be at root a quantum effect, separate from the physical body. Mind could control the body by consciously adjusting the c-variables to shift the underlying probabilities for various events. Moreover, if mind were separate from body, this subtle biasing of quantum probabilities might be accomplished either by flesh-and-blood people sitting next to you in a room, or by any free-floating mindlike essences: "spirits of the dead, beings from other dimensions, or dissociated fragments of living personalities." With the metaphase typewriter, even these ethereal quantum-mind-spirits could have their say.[44]

Herbert's device, forged from the latest that quantum theory and digital computing had to offer, was a 1970s gadget for an 1870s dream. Like the Victorian-era spirit mediums, Herbert sought to make contact with the other side—"the realm of mind, or spirit, or subquantum level, take your pick," as Sirag put it—and convey messages by convoluted alphabetic code: the table knockings and wall rappings of old replaced by radioactive sources and fancy electronics. Herbert assembled his apparatus in a cramped, out-of-the-way computer room nestled deep within the Medical Center at the University of California, San Francisco. A friend from Herbert's graduate-school days at Stanford had since joined the Medical Center staff, and he snuck Herbert and company into the facilities. In the computer room, Herbert had stashed a sample of the radioactive element thallium, first identified a century earlier by renowned chemist and outspoken spiritualist Sir William Crookes.[45]

Although not as stringent as today's safeguards, by the mid-1970s several barriers stood between would-be experimenters and radioactive materials like thallium. Herbert had to file a formal petition with the Department of Public Health of the State of California; just a few years earlier, authorities in the United States had banned the use of thallium in rat poison and pesticides precisely because of public health concerns. Rather than raise eyebrows among bureaucrats by describing his contraption in detail, Herbert wrote on his application merely that the radioactive thallium would serve as a "source of random pulses for statistical analysis."[46] He grew more expansive in an unpublished technical report on his contraption that same year. "It is probably no coincidence that thallium, our licensed source, is sandwiched in the periodic table of the elements between two of the traditional alchemical metals, mercury and lead." No wonder Crookes's element was so effective at producing "quantum anagrams" from the spirit world.[47]

Along with the radioactive thallium, Herbert had rigged up a Geiger counter and a fast-printing teletype machine. (It seems that Herbert could never fully escape his day job.) He loaded his own program onto the room-sized computer to convert time delays between radioactive decays into printed letters on the teletype. Radioactive decays are a prototypical quantum-mechanical phenomenon. Individual decay events—such as when this or that nucleus will decay—can never be predicted with certainty. Nuclei of a given type have an average rate at which they decay, related to the element's half-life; but individual nuclei from a large sample of a radioactive substance will decay at random times, scattered around the average value. Herbert zeroed in on that randomness. His device measured time delays between successive radioactive decays, and converted those time intervals to letters of the alphabet. If the gap between two radioactive decays was close to the average rate for thallium, then his metaphase typewriter would spit out a letter that appeared frequently in ordinary written English, such as *e* or *t*. If the time delay between successive radioactive decays departed further from the average rate, then Herbert's machine would produce less likely letters, such as *j* or *x*. Ever a stickler for accuracy, Herbert had obtained the statistics for English-language letter frequencies from an unclassified report by the National Security Agency.[48]

If ordinary quantum theory ruled the subatomic world, then the output from Herbert's machine should have been pure nonsense: a random jumble of letters spewed out in a row. But if Evan Harris Walker's musings were on track, and someone's (or something's) consciousness could skew the probabilities for radioactive decay—nudging individual events toward or away from an otherwise likely value—then that mind could control the sequence of letters tapped out by Herbert's teletype. Following along Walker's train of thought, Herbert reasoned that some conscious entity might be able to speak to the group by way of Herbert's machine. "A rather suspect communication channel you might think," teased Sirag in his article. "But then you haven't encountered the strange mind of Manny Hilbert."[49]

Herbert and company tried out the device several times. First they invited a series of self-proclaimed psychics to join them, asking the guests to use their conscious willpower to spell out a list of target words on the teletype. If consciousness consisted of Walker's hidden variables, Herbert contended, then the psychics might be able to use their minds to prod less-likely events into fruition (say, a longer-than-expected delay between radioactive decays), or pause likely events in their tracks. Proof would come from the string of letters rat-tat-tatted out on the teletype machine. Other times the group conducted séances around the machine, trying to make contact with recently departed colleagues who had known about the research before they died.[50]

The climax came in March 1974, when Herbert, Sirag, and about a dozen friends held a day-long séance to mark the one-hundredth anniversary of Harry Houdini's birth. They relished the irony: the famed magician had been an outspoken skeptic and debunker of spiritualism in his day. Yet Houdini, being the ultimate escape artist, had promised friends and family before he died that if there were any way to come back and communicate, he would. Now was his chance. In Sirag's fictional account, the metaphase typewriter whirred into action, spitting out the string of letters "anininfinitime," close enough to a recognizable phrase—"and in infinite time"—to convince the onlookers that in their brush with Houdini, his spirit had complained that their equipment was too slow to allow effective communication. During the actual demonstration, as Sirag recalls, they did not hear from Houdini, although the string "byjung" did crop up unexpectedly—just as a laboratory technician passed the room with a copy of *The Portable Jung* tucked into her pocket. Jungian synchronicity at its best. After the inevitable paper jams, celebratory drinking, and psychedelic drug use, the party disbanded. No hard conclusions to the mysteries of quantum mechanics, perhaps, but a good time was had by all.[51]

•

When Fundamental Fysiks Group cofounder George Weissmann arrived in Berkeley to study physics in 1971, he was, by his own lights, "a complete materialist." He had no truck with those who chased woolly spirits or pored over works by Eastern mystics. All that changed abruptly in 1974 when his father died, and George had what he can only describe as a "mystical experience" lasting several days. Looking back, he cites that event as the "awakening" he had needed. He wandered in and out of various Berkeley discussion groups, and worked his way through books like *Time, Space, and Knowledge*, a study of Tibetan thought by an American religous scholar. He delved more deeply into Buddhism, and he returned to books he had read as a teenager, including the writings of controversial French Jesuit priest and paleontologist Pierre Teilhard de Chardin. Writing during the early decades of the twentieth century, Teilhard pursued a notion of teleological evolution: all matter evolved in a goal-directed way toward greater and greater complexity. Consciousness emerged at critical stages of this complexification: it inhered in one form or another in seemingly inanimate objects like rocks and plants, and in higher, self-aware forms in humans. Teilhard posited a realm of shared consciousness, or "noosphere," extending beyond the minds of isolated thinkers. Immersing himself in Buddhist texts on the one hand and Teilhard's on the other, Weissmann recalls, made it "possible for me to think about quantum nonlocality." All the while, Weissmann had been interested in anomalies—tiny, seemingly inexplicable phenomena that might point to some hidden layers in the laws of physics. His twin interests in mysticism and anomalies pushed him ever more quickly into the realm of parapsychology.[52]

Elizabeth Rauscher, who founded the Fundamental Fysiks Group with Weissmann, had likewise caught the parapsychology bug. Like Sarfatti, her entrée into the world of psi came via the Stanford Research Institute. What had grabbed her was not the Geller studies, but a different set of experiments. One month after Uri Geller had arrived in Puthoff and Targ's psi lab, another unusual guest had appeared: a "spook" from the Central Intelligence Agency. The CIA, like other branches of the defense establishment, had begun to harbor fears of a "psi gap" vis-à-vis the Soviets, the consequences of which could prove as devastating (according to some) as the missile gap and the manpower gap. (Never mind that neither of those previous "gaps" had been real.) In July 1972, the Pentagon's Defense Intelligence Agency completed a lengthy classified report, entitled *Controlled Offensive Behavior: USSR*, detailing what was known about parapsychological research behind the Iron Curtain. "The Soviet Union is well aware of the benefits and applications of parapsychology research," declared the report's opening summary. "Many scientists, US and Soviet, feel that parapsychology can be harnessed to create conditions where one can alter or manipulate the minds of others. The major impetus behind the Soviet drive to harness the possible capabilities of telepathic communication, telekinetics, and bionics are said to come from the Soviet military and the KGB." And they were already off to a strong start: "Today, it is reported that the USSR has twenty or more centers for the study of parapsychological phenomena, with an annual budget estimated at 21 million dollars." With such a robust institutional base, the conclusion seemed inescapable: "Soviet knowledge in this field is superior to that of the US." Might the Soviet military and KGB be leaping ahead with new breakthroughs in telepathy, mind control, and psychokinesis?[53]

The CIA operative approached Puthoff to try to close the psi gap. Puthoff's prior experience in Naval Intelligence and the National Security Agency, combined with his new psi lab at the Stanford Research

Institute, made him an obvious target for CIA largesse. The agent hammered out an initial contract with Puthoff and Targ, and by October 1972 the first installment of $50,000 was in hand ($260,000 in 2010 dollars). Additional seed money came from the National Aeronautics and Space Administration (NASA), thanks in no small part to the urging of astronaut-telepathist Edgar Mitchell.[54]

With the infusion of cash, Puthoff and Targ rapidly expanded a side project that had been running alongside their Geller studies. The laser physicists had been working with several seers, some of whom had claimed prior psychic abilities and others who had not. The SRI scientists' goal was to investigate whether one person could receive telepathic messages or visual stimuli from another person, even if the "sending" person were far away from the "receiver." They dubbed the phenomenon "remote viewing."

A colleague at the Stanford Research Institute drew up a list of 100 locations in the San Francisco area, including swimming pools, children's playgrounds, a bicycle shed, specific benches on the Stanford University campus, a toll plaza, and so on. Each target location was within thirty minutes' driving distance from the SRI laboratory. The protocol that Puthoff and Targ reported included its share of spycraft spice: a member of the SRI upper management, "not otherwise associated with the experiment," drew up the secret list of target locations. He printed the name and address of each location on a separate card. Each card was then sealed in an envelope, each envelope assigned a number, and the whole stash locked in the division director's office safe. When the time came to conduct a remote-viewing test, the division director used a random-number generator to select a particular envelope from the pile in his safe. An outbound "target demarcation team" received the card from the division director, hopped in a car, and drove off to the specified location. Once they arrived at the site, their job was to stare intently at the specific object or location for fifteen minutes. Meanwhile, back in the laboratory, a test subject and an experimenter—neither of whom had any knowledge of the set of target locations, let alone the particular location toward which the outbound team was speeding—would wait for thirty minutes (to allow for the outbound team's travel time). Then the remote-viewing subject would begin to describe into a tape recorder any images or impressions that came to mind; the subject could also draw pictures. The experimenter who remained at the lab with the remote-viewing subject would ask questions to prompt further description or ask for clarifications. The subject's verbal descriptions were then transcribed, and the transcripts and drawings given to a panel of judges, along with a stack of photographs of the target locations that had been visited. The question became whether the judges would discern any statistically significant matches between the target locations and the stream-of-consciousness descriptions produced by the remote-viewing subjects.[55]

Puthoff and Targ reported some astounding results. After nine remote-viewing subjects had completed a total of fifty-one experiments, judges matched viewers' descriptions to photographs of the target locations at well above chance levels. In some cases, the odds appeared to be one in a million that the associations could have occurred merely by random chance.[56] The laser physicists managed to publish their findings in top-ranked scientific journals. They used appropriately scientific language to describe their results: "Information transmission under conditions of sensory shielding," for example, as they titled their 1974 article in the journal *Nature*; or "A perceptual channel for information transfer over kilometer distances" in their long 1976 article in the *Proceedings of the Institute of Electrical and Electronics Engineers*.[57] Journalists who caught a whiff of the research peppered their reports with juicier language, such as "mystic powers" and "supernatural phenomena." "If a man walked up to you on the street and told you that you had amazing mental powers that would enable you without using any equipment whatever to see through walls and watch things happening miles away," began one long article in the *San Francisco Chronicle*, "you would probably give him the fishy eye and walk away as quickly as possible"—and yet (the report continued) that was precisely what Puthoff and Targ seemed to be able to replicate in their laboratory at the prestigious Stanford Research Institute.[58]

Only decades later, after many of the early contracts and technical reports from the SRI remote-viewing work were declassified in the 1990s, did a fuller picture begin to emerge. The documents revealed an expensive and long-lived program, clandestinely funded by the Central Intelligence Agency, the Pentagon's Defense Intelligence Agency, and related national-security bureaus, to develop what some advocates jokingly called "ESPionage": the use of extrasensory perception (ESP) to peer into secret military establishments within the Soviet Union and elsewhere.[59]

Long before the national-security impetus behind remote viewing came out into the open, Elizabeth Rauscher had been fascinated by the local press reports on Puthoff and Targ's work. The SRI work seemed to herald new breakthroughs in the nonlocal nature of human perception. Rauscher decided she had to learn more about it. The headstrong self-starter, who used to gate-crash her way into the Lawrence Berkeley Laboratory as a high school student, decided there was no need to wait for an invitation to visit the SRI psi lab. She just showed up on Puthoff and Targ's doorstep one day. They tried to give her the brush-off until she showed them a long manuscript she had been working on, concerning theoretical efforts to explain nonlocality.[60]

Rauscher had been dabbling in relativity and cosmology since her return to graduate school. An autodidact in those fields, she got in touch with Princeton physicist and relativity specialist John Wheeler—the selfsame Wheeler of "participatory universe" fame with whom Jack Sarfatti was also enjoying an active correspondence at the time. Wheeler made frequent trips to NASA's Ames Research Center near San Francisco, and Rauscher often met with him while he was in town. They continued their discussions via letter, and he encouraged her forays into relativity and cosmology.[61] She had published a few short papers on obscure relativistic models and had begun to write a long monograph on the subject when she first heard about the Puthoff-Targ work on remote viewing.[62]

In the course of her work, Rauscher realized that one way to account for nonlocal effects—perhaps even to explain Bell's theorem, at a deeper level—would be to increase the number of dimensions of space and time. She began toying with a model in which the familiar coordinates of space and time were made complex: instead of a single dimension of time, for example, there would be two, a real component and an imaginary component. A similar doubling of the three dimensions of space (height, breadth, and depth) led to an eight-dimensional space-time rather than Einstein's four-dimensional version. The expanded space-time would contain new sets of shortest paths between here and there. What might look like a far spatial distance within a four-dimensional world might in fact have *no* space-time distance within the enlarged eight-dimensional universe. A long duration of time, as viewed within the four-dimensional slice, might take no time at all when viewed within the larger multidimensional system.[63]

When Rauscher got to those last features of her model, Puthoff and Targ stopped trying to shoo her out the door. For some time, they had been grasping for deep physical explanations that might account for their puzzling experimental results. They had routinely ended their early reports on remote viewing with a gesture toward Bell's theorem and quantum nonlocality, but they had not pursued the connection any further.[64] What they really needed was a house theorist—a consultant, expert in theoretical physics, who could work alongside them and focus on establishing some first-principles explanation, based on the laws of physics, that might explain the mysterious remote viewing phenomenon. Rauscher fit the bill. Her model explained, or at least could take into account, why their remote viewers seemed able to receive signals, instantaneously, across great distances; why the strength of those signals did not seem to fall off with distance; and even why some viewers seemed to receive signals from the future ("precognition"). Almost immediately, Puthoff and Targ arranged for Rauscher to serve as a paid consultant to their psi lab at the Stanford Research Institute. The extra consulting fees no doubt came in handy for the young mother trying to make ends meet on a graduate-student stipend.[65] (Fig. 4.6.)

Rauscher began her consulting work at the psi lab one year before she and Weissmann started the Fundamental Fysiks Group. By the time the Berkeley discussion group began she had participated in, or closely observed, dozens of remote-viewing experiments. In her mind, the experimental data on remote viewing seemed at least as statistically solid and repeatable as the one-in-a-million "golden events" that particle

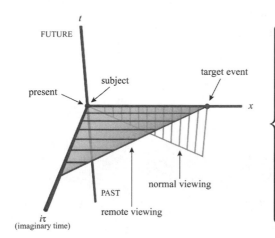

FIGURE 4.6. Elizabeth Rauscher's multidimensional approach to nonlocality and remote viewing. In this example, a subject sitting at the origin ($x = 0$ and $t = 0$) can receive signals instanteously from the "target event," separated in space from where she is sitting, if those signals travel through the imaginary-time dimension, τ. (Illustration by Alex Wellerstein, based on Rauscher [1979], 60.)

physicists chased with their huge accelerators. Even with her healthy dose of skepticism about the paranormal, she reasoned that "any subject (even if it doesn't exist) is a science, if the methodology of science is used to study it."[66] And so, as the first order of business for the Fundamental Fysiks Group, Rauscher, Sarfatti, Sirag, and Weissmann set out to replicate the Stanford Research Institute remote-viewing experiments. They dedicated all of June and July 1975 to the new experiments. In the end, they failed to find any statistically significant results, as they reported at that summer's annual meeting of the Parapsychology Association: independent judges only managed to match viewers' descriptions and sketches with photos of the target locations at chance levels. But they did find intriguing correlations all the same. One viewer produced surprisingly detailed descriptions of *different* targets, shifted from the intended target by a day or so. Perhaps, Rauscher and company suggested, this viewer had received precognitive visions of where the outbound observer would be going for the next session. All ample material that the Fundamental Fysiks Group pursued in follow-up sessions throughout the next year, including presentations by Puthoff and Targ themselves.[67]

Chapter 5

New Patrons, New Forums

The purpose of the PCRG [Physics/Consciousness Research Group] is to foster philosophical inquiry in quantum physics for the increased well-being of modern civilization's people, animals, and plants. PCRG recognizes that knowledge, in the form of critical inquiry, pursued in a context of love, is a path to spiritual wisdom. The ideal style of PCRG is Plato's Academy in Hellenic Athens rather than the Hellenistic Scholarship of the Great Library of Alexandria that rules the modern University.

—**Jack Sarfatti, 1977**

Like any intrepid explorers, members of the Fundamental Fysiks Group needed more than just passion to achieve their goals. They needed dependable base camps from which to mount their intellectual expeditions. Most of all, they needed cash. Given their mix of interests and the state of academe, hobbled by budget cuts and job losses, the young physicists needed to look beyond the usual physics institutions. They needed to carve out their own niche, a parallel universe with many of the trappings of the academic world but few of the constraints. With entrepreneurial flair, members of the group secured financial backing from some unusual patrons. Alongside CIA handlers and Pentagon officials, self-made millionaires with a hankering for quantum weirdness stepped in and kept the group afloat. With funds in hand the physicists forged new forums, safe spaces in which they could explore everything from the hidden byways of quantum theory to the nature of consciousness and the mysteries of the paranormal.

•

Jack Sarfatti's exuberant press releases about the Uri Geller tests, and the broader media coverage that members of the Fundamental Fysiks Group had begun to attract, invited some concerted pushback. Harold Puthoff, Russell Targ, and their psi lab at the Stanford Research Institute—whose tests of Uri Geller's psychic abilities and other paranormal phenomena such as remote viewing had captivated several members of the Fundamental Fysiks Group—did not suffer from any lack of spirited debunkers. Experimental psychologists, rather than other physicists, were quickest to jump into the fray. Some questioned the degree to which judges in the SRI remote-viewing experiments could have been swayed, consciously or unconsciously, by textual clues about the intended targets left in the transcripts. Others went after fine points of method, such as whether it was legitimate not to replace targets in the pool after one had been selected at random, thus steadily reducing the pool of potential targets with each viewing session. Such discussions trickled out with reasonable decorum, point-counterpoint, in major scientific journals such as *Nature*.[1]

Not everyone was content with the kid-gloves approach. Some came out swinging, comparing Puthoff's and Targ's research to Lysenko's decimation of Soviet genetics research after World War II; both, these critics argued, threatened to edge out legitimate science. Magician and parapsychology watchdog James "The Amazing" Randi—who had demonstrated to Jack Sarfatti how old-fashioned magic tricks could replicate many of Uri Geller's remarkable feats—took a more humorous approach. He dedicated an entire chapter of his popular book on "delusions" to the SRI psi lab's exploits, calling Puthoff and Targ "the Laurel and Hardy of psi." Thirty pages detailed what Randi considered to be Puthoff's and Targ's crimes against scientific method, statistical reasoning, and common sense.[2]

Physicist John Wheeler, whose ideas and correspondence had so inspired Sarfatti, Elizabeth Rauscher, and the others, likewise threw down the gauntlet. Wheeler prepared a talk for the January 1979 meeting of the American Association for the Advancement of Science (AAAS) to clarify why he thought his interpretation of quantum mechanics had been misappropriated in the parapsychological realm. To his dismay, he showed up at the meeting only to discover that his talk was scheduled in a session on science and consciousness alongside Harold Puthoff! Wheeler went on the warpath. First he appealed to the AAAS leadership to revoke the membership that had been granted, back in 1969, to the Parapsychological Association. The time had come, Wheeler charged, to "drive the pseudos out of the workshop of science." Then he trained his sights more narrowly upon Sarfatti, Rauscher, and company. Wheeler closed his AAAS lecture by admonishing, "Let no one use the Einstein-Podolsky-Rosen experiment to claim that information can be transmitted faster than light, or to postulate any so-called 'quantum-interconnectedness' between separate consciousnesses. Both are baseless. Both are mysticism. Both are moonshine." He elaborated in a long article in the *New York Review of Books*, "Quantum theory and quack theory," coauthored with longtime popular science writer Martin Gardner. (Jack Sarfatti latched onto Wheeler's charge of "moonshine," gleefully repeating to anyone who would listen that before the Manhattan Project, the great nuclear physicist Ernest Rutherford had famously rejected nuclear fission as "moonshine.")[3]

More than just editorialize, critics such as Randi, Gardner, and Wheeler began to organize. They formed groups like CSICOP (the Committee for the Scientific Investigation of Claims of the Paranormal) and ASTOP (the Austin Society to Oppose Pseudoscience; Wheeler having moved from Princeton to the University of Texas at Austin). Labeled a "scientific-vigilante organization" by some sociologists at the time, CSICOP attacked what its members considered New Age excesses. They conducted replication studies, founded a journal (the *Skeptical Inquirer*), and issued their own press releases, at times blurring the line between seemingly objective scientific body and self-interested lobbying group.[4]

And yet, as we now know, the joke was ultimately on the debunkers. Despite the thoroughgoing criticism and the overheated rhetoric, research on remote viewing continued unabated for more than twenty years, paid for with more than $20 million of taxpayer money (in 2010 dollars). The initial exploratory grant of $50,000 from the CIA, back in October 1972, snowballed over the years, with frequent inputs from the Defense Intelligence Agency and other branches of the Pentagon. While Wheeler pleaded with the American Association for the Advancement of Science to bar research like Puthoff and Targ's, the budget for their psi lab at the Stanford Research Institute swelled to nearly $1 million *per year* (about $3 million per year in 2010 dollars). Top-secret spin-offs sprang up around the country, usually established with Puthoff's help. No number of failed replications seemed to quell their backers' interest. When researchers at the Army's Aberdeen Proving Ground in Maryland conducted their own pilot study in 1978–79—having dished out $100,000 in consulting fees to Puthoff's SRI lab to get them going—they found no statistically significant results. But just like Rauscher and the Fundamental Fysiks Group members, the investigators at Aberdeen (including Evan Harris Walker, of consciousness-hidden-variables fame) had found enough surprising gems in the transcripts to keep at it. "The evaluation process is truly an art," concluded the secret Aberdeen report. "Our replication of the [SRI] protocol did not result in statistical significance," the report conceded, but "we learned a great deal about ourselves." And so the cash kept flowing.[5]

•

Puthoff and Targ thus managed to ride the Sputnik-era patronage machine into the New Age. Even after the Cold War bubble had burst, they still derived the bulk of their funds from defense-oriented agencies in the federal government—albeit agencies different from the ones that had funded most basic research in physics until then. Others became even more creative in their search for funds. They cultivated relationships with a different type of patron: eccentric enthusiasts with burning questions and money to spare. Here again members of the Fundamental Fysiks Group hearkened back to an earlier way of doing physics. Much like their quest to bring philosophy and big-thoughts speculation back into physicists' daily routines, their fund-raising efforts more closely resembled patterns from the 1920s and 1930s than from the Cold War years. During those earlier times, nearly all funding for physics research had come from private donors, philanthropical foundations, and local industries rather than from the federal government.[6] The hippie physicists thought it might be time to give that older funding model another try.

First to help out was Arthur Young, whose Institute for the Study of Consciousness in Berkeley helped to bring Saul-Paul Sirag, Nick Herbert, and Elizabeth Rauscher together. Young, a Princeton-educated engineer, had spent much of the 1930s and 1940s tinkering with ideas (and filing patents) for what would become the first commercially licensed helicopter. Jarred by the atomic bombings at the end of World War II—and financially secure thanks to his new helicopter design, which first took to the air in 1946—Young backed away from engineering and turned squarely to his other passions from undergraduate days: philosophy, Jungian psychoanalysis, and Eastern spirituality. By the time he opened his Berkeley institute in the summer of 1973, his intellectual journey had brought him squarely back to the mysteries of modern physics, some of which he had first dabbled with at Princeton. As he argued around that time, science could "best serve mankind and regenerate its search for truth" by merging the insights of quantum theory with "recent work in psychology, and perhaps ESP"; Heisenberg's uncertainty principle promised insights into the human condition, with its constant negotiation between "freedom and necessity."[7] He began to host outside speakers at the institute every Thursday night. Harold Puthoff and Russell Targ from the Stanford Research Institute became regulars, as did Berkeley physicist and Fundamental Fysiks Group charter member Henry Stapp. They were joined by philosophy professors and an eccentric computer scientist from a neighboring university. As the institute's live-in research assistant, Sirag began hosting a complementary discussion group, each Tuesday night, focused more specifically on physics and consciousness. They called themselves the "Consciousness Theory Group." Backed by Young's generous resources, the new group flourished. As Nick Herbert put it recently, Young's institute served as a "wonderful intellectual salon and sanctuary for the pursuit of the unusual, the extraordinary, and the marvelous."[8]

Within a few years, however, Sirag began to feel restless. Arthur Young seemed less inclined to entertain theories of physics and consciousness that deviated from his own. As it happened, a new wealthy patron appeared just in time: the toy manufacturer and paranormal enthusiast Henry Dakin. Dakin invited Sirag (who was dating Dakin's secretary at the time) to move into a house that Dakin owned across the street from his office in downtown San Francisco, and to continue the Consciousness Theory Group from there. In addition to Sirag, Herbert, and Rauscher, the group attracted experts in computer programming, visualizations of brain activity using electroencephalography (EEG), a biophysicist experimenting on the "psychic healing of bacteria," and more. One member had access to anatomy laboratories on Berkeley's campus and shuttled the group in after hours to examine dissected brains. Another, Charles MacDermid, an early pioneer in electronic music, spiced group meetings with his "weird vibrations," as Herbert put it. Backed now by Dakin's largesse—a place to hold meetings, and some extra cash to invite outside speakers such as hidden-variables theorists David Bohm and Evan Harris Walker—the Consciousness Theory Group aimed at nothing short of cracking the mystery of consciousness. And they were confident: they had a diverse mix of talents and they would stop at nothing in their quest. As Herbert explained, "We would take any drug (some of us), compose bizarre music, use EEG output in unusual ways, consort with psychics, Tarot [card] readers, tricksters, shamans, sex magicians and millionaire toy manufacturers (Henry Dakin)." Between Young's patronage and Dakin's, the group met twice a month for more than three years.[9]

Werner Erhard also became a generous patron. Fred Alan Wolf had met Erhard in London in May 1974, when Erhard asked if any physicists were in attendance during his May Lectures presentation; Wolf obliged during the intermission and was invited into the inner sanctum. The next month, back in Paris, Wolf brought Sarfatti to meet Erhard in the lobby of the Ritz Hotel. Once Sarfatti broke the awkward silence by insulting Erhard—using, as it happens, a slogan associated with Erhard's self-help program—he and Wolf were in.[10]

Werner Erhard (born Jack Rosenberg) had undergone a remarkable transformation just a few years earlier. Abandoning his wife, four children, and car-salesman job in Philadelphia, Rosenberg took off with a mistress (whom he later married) to forge a new life. After adopting his new alias, Erhard worked for a while in the encyclopedia business, quickly rising to managerial ranks. Soon he emerged as the enigmatic guru at the heart of the "human potential movement." His *est* workshops ("Erhard Seminars Training"), founded in 1971, had brought in nearly $3.4 million ($15 million in 2010 dollars) by the time he met with Wolf and Sarfatti just three years later. The group could already count at least a dozen "Sphere of Influence People" among its graduates, as internal *est* memos referred to them: famous entertainers (John Denver, Diana Ross, Valerie Harper, Roy Scheider, Cloris Leachman), astronauts (Edwin "Buzz" Aldrin), political advisors (Nixon's White House counsel John Dean), university presidents (Oberlin's Robert Fuller), Olympic athletes (skier Suzy Chaffee), and more. Far beyond this elite group, the *est* workshops garnered mass appeal. Within a few years, Erhard's organization could claim more than half a million graduates throughout the United States, each of whom had plunked down $250 for an intensive two-weekend, sixty-hour group-therapy session in a hotel ballroom.[11]

These days Erhard avers that *est* was distinct from the human potential movement, though journalists routinely categorized it as such at the time. As Erhard sees it, *est* was "a lot more rigorous in its thinking" than most of what passed for "human potential" back then. "It was a logical unfolding that brought people to insights that they found valuable in supporting themselves regarding the quality of their life, and their effectiveness in life."[12] Erhard and *est* quickly developed a vast and loyal following. Yet beginning in the mid-1970s, some critics began to allege that the methods employed in *est* sessions were "excessively confrontational." Three psychiatrists, one of whom had undergone the *est* training, cautioned in the pages of the *American Journal of Psychiatry* that *est* trainers "employ a confrontational, authoritarian model and often respond to disagreement from the participants with intimidation and ridicule."

It was reported that *est* trainers sometimes taunted and insulted participants in their effort to goad people into "getting it," Erhard's favorite phrase for a process by which someone could reevaluate deeply held beliefs and reassert control over his or her life.[13] *Newsweek* magazine reported Erhard's reaction to the latest critiques: "Erhard regards the report [in the *American Journal of Psychiatry*] as yet another failed attempt to find fault with his patented system of self-help. 'They've tried to dress *est* in other costumes like brainwashing and Fascism,' he says. 'Now it's psychosis-inducing. It's a legitimate process we have to go through, but none of the costumes fit.'"[14] Criticisms continued to swirl alongside glowing testimonials from some who had passed through the *est* training. All the while the rolls of *est* graduates continued to swell.[15]

Wolf and Sarfatti stumbled into Erhard's orbit with remarkably good timing. Ever since boyhood, Erhard had been fascinated by science, and by physics in particular. He had sought out popular treatments of physics during high school, intrigued by what he considered the "counterintuitive" features of modern physics. He kept wondering, "How did these people come up with that kind of an insight?" Years later, when he needed to adopt a new first name for his alias, he chose "Werner" after the fabled quantum physicist Werner Heisenberg.[16] With his newfound wealth, Erhard aspired to become a major benefactor for cutting-edge research. Just a year before meeting Wolf and Sarfatti, he had undertaken massive renovations to his San Francisco mansion, known as "Franklin House," which doubled as his living quarters and *est* headquarters. Throwing hundreds of thousands of dollars at the project, he aimed at nothing less than to establish it as "San Francisco's most dazzling salon." An investigative journalist—no fan of Erhard or *est*—concluded that "With a formal education that ended in high school, Erhard was determined to overcome his own intellectual shortcomings by surrounding himself with those whose very presence in his home would help to confirm his reputation as an enlightened source of big ideas."[17]

Once they were back in California from their European forays, Sarfatti and Wolf, joined by Saul-Paul Sirag, began working with Erhard's *est* trainers, coaching them on the new physics of Bell's theorem and nonlocality. Soon the consultants had a place of their own, when Erhard provided start-up funds for the physicists to establish the Physics/Consciousness Research Group, or PCRG. Sarfatti filed articles of incorporation with the state of California to establish the PCRG as a tax-exempt, nonprofit corporation. President and treasurer, Jack Sarfatti; vice president, Saul-Paul Sirag. As the corporate filing stipulated, "The specific and primary purposes of this corporation are to support new research, to publish scientific work, and to educate the general public on fundamental studies concerning the nature of consciousness in its relation to the laws of physics."[18] Erhard's charitable foundation donated about $5000 to help the group get started (more than $20,000 in 2010 dollars). With additional cash from a wealthy UFO enthusiast, the PCRG set up shop on two floors of an office building in San Francisco's tony Nob Hill neighborhood.[19]

Soon after the group's founding, Sirag explained that their goal was to "communicate the excitement and adventure of modern theoretical physics to the people in imaginative forms of communication." He was no stranger to those "imaginative forms of communication." The previous year he had composed a science-fiction opera, in which a physicist invents a means of time travel and confronts several paradoxes of causality and faster-than-light communication. (Charles MacDermid, the electronic music aficionado and Consciousness Theory Group member, composed the score.) Sirag was fishing around for some genuine physics notions to shore up his plot when he stumbled upon Sarfatti's forthcoming paper in *Psychoenergetic Systems* on quantum theory and Uri Geller's psychic powers, which Jack had written while on his European sabbatical. "Jack's paper appeared synchronistically," Sirag concluded, and he pasted some lines from it directly into the libretto.[20] Once they teamed up and began running the PCRG, they continued in a similar vein. They held public seminars on the new physics and composed ersatz curricular materials. One was Sarfatti's "Time Traveller's Handbook," a fictional prose-style account of themes similar to Sirag's opera, intended to educate and entertain. After all, Sarfatti proclaimed in the handbook, "Scientific speculation is exciting and a turn on." The handbook bristled with pop-culture allusions. Sarfatti couched his explanation of space-time diagrams of relativity, for example, in the language of Baba Ram Dass's 1971 best-seller, *Be Here Now*. (Born Richard Alpert, Ram Dass had taught psychology at Harvard alongside Timothy Leary until both were dismissed for experimenting on undergraduates with psychedelic drugs.) Sarfatti enlisted the Beatles' song "Being for the Benefit of Mr. Kite!" to clarify his distinction between classical and quantum physics. Physicist and quantum pioneer Max Born's description of instant, acausal quantum jumps came paired with Werner Erhard's own aphorism that "The only thing there is is instant enlightenment. It happens out of time, so it is really instantaneous."[21]

The novelist and playwright Robert Anton Wilson captured some of the flavor of the PCRG happenings in an article published in a Bay Area underground newspaper in 1976. Sarfatti, Sirag, and Nick Herbert had conducted a seminar on "Quantum physics and the transformation of consciousness" at Pajaro Dunes on Monterey Bay. Wilson reported that one participant had nicknamed the seminar a "quantum wonderland," although Wilson preferred his own labels of "a psychedelic Mensa" and "Zen physics seminar" instead. "This is some kind of epistemological Encounter Group, right?" another had asked. "As Physics/Consciousness organizer Dr. Jack Sarfatti said, sounding remarkably like Esalen or *est*, 'The observer is part of the data; maybe the observer even creates the data. There is no universe without you. You are essential,'" Wilson reported. A highlight for Wilson came when Nick Herbert played a recording of the sounds that came out of his metaphase typewriter. "We listened, I can assure you, as raptly as John Lilly ever listened to his dolphins," Wilson relayed. "The Physics/Consciousness Research Group is into encountering quantum reality totally—intellectually, emotionally, intuitively," he went on. "It accepts that the universe really is quite different from our traditional Aristotelian logic, Euclidean geometry and Newtonian causality," and hence that the universe of modern physics might best be described "in the metaphors of Zen, Taoism and Vedanta, or even in the language of parapsychology, ESP and shamanism." At last, "the climax of the seminar was decentralized and appropriately

Taoist," Wilson concluded, "as we split up into small bull-sessions to mull a bit on The Meaning Of It All."[22]

Some of the group's seminars met in the Nob Hill headquarters; others were held at local community colleges. They developed lecture series on the philosophy of quantum mechanics, possibilities for communication with extraterrestrials, physical models of the mind and brain, and "Science and religion in an uncertain quantum reality." Even the more traditional topics acquired a distinctive spin. Sarfatti's presentations on the development of quantum mechanics started out in a by-then standard way, recounting major developments by Albert Einstein, Niels Bohr, Werner Heisenberg, and others, and marched in short order to Puthoff, Targ, and remote viewing. The group launched a quarterly newsletter to keep subscribers up-to-date on the latest seminar offerings. They also began sending their pamphlets and preprints to a wide and eclectic group: everyone from the movie mogul Francis Ford Coppola and Uri Geller's associate, Andrija Puharich, to renowned physicists like John Wheeler and Richard Feynman and the editors of *Nature* and *Scientific American*.[23] All the while, the group's ties to Werner Erhard and his brand of self-improvement were never far from the surface. Flyers encouraged participants to use "the metaphors arising out of relativity, quantum physics, and the thermodynamics of information"—as detailed in the group's seminars and pamphlets—as "useful tools for you to create new ways of handling personal and professional relationships."[24]

For a time, Sarfatti enjoyed close interactions with Erhard. When a critic tried to tar the PCRG by association with Erhard, once Erhard's *est* workshops had begun to receive some negative publicity, Sarfatti shot back. "Mr. Erhard and I share a warm personal relationship," he made plain, "that enriches both of our lives independent of ideological and scientific beliefs."[25] In addition to the start-up funds provided by Erhard's charitable foundation, Sarfatti arranged for "several small grants from the personal account of Werner Erhard" to further the efforts of PCRG members, including a grant of $1500 to Fritjof Capra. A separate check for $2500 came from Erhard's foundation to support Saul-Paul Sirag's continuing research into physics and consciousness.[26]

Building on Erhard's generous support, the PCRG expanded its circle of donors. George Koopman, yet another eccentric entrepreneur, became one of the group's most significant backers. He had served as a military intelligence analyst during the Vietnam War. Some have alleged that when Koopman met members of the PCRG in the mid-1970s he was still working as an undercover agent for the Defense Intelligence Agency, covering what was known colloquially as the "nut desk"—that is, checking up on reports of UFOs and other occult or paranormal phenomena.[27] In response to Freedom of Information Act requests, neither the CIA nor the FBI would confirm or deny that Koopman had ever been on their payrolls; the National Security Agency did confirm that Koopman never worked for them. The Defense Intelligence Agency reported finding no records associating Koopman with the PCRG, but remained mum on whether Koopman had ever worked for the agency.[28] What is known for certain is that Koopman worked for a time making military training films as a contractor for the government. In fact, during the time he was sponsoring PCRG events, the FBI received a complaint against Koopman's filmmaking company, alleging that Koopman's firm had committed fraud against the U.S. government by acting on inside information from a local Air Force office. The tip, at least according to the complaint, had enabled Koopman's firm to lower its bid and hence squeeze out competition for a particular film project. After vetting the information provided by the FBI, the local assistant U.S. attorney declined to pursue the matter.[29]

Koopman's passion for filmmaking extended well beyond the occasional military training film. He coordinated stunts for the sleeper hit comedy *The Blues Brothers* (1980), starring John Belushi and Dan Aykroyd, including several car chases and the famous scene in which a police car fell onto the roof of a tall building, having been suspended by an (off-camera) helicopter.[30] Koopman liked to make things blast off as well as fall down. His next major venture was to found the American Rocket Company, a private firm specializing in low-cost delivery systems for launching payloads into space. In the midst of all those activities, Koopman became enamored of the Physics/Consciousness Research

Group. He financed one of the group's seminars at a ranch in Sonoma County during the summer of 1976, and participated in Berkeley discussion series as well, all while co-writing a book with psychedelics guru Timothy Leary.[31]

With cash flowing in from the likes of Erhard and Koopman, along with more modest donations from several other backers, the PCRG began to flourish. By its second year the organization had an annual operating budget of $35,000 (more than $130,000 in 2010 dollars). A big chunk went to rent, but most was spent on salaries for Sarfatti and Sirag and consulting fees to Nick Herbert, Fritjof Capra, Fred Alan Wolf, and others. The group's official profit and loss statement for 1976 showed additional loans ranging from $625 to $2500 paid to various group officers and consultants, over and above their salaries and consulting fees.[32] At least for a time, the physicists' gamble paid off: they began to thrive outside the usual funding model.

•

These overlapping discussion groups, experimental institutes, and public education forums merged in January 1976 for the first annual workshop on physics and consciousness at the Esalen Institute in Big Sur, California. Nestled in the cliffs overlooking the Pacific Ocean roughly 150 miles south of San Francisco, Esalen had served since its founding in 1962 as an incubator of the New Age movement. (Fig. 5.1.) Amid its large wooden buildings, famous hot-spring baths, and dramatic cliffside ocean views, Esalen had hosted informal workshops on everything from gestalt therapy and "transpersonal psychology" to the consciousness-expanding capacities of psychedelic drugs and Eastern mysticism. Critics dismissed the place as "a valhalla for frivolous self-absorptions," though even they agreed by the 1980s that much of what had seemed novel about Esalen's offerings had gradually seeped into the American cultural mainstream. One of the institute's founders, Michael Murphy, had long been fascinated (like Erhard) by the possibilities for "human potential" latent within modern science. Indeed, the cover of Esalen's first printed catalog, announcing its 1962 seminar series, fea-

FIGURE 5.1. The Esalen Institute in Big Sur, California, hosted a long-running workshop series on Bell's theorem, nonlocality, and their significance for human consciousness. (Courtesy Daniel Bianchetta.)

tured a fancy-looking calculus equation alongside a lotus flower, redolent with mystico-religious symbolism from Buddhist and Hindu traditions.[33]

A chance encounter had set Esalen's Michael Murphy onto the metaphysical track. Back in 1950, during his sophomore year at Stanford University, Murphy arrived at the wrong classroom, looking for a class on social psychology but landing instead in a course on comparative religion. The mesmerizing professor he stumbled upon opened Murphy's mind to the wide range of Eastern religions, especially Buddhism and Hinduism, and soon the young Murphy was hooked. The religious studies became more than an academic exercise; within months the former Episcopalian altar boy from a wealthy California family had forged a new personal faith, drawing equal inspiration from Indian Vedanta and Darwinian evolution. After a brief stint in the army during the mid-1950s—where he got in the habit of rising before morning reveille to meditate—he set off on a sixteen-month pilgrimage to a remote ashram in India. Ever the cultural hybrid, Murphy kept up his boyhood love of golf during his journey, and even taught his fellow adepts at the Indian commune how to play softball. Soon after he returned to the United States, he opened the Esalen Institute on a patch of coastal land that his family had owned for generations.[34]

By the early 1970s, Murphy was in regular contact with Werner Erhard, the two having come to symbolize the emerging California human potential movement. After donating early start-up funds to the PCRG, Erhard was also in contact with Sarfatti, Sirag, and several of the other physicists.[35] Murphy, in turn, invited Sarfatti to organize a month-long workshop at Esalen on "Physics and consciousness," open to invited participants during the weekdays and to the curious (and fee-paying) public on weekends. Murphy and Sarfatti announced the upcoming "Physics month" in the Esalen catalog. The sessions would focus on "some of the conceptual gaps and possibilities in theoretical physics and the relevance of modern physical thought for consciousness transformation on the planet." "One of the key questions," they clarified, "will concern the role of consciousness in the interpretation of quantum mechanics." Murphy, for one, hoped that the unique features of Esalen would stir some creative juices. "Perhaps a new kind of inspired physicist, experienced in the yogic modes of perception, must emerge to comprehend the further reaches of matter, space, and time." When it came to yogic modes of perception, Esalen was the place.[36]

Sarfatti set about lining up speakers. He tried to entice Caltech physicist and Nobel laureate Richard Feynman, who had already enjoyed a taste of what Esalen had to offer. (Feynman participated in a 1974 Esalen workshop led by John and Toni Lilly on sensory deprivation, out-of-body experiences, and some of their budding research on communication with dolphins.) "As director of the workshop," Sarfatti began, "I would very much appreciate your alive presence and hope that you will share your wisdom and genius with us!"[37] Feynman demurred. Despite his wide-ranging interests, Feynman had long been skeptical about philosophy. One of his many beloved anecdotes, told and retold late in life, centered on his frustration with a philosophy course through which he had suffered as an undergraduate. Even as he grappled with quantum theory in his own research, moreover, Feynman had consistently downplayed the kinds of philosophical engagement that Sarfatti and crew were seeking to pursue. Feynman had admonished his graduate students (and later the many readers of his influential textbooks) that the thorny matters of how to interpret the quantum formalism were all "in the nature of philosophical questions. They are not necessary for the further development of physics."[38] Thus when he received Sarfatti's invitation to the Esalen workshop, Feynman shot back a brief and characteristically humorous reply: "Due to the fact my doctor tells me I have labile blood pressure, I think it best that I do not attend because I know I would surely get involved in arguments."[39]

Others were more easily persuaded. Most members of the Fundamental Fysiks Group participated. Sarfatti also convinced David Finkelstein to attend. At the time Finkelstein was chair of the physics department of Yeshiva University in New York City; he would soon become director of Georgia Tech's School of Physics and editor of the *International Journal of Theoretical Physics*. Finkelstein had been toiling for a decade on an idiosyncratic approach to quantum gravity, long before that had become a mainstream topic among American physicists. Like so many others, Finkelstein had learned to leave such foundational topics alone during his graduate training at MIT in the early 1950s. That all changed during the mid-1960s when he volunteered two summers as a visiting professor at Tougaloo College, in Mississippi, as part of the Freedom Rides. During that time "I met some of the bravest people I know," Finkelstein recalled recently. "I felt so ashamed," he said: here were people laying everything on the line for something they believed in, putting themselves in harm's way to go knocking, door-to-door, to help register African Americans to vote. "They were doing such brave things. So I dropped everything," and put all his energies into what he had always wanted to do: develop a proper quantum-mechanical understanding of gravity. He had already derived a now-famous result about black holes. Now he pressed further. Did quantum mechanics demand its own formal logic structure? Could space-time itself be quantized?

By the late 1960s, his work had begun to appear in dribs and drabs; few took any notice. Finkelstein decided to beat the bushes. He sought out Feynman in 1975 and was not disappointed. "I had not had such a discussion before and do not expect one since," he wrote by way of thanks. "I imagined when I was reading the classic physics books that if I worked hard and had luck, some day I might find a place at the feast of reason, and that morning [at Caltech] I did."[40]

More feasts would soon follow. Sarfatti had been one of the few to notice Finkelstein's papers on "space-time code," and he invited Finkelstein to the Esalen workshop. By the time the invitation caught up with Finkelstein, he was already back in California for more face time with Feynman. The timing fit perfectly between Finkelstein's scheduled talks at Caltech and Berkeley. "I meant to spend one day at Esalen," Finkelstein recalled, "but wound up spending a week."[41]

Sarfatti expanded the speaker list beyond physicists. One prominent participant was Karl Pribram, the famous neurosurgeon and psychiatrist at Stanford University whose early work had clarified the structure and function of the human brain's limbic system and prefrontal cortex. At the time, Pribram was focusing on the question of consciousness from the vantage of neuroscience rather than quantum physics. Writing just days after the Esalen workshop had wrapped up, Pribram enthused to Werner Erhard about how much he had gotten from the experience. He had been "amazed at how little they [the physicists] knew about brain function," and was "pleased to find that they were enthralled by what has been accomplished" in the field. At the same time, "I was able to sharpen up many of my ideas on the possible configurations that the 'real' world might take. A goodly number of the ideas I have to work with come from physics and the interaction allowed me to express what I thought and to have misconceptions corrected." Most important to Pribram had been the format of the meeting. "The relaxed and informal atmosphere at Esalen leads to a kind of interchange which has become almost impossible anywhere else and I am grateful for being able to participate."[42]

Relaxed and informal it was. The Esalen workshop shaped up as half academic conference, half carnival. Speakers were slotted into two-hour sessions and offered the usual array of audiovisual equipment: overhead projector, 35-millimeter slide projector, blackboard. There the similarities to academe came to an end. Speakers wore crystals "as badge of office"; "large quartz and amethyst crystals were deployed around the room for beauty's sake and for their possible energy-transducing qualities." The goal was "to break the old scientific conference mold in which people standing at lecterns deliver formal papers to people sitting in chairs," Nick Herbert explained when preparing a follow-up workshop. "We wanted to become more mind-expanded, democratic, participatory, and delocalized, in the spirit of the New Physics. No problem at Esalen. There were no chairs to begin with, and the hot tubs, candles, and incense proved to be effective delocalization devices." When things got slow, people could always wander the grounds, get a massage, look out over the cliffs, or let their LSD trips take them where they may.[43] (Fig. 5.2.) Of course, some rules did apply. "Class space in the [hot-spring] baths must be reserved" through the central office, Esalen's conference staff reminded the organizers. Likewise, "If you use breathing methods in your workshops we ask

FIGURE 5.2. Nude couples on the balcony at the Esalen Institute, 1970. (Photograph by Arthur Schatz, reproduced by permission of Getty Images.)

that you inform all group members in advance of the contraindications involved." Most important: "Esalen policy excludes acting out of aggression in a way that might lead to physical injury. We ask that no group leader use coercion or pressure any person to participate in a way he or she does not choose." Good thing Feynman skipped after all.[44]

Esalen director Michael Murphy was so pleased with the month-long physics experiment that the workshops became a fixture. He invited Herbert and Sirag to organize a five-day version of the original workshop, focused on "Bell's theorem and the nature of reality." They obliged, and quickly set about planning the event. They shared their thoughts with Berkeley physicist and Fundamental Fysiks Group member Henry Stapp, who warned them (as Herbert recorded) that physicists might too easily slip into the "more comfortable territory of mathematical formalism, the fine details of proofs and experiments, rather than grapple with the difficult and unfamiliar task of constructing new realities consistent with Bell's discovery." "Stapp was exhorting us," Herbert concluded, "to shun the fleshpots of Egypt and to go forth and build new dwellings in the wilderness." That was precisely what Herbert and Sirag set out to do. Their first goal was to pare down the various derivations of Bell's theorem to their most basic form, removing extraneous assumptions. As Herbert noted, that task was made easier by the extensive analyses that had already been hashed out during Fundamental Fysiks Group discussions. "Could we extend this botany of possibilities at Esalen?" he wondered.[45]

Of course, as Stapp had warned them, Herbert, Sirag and their Esalen interlocutors did not have an easy time of it. Categorizing each others' candidates for the ultimate nature of reality was no simple matter. "Epithets" were "hurled at certain theories and theory-makers" during one "particularly disputatious session," which took on appropriately cosmic contours. "The weather seemed to track our moods," Herbert reported: "storm and heavy rain at the beginning (roads slide-blocked, power out at Esalen), later sparkling sunny with hundreds of lofting Monarch butterflies mocking physicists' airy notions of what can and cannot be beneath the ever-present world of phenomena." And so they went, hammering at the questions of Bell's theorem, nonlocality, and the nature of consciousness during their intense workshop.[46] Again Murphy was pleased, and Herbert's and Sirag's workshops became an annual event, meeting virtually every year until 1988—making them the longest-running seminar series in Esalen's history. In addition to Herbert and Sirag, John Clauser, Henry Stapp, and David Finkelstein became regulars, presenting their latest research each year on quantum nonlocality and the interpretation of quantum mechanics. (Fig. 5.3.) Fritjof Capra organized related workshops at Esalen on parallels between modern physics and Eastern mysticism.[47] Even Caltech's Richard Feynman got into the act. Though he had turned down Sarfatti's invitation to the original workshop in 1976, he reversed course and hosted his own (competing) Esalen workshop a few years later. The appeal of spending more time at Esalen—he had long admired the beautiful scenes along the rocky coast as well as in the naked co-ed hot-spring baths—seems to have trumped his allergy against dabbling with philosophical questions. Feynman's 1983 workshop, entitled "The quantum mechanical view of reality," featured in-depth discussions of Bell's theorem and the Einstein-Podolsky-Rosen paradox, interspersed "with sessions of primitive drum playing, yoga exercises, etc."[48]

•

The human-potential mavens Werner Erhard and Michael Murphy catalyzed further fund-raising efforts. The PCRG, for example, which Erhard had helped to get off the ground, announced soon afterward that it was raising funds to support John Clauser's next-generation nonlocality experiments. Securing such funding was no small matter: at just that time, a young physicist at Texas A&M, having become interested in Bell's theorem from Clauser, was hitting a brick wall in his efforts to get National Science Foundation support for his own version of Clauser's experiments. Repeatedly turned down from the usual funding sources, the Texas physicist, too, had to seek private support.[49] Meanwhile, Murphy and Nick Herbert convinced a wealthy participant in the Esalen workshops—Charles Brandon, one of the founders of shipping company Federal Express—to help support emerging research in the area. Brandon obliged, underwriting something they called "The Reality Foundation Prize." The first recipients were John Clauser and John Bell, who split the $6000 award money in 1982 (more than $13,000 in 2010 dollars). Bell was wary—he wrote to Clauser to inquire whether this was a "quack" group or not—and despite Clauser's reassurances, he declined to attend

FIGURE 5.3. Participants in the Esalen workshops on "Bell's theorem and the nature of reality," early 1980s. *Left*: Berkeley physicist Henry Stapp lectures on quantum nonlocality in Esalen's "big house." *Middle*: Georgia Tech physicist David Finkelstein explains his latest work on quantum logic to Esalen's codirector Michael Murphy. *Right*: Esalen's Michael Murphy ponders quantum reality. (Courtesy Nick Herbert.)

the award ceremony at Esalen. Herbert wrote to congratulate Bell on the award, assuring him that the champagne toasts drunk in Bell's honor had made the event "merry," but not undignified. To this day, meanwhile, Clauser's fancy plaque commemorating the Reality Foundation Prize hangs in his office at home.[50]

Though Bell turned them down, the curious little workshops at Esalen did attract other figures from Europe. The German physicist Dieter Zeh made his way from Heidelberg to Esalen for an intense meeting in 1983, for example, finding the Big Sur collective one of the few places in which he could discuss his ideas about the quantum measurement problem in sufficient detail. The group had discussed his work during previous workshops, so they were primed to dig into details with Zeh once he was able to join them. Nearly a decade passed before his "decoherence" interpretation of quantum measurement—now undeniably at the forefront of research—began to attract significant attention beyond the Esalen crowd.[51]

The French physicist Bernard d'Espagnat also seemed to take to the place. His 1971 book, *Conceptual Foundations of Quantum Mechanics*, had focused squarely on Bell's theorem and entanglement; he had also organized two influential summer workshops on the topic in Europe.[52] Yet he, too, felt the draw of Esalen, and accepted Nick Herbert's and Saul-Paul Sirag's invitation to participate in 1982. D'Espagnat's travel itinerary en route to Esalen illustrates how porous the boundaries between margin and mainstream could be. He piggybacked a short visit to John Wheeler's group on top of his trip to Esalen, asking Wheeler's secretary to coordinate directly with Sirag at Esalen to make all necessary arrangements.[53] On their way from Esalen to retrieve d'Espagnat from the local airport, Sirag and Herbert picked up a waterlogged hitchhiker, who had gotten doused in a big storm. Upon climbing into the backseat of their car, he announced that he was an armed robber, recently released from prison, who was making his way north to find food, clothes, and shelter. At the airport, d'Espagnat traded places with the hitchhiker in the car for the return trip. When Herbert and Sirag told them about their recent adventure, d'Espagnat replied (without missing a beat), "Who do you think is more dangerous—an armed robber or a theoretical physicist?" Once safely ensconced at Esalen, d'Espagnat quickly fell into a routine. He held "office hours" in the famous hot tubs: only once he and his interlocutors were reclining, naked, in the hot-spring baths would he discuss Bell's theorem and quantum nonlocality.[54]

Backed by money from unusual sources—most especially self-made millionaires and fixtures of California's New Age scene—the Berkeley physicists' Fundamental Fysiks Group, Physics/Consciousness Research Group, Consciousness Theory Group, and Esalen workshops became the only shows in town for puzzling through the implications of Bell's theorem and quantum entanglement. Berkeley's Henry Stapp closed a lengthy article on "Locality and reality" in 1980 by acknowledging the Esalen workshops for providing a space for him to work out his emerging ideas on entanglement and nonlocality. John Clauser, meanwhile—no fan of the Fundamental Fysiks Group's turn to parapsychology—exclaimed in a single breath in a recent interview that "those guys were a bunch of nuts, really," but that the group's "open discussion forum" was the only setting in which physicists could talk about the latest developments in quantum nonlocality. Clauser could likewise dismiss the annual Esalen workshops—"they would offer courses for mostly wealthy Los Angelinos and Bay Area folks who wanted to have their consciousness expanded"— and yet he rarely turned down an invitation to come speak about his latest work, year in and year out for the better part of a decade.[55] At a time when the vast majority of physicists still ignored or shunned the subtle business of how to make sense of quantum theory and its broader implications, members of the Fundamental Fysiks Group carved out flourishing, alternative spaces to keep the quest alive.

Chapter 6

Spreading (and Selling) the Word

> Packaging ideas as commodities, to be pushed in the same manner that other commodities are, lessens the impact they can make over time—unless that impact depends in part on scale of exposure or on linking specialists who are isolated by conventional disciplinary boundaries. In such cases, the market mode's capacity for promoting symbols as fad items may facilitate the process of new paradigm formation....A few of the counter-culture physicists have attempted a similar breakthrough of communication.
>
> **—Max Heirich, 1976**

Gathering around Esalen's hot tubs to discuss Bell's theorem was one thing. The "new physicists" faced the challenge of how to spread the fruits of their research beyond the Bay Area. At the time, physicists still could not hash out this material in the regular American physics journals. The longtime editor of the *Physical Review*—the mainstream workhorse of a journal, covering all topics in physics—actually banned articles on the interpretation of quantum mechanics. He went so far as to draw up a special instruction sheet to be mailed to referees of potentially offending submissions: referees were to reject all submissions on interpretive matters out of hand, unless the papers derived quantitative predictions for new experiments. As Fundamental Fysiks Group member and entanglement experimentalist John Clauser has pointed out, Bohr's famous response to the Einstein-Podolsky-Rosen paper, back in 1935, would hardly have qualified for publication under those strictures. Well into the 1970s, these policies shunted papers into unusual venues. Many went to the Italian journal *Nuovo Cimento*, which had adopted a more welcoming stance toward interpretive material. Others went to *Foundations of Physics*, the new journal founded in 1970 by two philosophically oriented émigré physicists working in the United States.[1]

Some of the most important papers, however, circulated by far more fragile means. Many appeared in a hand-typed, mimeographed newsletter, *Epistemological Letters*, including cutting-edge articles by John Bell himself. *Epistemological Letters* was produced by a private foundation in Switzerland and sent out to anyone who asked to be put on its mailing list.[2] Other papers circulated in crude photocopy form thanks to a larger-than-life character named Ira Einhorn.

Einhorn emerged as a darling of the New Left in the late 1960s. He led huge antiwar protests, hung out with famous Yippies like Abbie Hoffman and Jerry Rubin, and had a hand in some of the early environmentalist mass events, including the first Earth Day rally in 1970. While most of the high-profile hippies and radicals made their homes in San Francisco or New York City, Einhorn settled in his native Philadelphia. He became a local celebrity, the city's main conduit to the New Age. His ebullient intellectual energy charmed people from across Philadelphia's broad spectrum. College professors welcomed him and students flocked to him—especially to his freewheeling extension-school courses on psychedelic drugs. The city's down-and-out looked up to him as a community organizer who could get things done.

By the early 1970s, Einhorn had transformed himself into a freelance literary agent with an appetite for discussions about physics, consciousness, and the paranormal. He became one of Uri Geller's earliest promoters and served as a one-man distribution center for the Fundamental Fysiks Group's latest ideas. He also channeled several of them into the popular book market. Thanks to Einhorn's innovative networking, members of the Fundamental Fysiks Group performed an end-run around physicists' usual communication outlets. They managed to spread their message far and wide for the better part of a decade, until Einhorn's network came to a sudden and ignominious end.

•

Ira Einhorn grew up in Philadelphia, the oldest child in a proud, middle-class Jewish family. At the University of Pennsylvania in the late 1950s, he fell in love with the study of physics; he took physics courses his whole first year of college, intending to major in the subject. During his sophomore year, however, he came under the spell of a particularly inspiring mentor and his focus shifted to literature. Not long after Einhorn completed his undergraduate degree in 1961, that same mentor urged Einhorn to enroll in Penn's PhD program in literature. The professor had to twist some colleagues' arms to get Einhorn accepted, given Einhorn's spotty undergraduate record: Einhorn had a habit of failing to show up to classes that no longer held his interest. The professor prevailed and Einhorn began his graduate studies, only to drop out a year or two later. He read avidly on his own—he later boasted that he read a book a day, every day—but he had little patience for what he considered the staid, tweedy routines of academic life.[3]

One of the books that Einhorn encountered around that time was Thomas Kuhn's *The Structure of Scientific Revolutions*, first published in 1962. Kuhn's book would become a classic, easily one of the most influential books of the second half of the twentieth century. By the early 1980s Kuhn's book was the single most-cited book in all of the arts and humanities, eclipsing works by Freud, Chomsky, Derrida, Foucault, Wittgenstein, Heidegger, and, indeed, everyone else. One bemused observer noted that in the late 1970s, Harvard students were assigned Kuhn's book an average of two and a half times over their undergraduate careers: everyone from historians and sociologists to physicists, economists, and political scientists assigned the book in their classes.[4]

That runaway success lay far in the future when Einhorn discovered Kuhn's book. He became enamored of Kuhn's argument about the rise and fall of scientific worldviews. Einhorn was especially captivated by Kuhn's argument about anomalies: stubborn findings that fail to fit within prevailing scientific theories. Some of Kuhn's favorite examples included the accidental discovery of X-rays in 1895, and the unexpected detection of nuclear fission in a Berlin laboratory late in 1938. In both cases, the experimental data that would later be recognized as robust signals of major new phenomena had seemed, upon scientists' first encounter with them, to be little more than hiccups, minor deviations from the expected results that would presumably be assimilated or cleaned up down the road. Kuhn argued that when the collection of anomalies grows to a critical mass—when all those tiny blips and departures from expectations accumulate, and no accommodation with the reigning theory seems possible—they prompt a sudden "paradigm shift," reordering all our basic assumptions about how the world works.[5]

Einhorn was so taken with Kuhn's scheme that early in 1964 he put pen to paper and wrote to Kuhn directly. "Thank you for writing *The Structure of Scientific Revolutions*," Einhorn's letter began. "Its power and elegance somewhat reaffirm my already deteriorating faith in the ability of the academic world, as it is presently structured, to produce works of lasting significance in the humanities. A book such as yours makes one realize that there are still a few bright lights burning in the wasteland of modern humanistic thought." The recent graduate-school dropout—Einhorn was still a few months shy of his twenty-fourth birthday—went on to list a dozen further references ranging across art history, philosophy, psychology, and beyond, that Kuhn might wish to consult to help sharpen his thinking on the matter.[6]

This first exchange reveals several of Einhorn's enduring characteristics: no small amount of confidence in his own thinking; great passion for ideas about science and how today's scientific orthodoxy can become tomorrow's discarded paradigm; and a familiarity—with references and citations at his fingertips—with interesting ideas across a broad spectrum of subjects and disciplines. And there was more: Einhorn's abiding faith that people with common interests should communicate directly and informally. Just like the "Republic of Letters" during the Enlightenment, Einhorn believed that the postal service could knit together a network of like-minded thinkers.

Kuhn's response to the four-page, handwritten letter is equally striking. By the time he received Einhorn's letter, *Structure*—his second book—

had been out for two years. He was a full professor at Berkeley, having completed his doctoral and postdoctoral training at Harvard. "I probably need not tell you how much delight your good letter of January 16th has given me," Kuhn began, "but I do want to thank you for taking the trouble to write it. It is by all odds the most perceptive response I have yet received to my book, and it has helped my morale immeasurably." Kuhn thanked Einhorn for the many references, even asking for more information about an essay Einhorn had mentioned by Nelson Goodman, a renowned philosopher whose work would often, in later years, be compared with Kuhn's. And Kuhn agreed with Einhorn that Kuhn's protean notion of "paradigm," as introduced in *Structure*, still required "all sorts of additional work."[7] Ever perceptive, young Einhorn had put his finger on a major sticking point. A few years later a scholar isolated twenty-two distinct ways in which Kuhn used the term "paradigm" throughout *The Structure of Scientific Revolutions*: sometimes Kuhn used the term to denote a concept or theory, other times to denote a social structure such as a discipline or community, and still other times he seemed to use it to denote a method or laboratory practice. Kuhn endeavored to rectify the embarrassing conceptual muddle in later editions.[8]

Kuhn closed his letter to Einhorn by noting that he would soon be taking up a professorship at Princeton. With Einhorn living in nearby Philadelphia, Kuhn hoped they could arrange to meet in person. And so began a most curious relationship. Kuhn and Einhorn carried on a spirited correspondence for several years, interspersed with frequent personal visits. Sometimes Kuhn took the train to Philadelphia; other times Einhorn met him at his Princeton office and shared meals at his home. Kuhn dutifully sent drafts of his latest essays to Einhorn, who responded with detailed comments.[9] Kuhn was so impressed with the budding intellectual, in fact, that he recommended Einhorn for a prestigious fellowship in Harvard's Society of Fellows—the selfsame fellowship on which Kuhn had first begun working on *The Structure of Scientific Revolutions* a few years earlier.[10] (Despite Kuhn's letter, Einhorn did not secure the fellowship.) They traded ideas back and forth with ease. "Sorry to throw off half developed ideas this way," Einhorn wrote near the end of one of his letters, "but they are on my mind, and they might ring a bell in your mind. If not disregard them. Publication should be precise, conversation and letter writing a mess—how else can we learn?"[11] That notion—the unencumbered, free-spirited sharing of ideas by letter before they became ossified in print—would become an organizing principle of Einhorn's activities.

Kuhn agreed with Einhorn about the importance of informal correspondence, and their letters quickly reflected a growing camaraderie. "Professor Kuhn" became "Tom," "Mr. Einhorn" simply "Ira." At one point after Einhorn's peripatetic wanderings had kept him from writing to Kuhn for a few months, Kuhn greeted the next missive with cheer. "I had been wondering what you were up to and am correspondingly delighted to have your note. By all means let us get together," Kuhn enthused. "I do look forward to catching up on your activities." The feeling was mutual. Within days Einhorn replied to set up another meeting, emphasizing that "your enthusiasm is such a delight to experience."[12]

By that time, late in 1966, Einhorn had many activities on which to report for Kuhn. After dropping out of graduate school, Einhorn taught English for a while as a part-time instructor at Philadelphia's Temple University. Tiring of that—he still found academic life "sterile"—he began making frequent trips to California. The Esalen Institute in Big Sur became a home away from home. He plugged into the flowering counterculture movements there, experiencing firsthand the still-coagulating youth movement: equal parts antiwar protest, psychedelic drugs, Eastern mysticism, and communal living. He carried that jumble of experiences back with him to Philadelphia, along with a fast-growing network of tuned-in friends.[13]

In no time Einhorn had established himself as the "leading guru of Philadelphia's hippie community," as *Philadelphia Magazine* christened him in 1967.[14] He began teaching in a brand-new experimental program organized by the local chapter of the Students for a Democratic Society. Dubbed "Free University," and loosely affiliated with the University of Pennsylvania, the student-run alternative granted neither grades nor diplomas. Instructors volunteered to teach classes on any topic that interested them, and lessons continued as long as people showed up. Free U mixed the highbrow with the practical: courses on Nietzsche and esoteric literature were listed alongside workshops on how to avoid the draft, a pressing concern as the Vietnam War began to escalate. From the start, however, the most popular course was entitled simply "Evenings with Ira Einhorn." Drawing seventy students week after week, Einhorn used the course to ruminate on fast-paced societal changes. He gave updates on budding hippie communities from San Francisco's Haight-Ashbury district to Cambridge's Harvard Square.[15] He developed other courses for the Free University—likewise wildly popular—on psychedelic drugs and on "The World of Marshall McLuhan," whose theories about pop media and communication were just beginning to percolate. (Einhorn enjoyed quoting McLuhan's aphorisms, such as "Today's mysticism is tomorrow's science.") Einhorn sent his McLuhan syllabus to Kuhn, suggesting that McLuhan's work, though "wacky," might "cast oblique light" on Kuhn's own research topics. Organizers of Free University gave Einhorn's McLuhan course the pet name "Intro to Hippiedom."[16]

Building on his stand-out performances at Free U, Einhorn solidified his position as Philadelphia's head hippie. He became a kind of pied piper for the city's disenchanted youth. "Have rallied the kids in the drug scene around me," he reported at one point to Kuhn, "slowly trying to direct their energies to more constructive endeavors."[17] His course on psychedelics, which grew to include about 100 people per session, spawned other events around town. He organized a symposium on LSD at Temple University that drew 350 people, and he began to appear with increasing frequency on local radio and television programs to talk about the city's "psychedelic scene."[18]

In January 1967, the nation sat riveted before television coverage of the first "Be-In," held in San Francisco's Golden Gate Park. The day-long media spectacle aimed to unite the ardent New Left political protesters of Berkeley with the free-love and free-drug hippies of Haight-Ashbury. Einhorn recognized a good idea when he saw one, and he set about organizing Philadelphia's own Be-In. Having carefully secured all the necessary permits in advance, Einhorn's Be-In, held in a large city park that April, drew nearly 2000 people.[19] A local reporter marveled that Einhorn—"Philadelphia's best known social dropout"—always "manages to show up wherever there is trouble and has been credited with helping cool at least one potential riot situation at the University of Pennyslvania." Put simply, "He is the town Guru."[20]

A guru needs a fitting moniker. Since "Einhorn" means "one horn" in German, Einhorn began calling himself "the Unicorn." By the late 1960s the self-styled Unicorn had begun to mingle with the likes of Richard Alpert, the former Harvard psychology professor whose exploits with psychedelic drugs (including experiments on undergraduates) had cost him his job, along with that of his collaborator, Timothy Leary. When leading lights of the counterculture passed through Philadelphia, from Yippie antiwar protesters Jerry Rubin and Abbie Hoffman to poet Allen Ginsberg and composer John Cage, Einhorn served as their quasi-official host.[21] All the while he maintained contact with his "straight" interlocutors. "You sound even busier than I," replied Kuhn after one of Einhorn's updates. Other professors in the area, such as the eminent physicist Freeman Dyson at the Institute for Advanced Study in Princeton, came to know Einhorn as a "friendly and gracious host." Dyson developed great respect for Einhorn's courage in leading the antiwar protests.[22]

Einhorn's uncanny ability to interact with people from all lifestyles and persuasions came in handy early in 1970. Other groups had begun to plan big events for April 22, which they dubbed "Earth Day." They hoped that a mass gathering, modeled on the antiwar rallies and campus teach-ins, would help focus attention on environmental and ecological issues. The organizers of the Philadelphia event sought out Einhorn to land his counterculture constituency. They brokered a deal: Einhorn would help with arrangements in exchange for serving as master of ceremonies at the big event. When television crews swarmed to Philadelphia's Earth Day rally—one of the nation's largest, attracting tens of thousands of people to the same city park in which Einhorn had earlier hosted his Be-In—Einhorn dominated the media coverage. (Fig. 6.1.) One month later, *Philadelphia Magazine* devoted a feature-length article to the local

FIGURE 6.1. Ira Einhorn addressing the first Earth Day rally in Philadelphia, April 1970. (Courtesy Temple University Archives.)

celebrity on the occasion of his thirtieth birthday. Riding high, Einhorn mounted a semiserious campaign for mayor the following year.[23]

By that time, Einhorn's interests had broadened even beyond psychedelics or the budding environmental movement. His experiences with LSD and other drugs had heightened his curiosity about human consciousness. Einhorn read Andrija Puharich's *Beyond Telepathy* (1962) cover to cover; he considered it a must-read guide to the field. Puharich, originally trained as a medical doctor, had become an avid inventor of medical devices, holding dozens of patents for items like improved hearing aids. He had also devoted years of study to psychedelics, faith healing, and parapsychology, some of the work rumored to have been under the auspices of the CIA.[24] When a mutual friend offered to introduce Einhorn to Puharich in 1968, Einhorn leaped at the chance. They met at Puharich's parapsychology laboratory and personal residence in Ossining, New York, not far from New York City. They hit it off immediately and stayed in contact. Einhorn began making frequent visits to Ossining.[25]

Soon after they met, Puharich discovered Uri Geller, at the time still performing his mind-reading and spoon-bending feats in Tel Aviv nightclubs. Puharich whisked Geller to the United States, and Einhorn worked closely with Puharich behind the scenes to promote the scientific study of Geller's unusual abilities. Here, Einhorn was convinced, were crucial anomalies of just the sort Kuhn had described: psychic happenings and unexplained leaps of consciousness that just might topple the reigning paradigms of physics and psychology.[26]

•

Preparations for Earth Day brought Einhorn into contact with several of Philadelphia's leading businesspeople. An executive at General Electric, who served on the Earth Day planning committee, introduced Einhorn to a colleague, a vice president at the Pennsylvania branch of Bell Telephone. As with Kuhn before him, Einhorn and the Bell executive formed a fast friendship. Occasional lunches quickly blossomed into weekly dinners with Einhorn, the executive, and his wife at their suburban home.[27]

Einhorn proved his value to Bell Telephone soon enough. When the company planned to build a large switching station in a run-down neighborhood of South Philadelphia, and the locals pushed back against the corporate incursion into their territory, Einhorn's dining companion sought advice. The Unicorn stepped in and brokered a deal. After that, several high-level Bell executives, including the president of Pennsylvania Bell, began seeking out Einhorn's advice for how to improve community relations. As one of the executives explained, they turned to Ira for "help and counsel on what we might be doing wrong in various parts of the Philadelphia community."[28] For years, the Bell executives treated Einhorn as a highly prized management consultant, long before that role had become commonplace. Rarely would Einhorn's focus turn to this or that detail of corporate governance. Instead, he spun for his eager listeners a grander picture, a vision of an emerging networked society built around a new communitarian ethos. Einhorn cobbled together his message from his Esalen encounters, widespread reading, and interactions with like-minded thinkers such as Stewart Brand, who had just begun articulating a similar concept in his ragtag *Whole Earth Catalog*. The telephone executives couldn't get enough. Alongside the college kids and the acid freaks, the Bell executives treated Einhorn as a guru, their personal ambassador to the New Age. Soon the long-haired, potbellied, graduate-school dropout became a regular lunch companion to executive vice presidents, dining at one of Philadelphia's posh restaurants.[29]

Einhorn refused payment for these consultations. In lieu of cash, his Bell contacts paid him with services rendered. One of the first items Einhorn bartered was to get the Pennsylvania Bell leadership to pressure their colleagues at the world-famous corporate laboratory—Bell Labs, in nearby New Jersey—to study Uri Geller's psychic powers. The lab's scientists grudgingly agreed. Geller was ushered in under a thick cloak of secrecy; part of the deal was that no one would publicize the lab's involvement.[30] Though Geller managed to surprise some of the scientists and engineers with his abilities, little came of the 1972 meeting. Soon after that, Puharich landed Geller in the much more receptive psi lab at the Stanford Research Institute. Though the Bell Labs visit was a bust, Einhorn's close friend—the Bell Telephone vice president—gained much from the encounter. Along with Einhorn, the telephone executive began hanging out with Puharich and finding himself invited to parties with Geller.[31]

Beyond the Geller visit, Einhorn's friends at Bell Telephone provided a much more important service. The Unicorn's corporate contacts agreed to assume all costs and operations for his grand experiment in networking. Every few weeks, Einhorn sent a thick stack of papers to one of his associates at the telephone company—a smorgasbord of formal preprints, hastily typed press releases, clippings from newspapers and magazines, and informal musings of all kinds—along with a distribution list. The Bell executives would do the rest, making sure that the materials were photocopied and mailed out to whomever Einhorn had designated for a given package. With the corporate giant's help, Einhorn built just the sort of large-scale circulation system for informal ideas that he had envisioned in his early correspondence with Kuhn. He created, in effect, one of the first "listservs," or, as it later came to be known, "an internet before the internet," powered by photocopy machines, mimeographs, and postage stamps.[32]

Einhorn tended to his network with extraordinary care. Not just anything would go out. He handpicked the items and personally tailored the distribution list for each package, always striving to maximize the intellectual impact of particular ideas on targeted thinkers. Before long, his handlers at Bell had a collection of index cards with more than 300 names and addresses from which Einhorn would select recipients for a given mailing. The packages, featuring Bell's corporate logo on each envelope, traveled far and wide; by 1978 the list included recipients in more than twenty countries across North America, Western Europe, and even the Soviet bloc.[33] Einhorn, who proudly referred to himself as a "planetary enzyme," catalyzing intellectual reactions across the globe, cultivated a contact list that soon included everyone from famed anthropologist Margaret Mead to novelist and parapsychology enthusiast Arthur Koestler, futurist Alvin Toffler, inventor Arthur Young (founder of the Institute for the Study of Consciousness in Berkeley, at which members of the Fundamental Fysiks Group often crossed paths), and more. Business tycoons appeared on the rolls next to peace activists; heirs to billion-dollar fortunes alongside leading lights in the Esalen human-potential movement.[34] A British economist extolled Einhorn's network in the pages of Stewart Brand's *CoEvolution Quarterly* in 1979. The unusual network "circulates papers mutually between some of the most brilliant and original minds on the planet," she wrote. It had become "one of my best sources of epistomological [*sic*] speculation for the past ten years, even though I can not, for obvious reasons, suggest how to access or make inputs to it." "It was a wonderful arrangement," crowed Einhorn's original contact at Bell Telephone. "Ira had an absolutely incredible circle of friends."[35]

Not all recipients were equally appreciative. Several, like leading physicists Freeman Dyson and John Wheeler, never asked to be put on the mailing list in the first place. Dyson recalls sweeping nearly every package directly into the trash bin. "I found Einhorn interesting as a person but not as a scientist," he recently clarified.[36] The high-tech Diebold Corporation, on the other hand, took a rather different view. They commissioned an internal study of Einhorn's network in 1978 and invited Einhorn to corporate headquarters in New York for a follow-up discussion. The Diebold executives wondered whether networks like Einhorn's could help "sensitize" corporate management to "emerging social demands." Einhorn's great experiment in connectivity, and the ideas that flowed so freely within it, could acclimatize CEOs and upper management with "a better feeling" for "the mood of the time in which we live." By (in effect) eavesdropping on the latest developments "in a non-threatening environment," managers could anticipate the next big social and political rifts: a kind of early warning system before activists with bullhorns began shouting the same ideas across a picket line. Einhorn's network could even help the bottom line by fostering niche marketing.[37]

The latest ruminations by Fundamental Fysiks Group members Jack Sarfatti, Fred Alan Wolf, Elizabeth Rauscher, and Nick Herbert often topped Einhorn's lists. A journalist who was on the mailing list wrote about Einhorn's network late in 1976. He titled his article "Notes from the far-out physics underground" and highlighted recent work from the Fundamental Fysiks Group on the physics of psi as typical fare for the Unicorn's mailings. Sarfatti's original press releases about the laboratory tests of Uri Geller in London likewise circulated thanks to Einhorn's network.[38]

One item by Sarfatti and Wolf that Einhorn mailed out bore the straightforward-sounding title "A Dirac equation description of a quantized Kerr space-time"—from the sound of it, just the sort of material one might expect to read in a mainstream physics journal like the *Physical Review*. In case the title failed to tip off readers, someone added the helpful handwritten cue: "see pp. 4 and 5 on 'psi' effect." Thanks to the Unicorn preprint service, Sarfatti's and Wolf's latest brainwave went out to everyone from Uri Geller's main handler, Andrija Puharich, to physics-and-consciousness theorists Evan Harris Walker and Charles Musès, retired Air Force colonel and parapsychology advocate Tom Bearden,

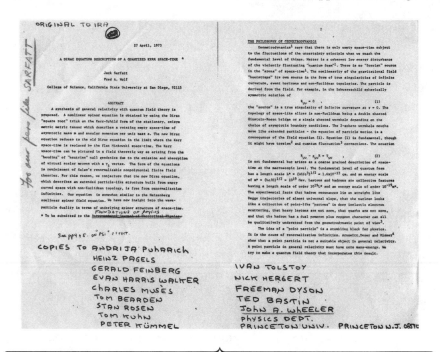

FIGURE 6.2. Preprint of a 1973 paper by Jack Sarfatti and Fred Alan Wolf, circulated by Ira Einhorn. The list of recipients appears at the bottom of the page, in Einhorn's handwriting. (John A. Wheeler papers, American Philosophical Society, Sarfatti folder, reproduced courtesy Fred Alan Wolf.)

Columbia University physics department chair Gerald Feinberg, Princeton physicists Freeman Dyson and John Wheeler, and Einhorn's old friend "Tom" Kuhn.[39] (Fig. 6.2.)

•

With his network reaching ever more people and his fame spilling far beyond Philadelphia, Einhorn's star rose higher still. He received an invitation to spend a semester at Harvard's Institute of Politics, tucked within the university's John F. Kennedy School of Government. (Einhorn finally got a fellowship at Harvard, a decade after Kuhn had tried to get him into Harvard's Society of Fellows.) The other fellows included public officials from state and local governments across the country, as well as a former

U.S. ambassador to South Korea.[40] Einhorn intoned at panel discussions on the current state of politics and culture, and he delighted in sizing up the Harvard population. "My new consciousness was reinforced by LSD, dope, and the loving I was doing," he explained to a young female reporter for the *Harvard Crimson*. "The kids here are incredibly bright, but everything is just so structured. There's simply no time to interact, relax, or enjoy yourself," he continued. Then he shot her a smile. "It's like you have to schedule your fucking."[41]

On campus and off, Einhorn parlayed his stature to further the cause of physics and consciousness. Together with Puharich, he organized a series of conferences on mind and matter, first at the University of Pennsylvania and later in Harvard's brand-new Science Center. He stepped up his visits to the West Coast as well, dropping in at Esalen and taking detours to the Bay Area to check in with members of the Fundamental Fysiks Group.[42]

Beyond the personal visits and network mailings, Einhorn helped the physicists reach a still larger audience. With his usual charm and persistence, he had forged an unusual alliance with Bill Whitehead, an editor at Anchor Books, a subsidiary of the large New York publishing firm Doubleday. Whitehead's task had been to build up the press's list in psychology. He interpreted that mandate broadly, tilting more and more toward parapsychology, the human potential movement, and Eastern mysticism. Whitehead's timing was impeccable. By tapping into the burgeoning New Age movement, many of his selections became commercial successes, and before long he had been promoted within Doubleday and then lured over to Dutton Books. At Dutton he rose to the rank of editor-in-chief, and Einhorn came along for the ride. In Einhorn, Whitehead had a first-class ticket to a pool of potential authors; and Einhorn, in turn, exercised outsized influence on what books made it into print.[43]

Among his first successes, Einhorn convinced Whitehead to republish Puharich's *Beyond Telepathy*, the book that had first captured Einhorn's imagination in the early 1960s. Not everyone had been as enamored as Einhorn, and Puharich's book had gone out of print. No problem: a reprint of *Beyond Telepathy*—complete with a new preface by Einhorn himself—came out in Whitehead's series in 1973, followed, one year later, by Puharich's rambling *Uri: A Journal of the Mystery of Uri Geller*. In *Uri*, Puharich expanded upon the by-then familiar Geller feats. Not just mind reading and spoon bending; now there were tales of extraterrestrial contacts and the teleportation of Puharich's briefcase from the Ossining ranch to Geller's Tel Aviv apartment. Puharich acknowledged in the opening pages that "Ira Einhorn's imagination helped to formulate this book and to get it to the attention of publishers," and that editor Bill Whitehead's "cool judgment and courage got the book published."[44]

Soon Einhorn was getting other colleagues in on the act. Bob Toben, a high-school pal of Fundamental Fysiks Group member Fred Alan Wolf's, had become intrigued by Uri Geller and reports of related paranormal phenomena. He wrote a short article about the latest mysteries for an obscure underground journal that caught Einhorn's eye. Soon the Unicorn had set up Toben with his own book contract in Whitehead's series. Toben had little scientific background, so he teamed up with Wolf and Jack Sarfatti to write a popular book speculating on how modern physics might account for Geller's psychic fireworks. They began brainstorming together in California a few months before Sarfatti and Wolf set off for their European foray. Toben flew to Europe to continue working with them. Much of the book was planned out in a favorite Parisian café, while Sarfatti was in "an altered state of consciousness." Toben then put the finishing touches on the manuscript at Puharich's Ossining estate, Einhorn in tow. With Einhorn's help, *Space-Time and Beyond: Toward an Explanation of the Unexplainable* appeared in 1974. A curious hybrid, the book consisted of a hundred pages of pen-and-ink cartoons by Toben offering a hip, New Age guide for the perplexed, followed by dense scientific appendices with quotations cobbled from Sarfatti's and Wolf's favorite sources, such as physicists Eugene Wigner and John Wheeler. And it sold; Einhorn had gauged the market well. The first edition sold about 50,000 copies. Translations appeared in German and Japanese, and a second English edition, brought out by Bantam a few years later, sold handsomely as well.[45]

Space-Time and Beyond presaged a wave of popular books. That same year the Esalen Institute Publishing Program, in conjunction with Viking Press, published Lawrence LeShan's *The Medium, the Mystic, and the Physicist: Toward a General Theory of the Paranormal*. LeShan, trained as a psychologist and deep into parapsychology by the late 1960s, had already been in contact with Bob Toben before their books came out.[46] While working in Europe on *Space-Time and Beyond*, meanwhile, Sarfatti met up with Fritjof Capra, another physicist interested in the grand mysteries of quantum theory and their possible ties to mysticism. Capra had no connection yet to Einhorn or Whitehead; he would enter Einhorn's network later, after moving to Berkeley and joining the Fundamental Fysiks Group. In the meantime, Capra endured dozens of publishers' rejections before a small press took a gamble on his book *The Tao of Physics*.[47]

The genre really only took off once the Fundamental Fysiks Group was up and running. The members' raw enthusiasm, combined with Einhorn's bookselling sense, made a potent combination. Berkeley physicist Henry Stapp remembers a party at his house in 1975 for the Fundamental Fysiks Group, following one of his presentations. As the party stretched late into the night, Stapp looked around his backyard. He overheard several clusters of people hatching plans for the best-selling books they intended to write; the mood, Stapp recalls, was one of exuberance. In short order, they began making good on those plans.[48]

Several of the early books took shape in Arthur Young's Institute for the Study of Consciousness in Berkeley, which had served for years as a frequent stomping ground and sometime residence for members of the Fundamental Fysiks Group. Young's institute had also become a favorite destination for Einhorn during his California trips; Young, in turn, became an early member of Einhorn's network. When Young, a former aeronautical engineer, began working on two books about his own theories about physics and consciousness, Einhorn lent a hand. He edited several drafts of Young's *The Reflexive Universe: Evolution of Consciousness* and *The Geometry of Meaning*, both of which appeared in 1976. (Young thanked Einhorn in his acknowledgments, alongside Young's onetime protégé and tenant, Fundamental Fysiks Group member Saul-Paul Sirag.)[49] Not long after that, Sirag contributed his own chapter on physics and consciousness to George Leonard's *The Silent Pulse*, published by Einhorn's editor-friend Bill Whitehead. Like Young, Leonard—an Esalen fixture and proponent of "humanistic psychology"—had been a member of Einhorn's network for years.[50]

Jeffrey Mishlove crossed paths with Einhorn at Young's institute; they enjoyed a game of Go whenever the Unicorn was in town. Mishlove, a Berkeley graduate student, first connected with Sirag and Sarfatti at Young's institute as well, before joining the Fundamental Fysiks Group discussions. His best-selling book, *The Roots of Consciousness: Psychic Liberation through History, Science, and Experience*, appeared in 1975, complete with a lengthy appendix by Jack Sarfatti on "the physical roots of consciousness." The book began as a kind of proof-of-principle. Mishlove was hard at work on what would become the first doctorate in parapsychology awarded in the United States. Before his graduate-school committee would allow him to jump into a dissertation on the topic, they required him to write an extensive report on the history and current status of parapsychology. The enterprising Mishlove—who was also running his own local radio show about the paranormal and help-ing arrange large events in the Bay Area, such as Uri Geller's famous performance in 1973 before a standing-room-only crowd of 1500 people in Berkeley's Zellerbach Auditorium—turned his homework assign-ment into a major publishing opportunity.[51]

Then came Gary Zukav's book, *The Dancing Wu Li Masters*. Zukav had been Jack Sarfatti's roommate in North Beach, San Francisco, and he frequently tagged along with Jack to meetings of the Fundamental Fysiks Group and the Physics/Consciousness Research Group.[52] Sarfatti also invited Zukav to attend the first "Physics and consciousness" workshop at Esalen in January 1976. The meeting left a deep impression on Zukav; it also introduced him to even more tuned-in physicists. Zukav crafted his entire book around discussions at the Esalen workshop, entitling the first chapter "Big week at Big Sur." As he wrote in the opening pages, "I had spoken often to Jack Sarfatti, who is the physicist director of the Physics/Consciousness Research Group, about the possibility of writing a book, unencumbered with technicalities and mathematics, to explain the exciting insights that motivate current physics. So when he invited me to a conference on physics that he and Michael Murphy were arranging at the Esalen Institute, I accepted with a purpose." David Finkelstein, another physicist whom Zukav met at the Esalen workshop, contributed a preface; Zukav acknowledged Sarfatti and Finkelstein as the "godfathers of this book." He went on to thank most members of the Fundamental Fysiks Group for further help: Henry Stapp, Elizabeth Rauscher, John Clauser, George Weissmann, Fred Alan Wolf, Fritjof Capra, Saul-Paul Sirag, and Nick Herbert. But the real star of the book—at least in the first edition—remained Sarfatti. The entire discussion, ranging over relativity, quantum theory, Bell's theorem, and nonlocality, marched toward a concluding chapter that focused on Sarfatti's still-fresh ideas about quantum entanglement and psychic phenomena.[53]

The book's unusual title came from Esalen's resident T'ai Chi master, Al Huang. At dinner one evening, early in the month-long physics workshop, Huang had explained that the Chinese word for physics was "Wu Li," which could be translated literally as "patterns of organic energy." The evocative phrase captured everyone's imagination. Huang had gone on to explain that the Chinese characters making up "Wu Li" could also mean a variety of notions, depending on the spoken inflection or intonation. The term could be translated as "my way," "nonsense," "I clutch my ideas," or "enlightenment," each connoting a more individualistic, epistemic bent than the English word "physics." The physicists sitting around the table began to chime in: the richness of the term seemed an eerie match for the direction their own research had taken them, from quantum physics to the nature of consciousness. Zukav now had the outline for his book. He divided his discussion into sections, each labeled by one of the meanings of "Wu Li."[54]

Despite the similarity in title to Fritjof Capra's iconic book *The Tao of Physics*, Zukav's point in *Dancing Wu Li Masters* was not to draw out parallels between modern physics and insights from ancient Eastern religions such as Buddhism or Hinduism. Rather, Zukav focused on the development of quantum physics and relativity in the West. Taking inspiration from the T'ai Chi master's lesson about the many meanings of "Wu

Li," Zukav introduced readers to a particular view of how to interpret the startling discoveries. He dwelled upon suggestive statements from the likes of Niels Bohr and Werner Heisenberg about the importance of the role of the observer in quantum mechanics, since our choices of what to measure and how to interact with quantum systems change the very systems under study. He drove the point home by quoting a long passage from John Wheeler's still-obscure conference paper on the shift from "observers" to "participators," which Jack Sarfatti had found so inspiring.[55]

Zukav went further, writing that the emphasis upon the observer (or participator) meant that "Physics has become a branch of psychology, or perhaps the other way around." He argued that quantum physics had at last shattered the illusion of objectivity:

> "The exact sciences" no longer study an objective reality that runs its course regardless of our interest in it or not, leaving us to fare as best we can while it goes its predetermined way. Science, at the level of subatomic events, is no longer exact, the distinction between objective and subjective has vanished, and the portals through which the universe manifests itself are, as we once knew a long time ago, those impotent, passive witnesses to its unfolding, the "I"s, of which we, insignificant we, are examples. The Cogs in the Machine have become the Creators of the Universe.

Hence the significance of translations such as "my way" and "enlightenment" for "Wu Li": "If the new physics has led us anywhere, it is back to ourselves, which, of course, is the only place that we could go." Indeed, enthused Zukav, "Do not be surprised if physics curricula of the twenty-first century include classes in meditation."[56]

Like a spider suspended in amber, the first edition of Zukav's book captured a moment in the ferment of the Fundamental Fysiks Group and Physics/Consciousness Research Group discussions. When introducing readers to the notion from relativity that space and time form a single, united thing called "space-time," Zukav followed Sarfatti's lead and likened the relativistic concept to Baba Ram Dass's spiritual manual *Be Here Now*. Later chapters pursued Sarfatti's interpretation of Bell's theorem and quantum entanglement. Zukav described Sarfatti's latest schemes for how one might control Bell-styled nonlocality and use it to transmit signals faster than light. Zukav likewise emphasized possible links between Bell's theorem and parapsychological phenomena such as the remote-viewing results of Stanford Research Institute physicists Harold Puthoff and Russell Targ.[57]

In fact, the entire framework for the book followed a Physics/Consciousness Research Group model. Just two weeks after the Esalen workshop that Zukav had attended, Sarfatti submitted a follow-up grant proposal to Werner Erhard's charitable foundation, seeking further funding for the PCRG. He included a packet of photographs from the Esalen meeting, emphasizing on the first page that "This is the kind of experience supported by your donations." One photo showed Sarfatti and T'ai Chi master Al Huang sitting together in deep repose. The caption read: "Sarfatti and T'ai Chi Master, Al Huang, doing Wu-Li. Wu-Li is the Chinese word for physics. It also means pattern of organic movement. It also means 'nonsense,' 'my way,' 'enlightenment,' and 'I clutch my ideas,' depending on the pronunciation of 'li.'"[58]

Sarfatti had been planning his own follow-up book. In a proposal that he drew up just as Zukav's book was hitting the shelves, Sarfatti emphasized that books like his own *Space-Time and Beyond*, Capra's *Tao of Physics*, and Zukav's *Dancing Wu Li Masters* had created a robust market for discussions of physics and consciousness.[59] But Sarfatti's latest book never materialized. Instead, Zukav's book hit something of a speed bump, revealing the limits as well as the promise of communicating ideas in the form of flashy trade-press books. The earlier books had been relatively small-scale affairs that happened to do well commercially—sleeper hits or cult classics rather than major publishing events. (None of the earlier books, for example, received reviews in the *New York Times* upon first publication.) Their success raised the stakes. Zukav's book received a major launch in 1979 and that brought out the critics. Several reviewers were quick to attack what they considered the book's scientific infelicities, heaping scorn upon Zukav's main informants. One physicist, reviewing the book in *Physics Today*, complained that Zukav had been too heavily influenced by "the 'Physics/Consciousness' movement of northern California, and its leading spokesman, Jack Sarfatti." A *New York Times* reviewer likewise fumed that there was something "truly insidious about this tract-posing-as-primer," parroting, as it did, "the dubious notions of certain renegade physicists."[60]

Zukav snapped into crisis mode, rewriting several sections of the book before its second printing. The result: most of the references to Sarfatti hit the cutting-room floor, and Zukav dialed back the discussions of quantum-enabled telepathy and clairvoyance. All reference to the Physics/Consciousness Research Group disappeared; Zukav wrote simply that "a friend" (otherwise unnamed) had brought him to the Esalen workshop. The heavily edited closing chapter hewed more closely to Henry Stapp's interpretation of Bell's theorem (which made little room for ESP or clairvoyance), moving Sarfatti's unorthodox ideas—followed by Stapp's critique—to a footnote.[61] Naturally Sarfatti felt betrayed, and not only for what he considered an Orwellian rewriting of history. Sarfatti accused Zukav of reneging on their earlier deal: Sarfatti said the deal had been for him to receive 10 percent of the royalties in exchange for his extensive coaching. Instead, Sarfatti claimed that Zukav used the money to pay for the last-minute revisions, including the expense of producing new plates for the second printing.[62]

Although Zukav's friendship with Sarfatti came to an abrupt end, the quick fix worked: *The Dancing Wu Li Masters* became a break-out success. Within its first four years the book went through nine printings; a paperback edition quickly sold another quarter million copies. The amended version received critical acclaim as well, sharing an American Book Award in 1980 with that other enduring favorite, Douglas Hofstadter's *Gödel, Escher, Bach*. The prominent publishing house HarperCollins brought out a paperback edition of Zukav's book in 2001 as part of its Perennial Classics series.[63]

•

Einhorn and the Fundamental Fysiks Group thus forged some impressive, alternative tracks for putting their ideas into circulation. Einhorn's network allowed Sarfatti, Wolf, Herbert, Rauscher and the others to sidestep physicists' usual communication outlets. By easing their way into the commercial book-publishing world, Einhorn further spread word of the "new physics"—and its potential for parapsychology and an expanded worldview—to audiences far larger, and far more diverse, than physicists' ordinary routines would have allowed. Indeed, Einhorn and the Fundamental Fysiks Group helped seed what one publisher called a new "popular metaphysics."[64] These were the first broad-market books devoted to the interpretation of quantum theory in more than two generations. Despite a surge of popular science writing in the United States after World War II, virtually no popular books (at least in English) had tackled the quantum puzzles and paradoxes that had so animated earlier physicists and popularizers like Einstein, Bohr, Heisenberg, and others in the 1930s and 1940s. Einhorn and the Fundamental Fysiks Group injected new energy and new source material into a dormant domain. In turn, books like *Space-Time and Beyond*, *The Tao of Physics*, and *The Dancing Wu Li Masters* inspired dozens of imitators during the 1980s and 1990s, jumpstarting a now-flourishing market for popular physics books on quantum theory and its mysteries.[65]

The new visibility invited concerted pushback. As Zukav and Sarfatti learned the hard way, commercial publishing operated under its own rules and constraints. The physicists could get the word out, but not all messages would flourish under the informal strictures of the marketplace. Einhorn's network also proved to be less robust than it had first appeared. In the end, the network depended on the whimsy of an idiosyncratic individual and the goodwill of a few highly placed corporate executives; it had little secure institutional backing. Einhorn's network ended as abruptly as it had begun, once Einhorn was no longer in a position to carry it on.

The first hint that trouble might be brewing came in November 1977.

Einhorn had worked with Andrija Puharich to organize an international conference on quantum mechanics and parapsychology, building on the successes of their conferences at Penn and Harvard. With funding from a private foundation based in London, the meeting was to be held in Reykjavík, Iceland, and feature presentations from many in what Einhorn jokingly called his "psychic mafia," including Harold Puthoff and Russell Targ from the Stanford Research Institute psi lab, Elizabeth Rauscher from the Fundamental Fysiks Group, and Evan Harris Walker from the Army's Aberdeen Proving Ground. The meeting went ahead as planned, but at the last moment Einhorn failed to show up. His longtime girlfriend had just left him, he said, and his behavior had become a bit erratic.[66]

In fact, his girlfriend, Holly Maddux, had been missing since early September. Even when authorities feared the worst—that Maddux had been the victim of foul play—few could imagine that peace-loving Einhorn, the New Age guru of Philadelphia, could have had anything to do with Maddux's disappearance. Maddux's family, back in Texas, was less convinced. They hired a retired FBI agent as a private detective to look more closely into Einhorn. Eighteen months later, police executed a search warrant at the Unicorn's apartment, where they found the victim's rotting corpse stuffed into a trunk. Her skull had suffered multiple fractures; she had been beaten beyond recognition.[67] (Fig. 6.3.)

FIGURE 6.3. Investigators remove a trunk containing Holly Maddux's remains from Ira Einhorn's apartment in Philadelphia in March 1979. (Associated Press/Temple University Archives.)

The murder case shocked friends and foes alike. The front-page headline in the *Philadelphia Daily News*—"'Hippie Guru' held in trunk slaying"—eclipsed the other news of the day: the Three Mile Island nuclear reactor meltdown in nearby Harrisburg, Pennsylvania. Einhorn's lawyer for the arraignment hearing was a young Philadelphia defense attorney named Arlen Specter, who was about to embark on his successful campaign for the United States Senate. Specter convinced the judge to release Einhorn on just a fraction of the bail that the prosecutor had demanded. Central to Specter's case was a parade of character witnesses, including Einhorn's benefactors from Bell Telephone, who proclaimed that the peace-loving, intellectual, environmental-activist Einhorn could not possibly have been guilty of so hideous a crime. At the very least, Specter continued, Einhorn, Philadelphia's native son, posed little flight risk.[68]

Specter's persuasion worked. Despite the murder charges and the gruesome evidence retrieved from Einhorn's apartment, the judge set bail at just $40,000 (about $120,000 in 2010 dollars). Pennsylvania law required that only 10 percent of the total, a mere $4000, be paid up front. Even that modest sum hovered a good bit above what Einhorn could afford. A wealthy heiress in Montreal, who had been an avid follower of Einhorn's network mailings, stepped in and put up the cash after conferring with Fundamental Fysiks Group organizer Elizabeth Rauscher (who, like most of Einhorn's confederates, could not imagine that Einhorn was guilty).[69]

Several of Einhorn's friends at Pennsylvania Bell privately believed that Einhorn must be innocent. Nonetheless, the telephone company immediately halted all operations of Einhorn's network, fearing bad publicity.[70] Just like that, Einhorn's international network—the main conduit for preprints and press releases by members of the Fundamental Fysiks Group—fell silent. Others tried to rig up their own replacements. Within a year, Jack Sarfatti had begun mailing items to what he dubbed his "Quantum Communications Network," a list including fifty-seven names and organizations, many of whom, as in Einhorn's original network, had not asked to be included. Yet the later imitations never achieved Einhorn's reach or scale.[71]

Einhorn, meanwhile, chose to interpret the phrase "released on bail" rather liberally. He made frequent trips outside the Philadelphia area, first to California to check in at Esalen and then to hang out in San Francisco for a while. He met up with Jack Sarfatti in Sarfatti's unofficial headquarters (Caffe Trieste, North Beach). After hearing him out, Sarfatti organized a meeting at which Einhorn could tell his side of the story. Several members of the Fundamental Fysiks Group attended. (Tongue in cheek, Sarfatti described Einhorn's trip as "Ira's confessional passage to the Monastery of Esalen in Big Sur where he was to be absolved of all guilt.") The main impression Einhorn left on his listeners was his outward calm. "He certainly didn't *act* as if he did it," Sarfatti later told a journalist.[72]

A few weeks after the California jaunt, Einhorn actually left the country, traveling to Montreal to visit the heiress who had posted his bail. These short forays proved to be test runs for what was to come. Just weeks before his murder trial was set to begin, Einhorn and his new girlfriend fled to London.[73] Under a series of assumed names, he managed to elude authorities for nearly two decades, ultimately settling in the south of France. While he was on the lam, the Philadelphia district attorney's office went ahead with its murder trial in 1993. In one more bizarre twist to an already bizarre case, Einhorn was convicted in absentia of the murder charges, a rarity in U.S. law.[74]

In the late 1990s, Einhorn's luck began to run out. FBI agents and Interpol colleagues tracked him down.[75] Even then it seemed Einhorn might receive a reprieve: France refused to extradite the fugitive, citing a policy of not returning individuals to nations in which they might face the death penalty. The diplomatic crisis reached the highest echelons. Thirty-five members of the U.S. Congress signed a letter to French president Jacques Chirac urging Einhorn's extradition; Secretary of State Madeleine Albright and Attorney General Janet Reno personally intervened with their French counterparts.[76] The Pennsylvania state legislature, meanwhile, passed a bill specifically in reference to Einhorn's case that allowed defendants who had been convicted in absentia to request a new trial. The bill, officially titled the "Post Conviction Relief Act" but known informally as the

"Einhorn law," quickly attracted critics, who challenged the constitution-ality of such a legislative incursion into the judicial branch's domain.[77]

All the while Einhorn enjoyed a peaceful life in the French countryside. His country home, he told a visiting journalist, betrayed a "touch of Eden." He even got to partake in the wonders of the Internet, networking at light speed rather than relying on the old mechanisms of photocopier and postage stamp. He created an email account—his address, User886114 @aol.com, appropriately anonymous—and began corresponding daily with members of his former network. He bragged to a reporter in 1999 that he had received about 9000 emails from Jack Sarfatti alone.[78] (Those who know Sarfatti will recognize that as a low-ball figure.)

After four years of wrangling, the French released Einhorn to U.S. authorities in July 2001. For one thing, the Philadelphia district attor-ney's office clarified that at the time Einhorn was arrested, the State of Pennsylvania did not have the death penalty on its books. (It was added later.) That meant that Einhorn could not face the ultimate punishment. Moreover, the Philadelphia courts agreed to conduct a new trial of Ein-horn, rather than let the in absentia verdict stand.[79]

The new trial began in autumn 2002, capping a case with more than its share of legal curiosities. Einhorn claimed that the prosecution had tampered with witnesses and withheld evidence during the 1993 trial, although no independent corroboration of these allegations surfaced.[80] In a separate development, a Philadelphia jury awarded a record-break-ing $907 million verdict in a wrongful death suit brought against Einhorn in civil court by Maddux's family, after he had surfaced in France but before extradition had been secured.[81] (The family wanted to ensure that Einhorn could not profit from selling his story.) Meanwhile, the transcript of Einhorn's 1979 arraignment hearing—at which a dozen witnesses had testified to Einhorn's upstanding character, including two Bell Tele-phone executives who described, for an incredulous judge, their role in facilitating Einhorn's network—vanished into thin air. City and county clerks continue to point fingers. Einhorn claims that the disappearing transcript is part of a larger government conspiracy against him.[82]

The trial in 2002 electrified Philadelphia. Hundreds of people flocked to the courthouse to catch a glimpse of the aging guru; some camped out all night to secure a ticket to one of the few seats in the crowded courtroom.[83] The prosecutors argued (as they had maintained ever since Maddux's body was discovered in Einhorn's apartment) that Einhorn killed Maddux in a jealous rage: she had just met a new man and had returned to the apartment to pick up her last items before moving out. Einhorn countered with his own theory which, like the prosecution's, had changed little since 1979. As far as Einhorn was concerned, he had been framed for Maddux's murder by the CIA, the KGB, or both. His work on Uri Geller, which by the late 1970s had expanded into all manner of para-psychological topics, was getting too close to top-secret psychic warfare plans, he maintained. His network, distributing the latest psi theories from members of the Fundamental Fysiks Group, had to be stopped. Ein-horn argued that dark forces—deep-cover intelligence agents working for one or another of the world's superpowers—had killed Holly Maddux in order to silence him.[84]

Despite their early support, few (if any) members of the Fundamental Fysiks Group lend Einhorn's conspiracy theory much credence today. The Swedish woman whom Einhorn had married while on the run pleaded with Sarfatti and others to testify on Einhorn's behalf at the new trial, but Sarfatti kept his distance.[85] In October 2002, the jury deliberated for two hours before finding Einhorn guilty of first-degree murder. The presid-ing judge dismissed Einhorn, who had once inspired world-renowned scholars, multinational corporations, and major publishing firms, as "an intellectual dilettante who preyed on the uninitiated, uninformed, unsus-pecting, and inexperienced people." Today Einhorn sits in a Pennsylvania prison, serving a life sentence with no possibility for parole. He spends his time honing a treatise on what he considers to be a grave miscarriage of justice.[86] Most observers agree, however, that Ira Einhorn killed more than just the physicists' special preprint network.

Chapter 7

Zen and the Art of Textbook Publishing

Capra is an engaging teacher. . . . Here is the kind of physics one should have had in high school, instead of the grey formulae which hung in the school laboratory like eminent ancestors whom one could never live up to, let alone surpass.

—A. Dull, 1978

embers of the Fundamental Fysiks Group, catalyzed by Ira Einhorn and his contacts at major publishing firms, helped to launch a new type of popular book in the 1970s: accessible books that compared striking features of modern physics, such as Bell's theorem and nonlocality, with staples of the counterculture and New Age revivals, from parapsychology to Eastern mysticism. Some of the books enjoyed critical acclaim, and many achieved commercial success. But the books served to do more than introduce members of the Fundamental Fysiks Group and their enthusiasms to new audiences. Several operated in multiple registers, blurring genres that usually remained distinct: both popular book for the masses and textbook for science students. Fritjof Capra's *The Tao of Physics*, first published in 1975, remains the most emblematic and successful of the group's efforts in this domain. Capra's *Tao* exemplifies the hybrid nature of the new books, and the diverse roles they came to play.

Capra had met key members of the Fundamental Fysiks Group in the course of his writing, including Elizabeth Rauscher, George Weissmann, Fred Alan Wolf, and Jack Sarfatti. (In fact, after Sarfatti and Capra met in London via David Bohm in the fall of 1973, Sarfatti stayed with Capra and his parents in Innsbruck during a brief holiday. They crossed paths again that spring at Abdus Salam's International Centre for Theoretical Physics in Italy.) Capra returned to Berkeley just as his book was coming out, long before it had become the runaway best-seller we know it as today. Still relatively unknown, Capra spoke about his book and its larger themes—the multiple parallels, as Capra saw them, between the world-views of modern physics and Eastern mysticism—at the first sessions of the Fundamental Fysiks Group in May 1975. Capra helped to organize a follow-up series of discussions about the book and its larger themes at the Lawrence Berkeley Laboratory the next year, filled with regular members of the Fundamental Fysiks Group. He became a core member of the group, participating as well in various Physics/Consciousness Research Group activities and Esalen workshops.[1]

By that time the book, like its author, had already traveled a long route. Austrian-born Capra completed his PhD in theoretical particle physics at the University of Vienna in 1966, and moved on to a postdoctoral fellowship in Paris. There he witnessed the student uprisings and general strikes of May 1968, scenes that left a deep impression on him. He also met a senior physicist from the University of California at Santa Cruz who was spending some sabbatical time in Paris. The professor invited Capra to Santa Cruz for a follow-up postdoctoral fellowship, which Capra gladly accepted. He arrived in Santa Cruz in September 1968.[2]

Capra broadened his horizons on many fronts in California. As he later wrote, he led "a somewhat schizophrenic life" in Santa Cruz: hard-working quantum physicist by day, tuned-in hippie by night. He continued his political education, already stoked by the Paris of 1968: he went to lectures and rallies by the Black Panthers; he protested against the war in Vietnam. He took in "the rock festivals, the psychedelics, the new sexual freedom, the communal living" that had become de rigueur among the Santa Cruz counterculture set. He also began exploring Eastern religions and mysticism—an interest originally sparked by his filmmaker brother—by reading essays and attending lectures by Alan Watts, a local expert on all things Eastern who had assisted Esalen's Michael Murphy in his continuing investigations of Buddhism, Hinduism, Taoism, and the rest.[3]

In the midst of these explorations, Capra had a powerful experience on the beach at Santa Cruz during the summer of 1969. Watching the ocean waves roll in and out, he fell into a kind of trance. As he later described it, the physical processes all around him took on a new immediacy: the vibrations of atoms and molecules in the sand, rocks, and water; the showers of high-energy cosmic rays striking the atmosphere from outer space; all these were more than the formulas and graphs he had studied in the classroom. He felt them in a new, visceral way. They were, he gleaned, the Dance of Shiva from Hindu mythology. He began to notice similar parallels between cutting-edge quantum theory and central tenets of Eastern thought: the emphasis upon wholeness or interconnectedness, for example, or upon dynamic interactions rather than static entities.[4]

In December 1970, his visa about to expire, Capra returned to Europe. With no new job lined up, he began to check in with some of his contacts to see if he might find some steady position. He wandered into the theoretical physics division at London's Imperial College, whose leader he had met in California. The physicist had no fellowships to offer—finances had become as difficult for British physicists as for their American colleagues by that time—but with the financial downturn there were at least some empty desks around. And so Capra set up shop at Imperial: no position, no income, but a tiny corner of office space he could call his own.[5]

His financial situation quickly grew dim. He took on private tutoring jobs; he did some freelance work writing abstracts of recent physics articles for the *Physikalische Berichte*. When he could spare the time, he delved more deeply into his readings of Eastern texts, inspired as much by Alan Watts's teachings as by his own mystical experience on the beach. And he hatched a plan to put some of his hard-won physics knowledge to use: he would write a textbook on his beloved subject of quantum physics. If he could write the book quickly enough, he reasoned—and if he could get a major textbook publisher interested in the project—he might pull out of his financial tailspin. Not only that, the textbook might make him a more attractive candidate for a teaching position down the road.[6]

By November 1972 he had drawn up an outline for the book and begun drafting chapters. He reached out to another contact for advice: MIT's Victor Weisskopf, whom he had met at a recent summer school in Italy. Weisskopf, like Capra a native of Vienna, was by that time a grand old man of the profession. He had recently completed a term as director general of CERN, the multinational high-energy physics laboratory in Geneva, where John Bell made his career. By the time Capra sought Weisskopf's advice, the elder physicist was well into a sideline career as a success- ful popular-science writer. He had also published a highly influential textbook on nuclear physics—a book that held the honor, Weisskopf was always happy to recount, of having been the book most frequently stolen from the MIT libraries. Weisskopf had suggested the idea to Capra of writing a textbook when they met at the summer school. Capra sent his chapter outline to Weisskopf, hoping for some further encouragement. He also hoped his senior colleague would use his contacts to help line up a publisher and secure an advance payment in anticipation of future royalties.[7]

Back and forth their letters flew: Weisskopf commenting on Capra's proposal, and soon on individual chapter drafts; Capra thanking him for his comments but pressing again and again for more tangible forms of support. "As you know, the problem of financial support has become vital for me," Capra responded in January 1973, "and I wonder whether I could approach a publisher for a contract" at that stage of the project. If so, which publisher would Weisskopf recommend, and would Weiss- kopf mind contacting the press directly to recommend the book? "I am sorry to bother you with these problems, but I have indeed very little time to work on the book at the moment, because I am not supported by anybody and have to make my living with much less creative work." Weisskopf's responses—asking for more drafts and sending along fur- ther comments—sidestepped the issues Capra found most pressing. Capra reiterated his urgent need to line up a publisher and get some financial support.[8]

A few weeks (and chapter drafts) later, Weisskopf addressed Capra's main concern. "I like your style and find many things well expressed," he began. "I would again like to encourage you to go ahead and finish the manuscript." But, Weisskopf advised, Capra should wait before approaching a publisher until he had a complete manuscript in hand. He should also understand that few publishers offered advances for textbooks anymore. "I understand your need for financial support but I suppose you are aware of the fact that a book like this is not going to bring in much money because of the nature of the subject. The best that one can hope is something like $1 thousand the first year and less thereafter." Writing a textbook, Weisskopf counseled, might be a noble endeavor, but it was a lousy get-rich-quick scheme.[9]

Just at that moment, Capra received his invitation to visit Berkeley and give some talks to Geoffrey Chew's group. (Capra had sent some early essays comparing Chew's core notion of particle physics—a self-consistent particle "bootstrap"—to central doctrines of Buddhist thought. Chew had passed these essays on to his graduate students Elizabeth Rauscher and George Weissmann, who in turn encouraged Chew to invite Capra to visit.) While back in California, Capra also checked in with his former postdoctoral advisor at Santa Cruz. They talked about Capra's parallel projects: continuing his exploration of Eastern mystical thought, and pushing forward on his textbook project. The Santa Cruz physicist—"a rather hard-headed and pragmatic physicist" in Capra's estimation, hardly one drawn to the woolly countercultural currents swirling around him—encouraged Capra to combine his interests and change gears with his book project. Rather than write a physics textbook, why not refocus the book to explore the parallels between physics and Eastern thought that had so intrigued Capra since his transcendental experience on the beach? Coming on the heels of Weisskopf's realistic cautions about how well a textbook might sell, Capra took his former advisor's advice. Upon his return to London, Capra began composing new chapters on Eastern mysticism—one each on Hinduism, Buddhism, Confucianism, Taoism, and Zen—and interleaving them with the textbook chapters he had already written.[10]

Capra found the new plan inspiring, and set about trying to interest publishers in the project. A dozen rejections later, a small London-based publishing house agreed to take a gamble on it, even offering Capra the long-sought, if modest, advance payment that allowed him to finish writing it up. Completed manuscript in hand, Capra next managed to interest a tiny American publisher to bring out an edition in the United States: Shambhala Press, then just five years old, which had been founded in Berkeley to publish books on Eastern mysticism and spirituality. *The Tao of Physics* thus appeared simultaneously in Britain and the United States in 1975.[11]

A few months later Capra met Weisskopf in person again, at a conference in California, and Capra presented to Weisskopf a copy of the book. Weisskopf read most of it on his plane flight back to Massachusetts and he "liked it very much," he reported back. "It is very hard for me to judge whether you have succeeded in your task," Weisskopf continued, "since it addresses itself to a very specific kind of public than you find here in the East." Translation: we have no hippies at MIT. "I do believe, however, that it is a good book and that there will be many people who will have a better idea of physics after they have read it." Weisskopf shared his concern that some readers might be "scared off by the 'Tao' side of the deliberation," but conceded that "you can't make it right for everybody." He closed on a brighter note: "I wish you all luck and wonder how the sales will go."[12]

They went well. The first edition from Shambhala—20,000 copies—sold out in just over a year. Bantam brought out a pocket-sized edition in 1977 as part of its New Age series, with an initial printing of 150,000 copies. By 1983 half a million copies were in print, with additional editions prepared in a host of foreign languages. Twenty-five years later the book had achieved true blockbuster status: forty-three editions, including twenty-three translations—everything from German, Dutch, French, Portuguese, Greek, Romanian, Bulgarian, and Macedonian to Farsi, Hebrew, Chinese, Japanese, and Korean—with millions of copies sold worldwide.[13]

Many factors seem to have combined to jolt the book into the sales stratosphere. For one thing, Capra enjoyed a firm command of the physics; he had been well trained. The fact that the physics-heavy portions of his book had begun as drafts for a textbook—and that those sections had benefited from careful readings by a towering physicist like Viki Weisskopf—surely helped Capra clarify just how he wanted to present difficult concepts such as Heisenberg's uncertainty principle or quantum nonlocality. Moreover, his incursions into Eastern thought, while sometimes belittled by specialists in religious studies, nonetheless sprang from a genuine earnestness.[14] Capra had become a seeker, reading everything he could get his hands on. By the time he finished the book, he had spent years experimenting with alternate modes of encountering the world, always pushing to absorb the insights of the ancient mystical traditions. And then there was his impeccable timing. With the New Age rage in full force by the mid-1970s, conditions were ripe for a book like *The Tao of Physics*. Capra's book capitalized on a tremendous, diffuse, untapped thirst, a widely shared striving to find some meaning in the universe that might transcend the mundane affairs of the here and now. The market for Capra's book had been teeming like a huge pot of water just on the verge of boiling. *The Tao of Physics* became a catalyst, triggering an enormous reaction.

●

When Capra set out to promote the book, he seemed straight out of central casting. "Tall and slim with curly brown hair skirting the nape of his neck," cooed one *Washington Post* reporter, "Capra, with California tan, shoulder bag, and a Yin Yang button pinned to his casual jacket, seems more a purveyor of some new self-awareness scheme than a physicist." (Fig. 7.1.) It soon became clear, however, that Capra was more than just a pretty face. He was on a mission not just to explore the foundations of modern physics, but to alter the very fabric of Western civilization; "a cultural revolution in the true sense of the word," as he put it in the book's epilogue. As he saw it, modern physics had undergone a tremendous sea change in its understanding of reality, and yet most physicists—let alone the broader public—had failed to appreciate the consequences. The "mechanistic, fragmented world view" of classical physics had been toppled by quantum mechanics and relativity, but Western society still

FIGURE 7.1. Fritjof Capra discussing *The Tao of Physics* in November 1977. (Photograph by Roger Ressmeyer, reproduced by permission of Corbis Images.)

carried on as if Einstein, Bohr, Bohm, and Bell had never lifted a pencil. "The world view implied by modern physics is inconsistent with our present society, which does not reflect the harmonious interrelatedness we observe in nature," he explained. A proper understanding of what modern physics had achieved—especially its "philosophical, cultural, and spiritual implications"—could help restore the balance before it was too late.[15]

The new worldview of modern physics was not just out of step with Western traditions, Capra concluded; it had rediscovered the age-old sutras of Buddhism, Vedas of Hinduism, and *I Ching* of ancient Chinese thought. "The further we penetrate into the submicroscopic world, the more we shall realize how the modern physicist, like the Eastern mystic, has come to see the world as a system of inseparable, interacting, and ever-moving components with man being an integral part of this system." Only in the twentieth century, in the wake of the great revolutions of modern physics, were physicists beginning to throw off the yoke of Cartesian dualism, the assumption that had reigned since the seventeenth century that mind and matter occupied separate realms, firmly cut off from each other. Capra acknowledged that much good work had followed upon postulates of René Descartes and Isaac Newton such as reductionism, the dictum to study nature by breaking things down into their smallest parts and focusing on the mechanisms by which parts interact to become wholes. But the centuries-long trend of Western science had not been without its costs. "Our tendency to divide the perceived world into individual and separate things and to experience ourselves as isolated egos in this world," Capra contended, had long been understood in Eastern traditions as a mere "illusion which comes from our measuring and categorizing mentality."[16] Western observers' impressions of the physical world as pointillist and fundamentally cleaved off from human consciousness arose not from the nature of reality per se, but from the mental filters and habits we happened to have imposed upon our investigations. Three centuries after Newton and Descartes, quantum physicists had only just learned that "we can never speak about nature without, at the same time, speaking about ourselves"—a deep insight that Capra considered comparable to age-old Hindu, Buddhist, and Taoist teachings.[17]

In the remainder of his book, Capra marched through a series of further parallels between the latest lessons of quantum physics and the venerable mantras of Eastern mysticism. First and foremost was what he saw as the "organicism" or holism implied by quantum interconnectedness: ultimately the quantum world is not divisible into separate parts, but is woven into one seamless whole. He drew on several of the Fundamental Fysiks Group's favorite physicists, among them Henry Stapp, David Bohm, and John Wheeler. Like group member Jack Sarfatti, Capra emphasized Wheeler's shift from "observer" to "participator," even including a long quotation from one of Wheeler's little-noticed conference talks on the theme. "The idea of 'participation instead of observation,'" Capra noted, "has been formulated in modern physics only recently, but it is an idea which is well known to any student of mysticism," which, after all, has always required "full participation with one's whole being."[18]

Capra also saw deep parallels between the koans, or riddles, of Buddhist thought, the constant interplay of opposites in Taoism, and the paradoxes of quantum theory. Bohr's complementarity called on physicists to transcend what appeared to be opposites: neither wave nor particle but both. Although "this notion of complementarity has become an essential part of the way physicists think about nature," Capra explained, the physicists had come late to the party: "in fact, the notion of complementarity proved to be extremely useful 2,500 years ago," when the Chinese sages promoted the dialectic of yin and yang to the center of their cosmos. Little wonder, Capra concluded, that Bohr adopted the yin-yang symbol for his family coat of arms. Einstein's relativity, meanwhile, with its interconversion of matter and energy via $E = mc^2$, echoed the Eastern emphasis upon dynamism and flow: the universe caught in a never-ending dance, rather than being a collection of static objects. The merging of space and time into a unified space-time likewise brought the physicists' cosmic picture into line with long-standing Eastern intuitions.[19]

More than just the conclusions of physics and mysticism seemed to line up. Capra saw deep similarities in their methods—their *Tao*, or "way"—as well. He insisted that both sides were, at root, empirical. Physicists could formulate no lasting knowledge without careful observations and experiments. So, too, did close and careful observations form the backbone of mystical knowledge, observations or intuitions gleaned by the mystics during meditation or other altered states of consciousness. Mystics achieve this "direct insight," Capra explained, by "watching rather than thinking." The centrality of experience to all Eastern traditions thus gave them "a strong empirical character," closely "parallel to the firm basis of scientific knowledge on experiment." In both cases, knowledge rests on observations that transcend the ordinary world of appearances, pushing beyond our usual senses: to the impossibly small realm of atoms and particles, or to the inner space of altered consciousness.[20]

The Tao of Physics succeeded in that rare category, the crossover hit. It held broad appeal for hundreds of thousands of readers who were not physicists, or academics of any sort. Looking back a few years after its original publication, one reviewer marveled that Capra's book had sold "amazingly well, not only to the usual Shambhala devotees of Eastern religion but also to engineers, Caltech grad students and people of the general population who, a few years later, would be reading Carl Sagan." Reviewers routinely touted Capra's clear expository style. "He has a pleasing way of raising and answering questions," proclaimed one. "The book is very exciting, and Capra is clearly in earnest," cheered another; "I do not know of a better general introduction to the major concepts of modern physics." Another emphasized that Capra "writes fluently, quotes charming *haikus* and Zen *koans*"; his presentation never suffered from the "ponderous" and "abstract" style so often adopted in popular treatments of modern physics. More than just a good read, the book received a great deal of serious, scholarly attention. Academic journals specializing in philosophy, history, and sociology carried reviews. The journal *Theoria to Theory* published a lengthy review section on the book, with detailed comments from three specialists in philosophy and religious studies. Sociologists and philosophers of science likewise devoted substantial articles to the book, picking through the claimed parallels and subjecting each to sustained critique.[21]

Some scholars accepted the basic approach that Capra followed—comparing quotations from physicists and mystics to illustrate the parallels between them—but complained that Capra overlooked equally good parallels a good bit closer to home. Why focus only on Eastern mystics like Lao Tzu, some reviewers asked, while ignoring all the influential adherents of Western mystical traditions, from the pre-Socratics through the neo-Platonists? Surely pronouncements by ancient Greek philosophers like Parmenides or Anaximander about the nature of matter and the essence of change could stack up equally well with telltale quotations from quantum physicists. Where better than in Plato's *Timaeus*, with its notion of a "womb of becoming," could one find parallels to modern physicists' description of the vacuum state? Immanuel Kant's meticulous parsing of the worlds of appearance and existence, moreover, seemed just as robust an anticipation of quantum physicists' emphasis upon the active role of the observer as the Hindu concept of "maya."[22]

Others were less sanguine about the "parallelist" approach itself. What controls or provisions were in place to guard against cherry-picking juicy quotations out of context, with no justification of those quotations' representativeness? How did analysts like Capra handle subtle nuances of translation? After all, both Sanskrit aphorisms and mathematical equations had to be rendered into a common language (in this case, English) before they could be compared. Quotations might not only be ripped out of context but also out of their original vocabularies, introducing all manner of distortions.[23]

Capra was also far too cavalier a "lumper" for many of these analytical "splitters." Did it really make sense to dump Hinduism, Buddhism, Taoism, Confucianism, and more into a single Eastern worldview? Several reviewers found that to be as dubious and unhelpful as papering over all the distinctions between Judaism, Christianity, and Islam and proclaiming a unified Western worldview. Meanwhile, some reviewers were quick to point out that many of Capra's favorite examples from modern physics were far from accepted as mainstream. Physicists still debated many of the points under consideration, even as Capra tended to "muddle" together competing—even contradictory—interpretations into a single worldview of modern physics. The effect, some complained, produced a "schizophrenic" oscillation between "different, often conflicting, inter-pretations of physics." Or, as another reviewer put it, Capra "dissolves the precise meaning of technical terms in order to establish analogies; and he plays down discrepancies between points of view when they threaten his parallels." Same for Capra's treatment of "empiricism": are mystics' intuitions, gleaned individually during altered states of consciousness, really comparable to physicists' experiments, which, at least in principle, could be conducted in public and replicated by others?[24]

Finally, what was the actual point? Some accused Capra of circular reasoning: physics concepts were supposed to be bolstered by their resemblance to ancient wisdom; but Capra used the template of ideas from modern physics to pick out and accentuate particular ideas from the mystics. So which was giving credence to which? Capra likewise seemed to flip-flop between the argument that the two traditions yielded separate but complementary visions of reality, and the suggestion that one tradition confirmed or validated the other. What, meanwhile, might account for the similarities that did exist? As one philosopher pointed out, the mere fact that two people said things that sounded alike in no way meant that what they said was true. In the end might the similarities, striking as they may seem when stacked up, one after the other, in Capra's charming and instructive book, tell us only about "basic tendencies of the human mind, or perhaps similarities in metaphor," rather than how the world really works? The similarities could have an even more mundane explanation, suggested a skeptical sociologist: "contamination" over time, since, by the twentieth century, both physicists and mystics had heard a lot about each other. Perhaps there had been conscious or unconscious borrowings of terms. Some of these borrowings were clearly intended to be ironic or humorous, as in physicist Murray Gell-Mann's "eightfold way" of symmetries among nuclear particles, echoing the famous Buddhist path to nirvana. Those types of parallels, critics maintained, reflected at most a matter of semantics but nothing of substance.[25]

In the end, all of the learned rebukes, steeped as they were in philosophy, theology, and sociology, served to reinforce a most remarkable point: Capra's little paperback had clearly hit a nerve. No other popularizations received such serious and sustained scrutiny—scrutiny that continued to fill academic journals fully fifteen years after *The Tao of Physics* had first appeared in print.

•

Perhaps the most surprising response of all, however, came from scientists. Some certainly responded as we might expect, downplaying the book as mere popularization and dismissing the countercultural overtones as just so much Zeitgeist pap. Famed biochemist and science writer Isaac Asimov, for one, bewailed the "genuflections" to all things Eastern made by "rational minds who have lost their nerve." Jeremy Bernstein, Harvard-trained physicist and staff writer for *The New Yorker*, went further. He concluded his review of Capra's book, "I agree with Capra when he writes, 'Science does not need mysticism and mysticism does not need science but man needs both.' What no one needs, in my opinion, is this superficial and profoundly misleading book."[26]

These predictable responses, however, were by no means the norm. Mysticism aside, Capra offered a vision around which many physicists could rally. In his opening chapter he had noted the "widespread dissatisfaction" and "marked anti-scientific attitude" of so many people in the West, especially among the youth: "They tend to see science, and physics in particular, as an unimaginative, narrow-minded discipline which is responsible for all the evils of modern technology." Capra declared, "This book aims at improving the image of science"; the insights and joys of modern physics extended far beyond mere technology. Indeed, "physics can be a path with a heart, a way to spiritual knowledge and self-realization." Few reviewers missed the point. *Physics Today* ran a review of the book by a Cornell astrophysicist. The review began by citing the profession's litany of woes: the "anti-scientific sentiment" of the age, which distressed Capra and his critics alike, "manifests itself on all levels of our society, from a decrease in funding for basic research to a turning to Eastern mysticism and various forms of occultism." Not an auspicious start for the volume under consideration. Yet the reviewer judged Capra's book to be a great success. For one thing, the book got the physics right. (On this point, reviewers had far fewer complaints than with Gary Zukav's otherwise-similar *Dancing Wu Li Masters*. Unlike the highly trained Capra, Zukav had no previous background in physics.) Even more important: *The Tao of Physics* integrated "the abstract, rational world view of science with the immediate, feeling-oriented vision of the mystic so attractive to many of our best students."[27]

The reviewer's comments proved more than a passing observation. Just as the review was going to press, Capra was busy teaching a new undergraduate course at Berkeley based on his book. He reported proudly to MIT's Victor Weisskopf that one-third of the students were science majors, eager to learn about the foundations of modern physics: just the sort of philosophical material they were not receiving in their other physics classes. Soon the *American Journal of Physics*, devoted to pedagogical innovations in the teaching of physics, began carrying articles on how best—not whether—to use *The Tao of Physics* in the classroom. One early adopter began by citing the huge market success of Capra's book. "This leads naturally to the question," he continued, "how can a physicist utilize this interest by offering a course using Capra's book?" A follow-up article commented matter-of-factly:

> Anyone involved in physics education is likely to be asked to comment on parallelism [between modern physics and Eastern mysticism] at some stage. It would be easy to dismiss such ideas entirely, and in so doing possibly risk alienating a new-found interest among students. This field has the potential of appealing to the imagination and should perhaps be carefully explored and maybe even "exploited."

With budgets falling and enrollments crashing, physicists could ill afford to turn their noses up at anything that might bring students back into their classrooms.[28]

An early article in the *American Journal of Physics* illustrated one way forward. It described a successful course ("Zen of Physics") that David Harrison, a physics professor at the University of Toronto, had offered two years in a row. Like Capra, Harrison had already been intrigued by parallels between quantum physics and other intellectual traditions, ranging from the ancient Greeks to Eastern religions. He was also plugged into the New Age community in and around Toronto. He conceived the course, he later recalled, as one way to bring these many facets of his life together. The new course felt like a liberation. For one thing, the intimate classroom setting allowed for in-depth discussion. Essays and "special individual projects" replaced problem sets and (he argued) fostered greater enlightenment and enthusiasm among the students— so much so that Harrison answered popular demand by offering a more advanced follow-up course, playfully dubbed "Son of Zen of Physics." Within a few years, the original course had become so popular that he had to break it up into smaller tutorials (capped at twenty-five students each) to accommodate the expanding enrollments while retaining the discussion-based approach. He developed novel assignments as well, such as this one:

In 1969 [Lawrence] LeShan devised an interesting test. He collected 62 quotations, some from modern theoretical physicists
and some from mystics. Below, we have randomly selected 20 of
these quotations. . . . Beside each statement clearly mark a *P* if
you believe it was made by a physicist and an *M* if you believe it
was made by a mystic.

Examples: "It is the mind which gives to things their quality, their foundation, and their being"; or "Thus the material world constitutes the whole
world of appearance, but not the whole world of reality; we may think of
it as forming only a cross section of the world of reality."[29]

To Harrison's delight, he soon found his course filling up with physics
majors alongside nonscience students, just as Capra had found in his
course at Berkeley. A few years later he published another article in the
American Journal of Physics detailing one of the lesson plans he had developed for the course, on Bell's theorem and quantum nonlocality. He noted
that standard physics curricula and textbooks still had not incorporated
Bell's theorem. (Indeed they hadn't: the first graduate-level textbook to
discuss Bell's theorem appeared in 1985; the topic didn't appear in undergraduate textbooks until even later.) Consequently, he had to seek extra
help to make sense of the material for himself. "Tuned-in" friends in
his local countercultural circles told him about the Fundamental Fysiks
Group, and he launched into a long and spirited correspondence with
group member Nick Herbert, whom he thanked extensively for explaining
Bell's theorem to him in detail. Only through his *Tao*-inspired course, in
turn, did physics students learn about such major, foundational topics.
Supplemental materials like his latest article filled a real curricular hole.[30]

By no means did courses like "Zen of Physics" swamp ordinary physics
department offerings. All the same, the courses, and Capra's book, clearly
left their mark. As late as 1990, two scholars noted that university physics courses throughout the United States still routinely listed *The Tao of
Physics* on their syllabi as a "helpful reference."[31] The book's presence in
the classroom became so common that some physics teachers began to
push back. They criticized what seemed to be loose analogies or paral-

lels between physics and mysticism, which they feared were as likely to confound students as enlighten them. Whether or not Capra's parallels could stand on their own, did they really hold special pedagogical value? After all, to truly understand mysticism required the same kind of discipline, training, and time that understanding modern physics did; why becloud one difficult subject with allusions to another? Debates broke out in the pages of the *American Journal of Physics*. At one point Harrison, the "Zen of physics" professor, shot back: "It should also be emphasized that most of these students would not have taken an offering by the Physics Department if it were not this one." Or, as he put it more recently, he was among the best in his department at delivering "bums in the seats."[32]

In a roundabout way, Capra thus fulfilled his original goal: he wrote a successful textbook after all. Physicists across the continent eagerly snatched up *The Tao of Physics*. In their classrooms, the book helped demonstrate to disaffected students—or so physicists hoped—that they, too, were "with it." Along the way, the book inspired some of the first lesson plans on topics like Bell's theorem. Capra's book thus smuggled some attention to interpretive, foundational issues back into the classroom.[33]

Chapter 8

Fringe?!

> At the time of the first conference, it was agreed that there would be no publicity for the conferences from Est. Est has adhered strictly to this agreement. Although they are not publicized by Est, the conferences are not secret; they are widely known within the physics community.
>
> **—Sidney Coleman, 1981**

The Fundamental Fysiks Group's dogged efforts to meld quantum entanglement with parapsychology and Eastern mysticism set the group's members apart from the physics mainstream, but their self-made universe was not closed unto itself. Despite the intuitions (or even hopes) of scientists and philosophers, no clear demarcation separated the Fundamental Fysiks Group from "real" physics. In fact, highly successful physicists often sought to cross paths, figuratively and literally, with members of the Fundamental Fysiks Group.

•

When pressed to give an opinion on possible connections between parapsychology and his own work on quantum entanglement, John Bell refused to dismiss the matter out of hand. In a charming letter to a longtime parapsychology researcher, Bell wrote that his experiences as a young physics student tempered his judgment of psi research. Critics of parapsychology often complained about the small number of times that researchers had successfully repeated a claimed psi effect. But perhaps that was not too different from Bell's own frustrations in his student laboratory in Northern Ireland, he wrote, where he had failed miserably to reproduce the well-known laws of electric attraction and repulsion. He had "formed the opinion that electrostatics could never have been

convincingly discovered in my home country—because of the damp."
Perhaps a similar confounding factor masked exotic parapsychological
phenomena as well, making them difficult to reproduce at will. In any
case, Bell closed, good scientists should certainly keep an open mind:
physicists had been surprised by seemingly impossible phenomena several times before.[1]

Yale's eminent physicist Henry Margenau reasoned along similar lines.
Like Bell, Margenau was no stranger to the mysteries of quantum theory.
He had published one of the earliest responses to Einstein's famous paper
with Podolsky and Rosen on "spooky actions at a distance" in quantum
theory, all the way back in 1936. He had gone on to write successful physics textbooks and treatises on the philosophy of science.[2] Margenau also
launched the journal *Foundations of Physics*, in which several members of
the Fundamental Fysiks Group published their work. In the midst of his
long and distinguished career, Margenau delivered a keynote address
to the American Society for Psychical Research. He likened the current
state of psi research to the early days of radioactivity, when a few stubborn signals—clicks in Geiger counters—could neither be easily reproduced nor convincingly explained.[3] In 1978 Margenau teamed up with
psychologist Lawrence LeShan, just a few years after LeShan published
his book *The Medium, the Mystic, and the Physicist*. Together they wrote a
short submission to the top-flight journal *Science* urging the scientific
study of extrasensory perception (ESP).[4]

Like members of the Fundamental Fysiks Group, Margenau highlighted quantum entanglement as a likely means of reconciling ESP with
action at a distance. As he and LeShan wrote in their piece for *Science*,
"ESP is not stranger than some of the discussions" that had recently
emerged in quantum theory, such as Bell's theorem. Marching through
one example after another, they found no contradiction between well-
tested scientific laws and ESP. "We *can* find contradictions between ESP
and our culturally accepted view of reality, but not—as many of us have
believed—between ESP and the scientific laws that have been so labori-
ously developed." In the absence of those contradictions, they suggested,
"it may be advisable to look more carefully at reports of these strange

and uncomfortable phenomena which come to us from trained scientists and fulfill the basic rules of scientific research."[5] What happened next left Margenau fuming. Not only did *Science* fail to publish Margenau and LeShan's letter to the editor but the editors also neglected even to acknowledge receipt of the submission. Nine months later Margenau sent a sharp letter complaining about the conspicuous lack of response. "Is this your policy in dealing with matters of this sort?"[6] A month later Margenau was still outraged by "these violations of every canon of editorial courtesy." Their letter about ESP never did appear in print.[7]

Other leading lights interacted directly with members of the Fundamental Fysiks Group on topics of shared interest. Nobel laureate Eugene Wigner, John Wheeler's longtime colleague at Princeton, had introduced the notion that human consciousness might be necessary to collapse a quantum wavefunction, reducing its plethora of possibilities to one actual outcome. Wigner's idea had provided early inspiration to members of the Fundamental Fysiks Group, as they groped toward a physical account of parapsychological effects. Wigner, in turn, commented generously—in public and in print—on Elizabeth Rauscher's working explanation for the remote-viewing results that Harold Puthoff and Russell Targ had recorded in their psi lab at the Stanford Research Institute. Wigner concurred with Rauscher that Bell's theorem seemed relevant, and encouraged her work on multidimensional space-times and psi phenomena.[8]

Wigner's former student Abner Shimony dabbled with similar material. One year before he teamed up with his friend and collaborator John Clauser to write a definitive review article on Bell's theorem and its experimental tests, Shimony published a rather different article in Margenau's journal, *Foundations of Physics*. Together with three of his graduate students, Shimony put Wigner's idea about consciousness and quantum theory to the test. Recall that, to Wigner, an electron being put through its paces between the poles of two magnets would only assume a definite value for its spin—either spin up or spin down—once a human observer had bothered to look at the array of detectors. The act of trying to observe the electron's spin, in Wigner's view, forced the electron to assume only one of its two possible outcomes. If Wigner's idea were correct, Shimony reasoned, then one should be able to reproduce telepathy in the laboratory. Shimony sat one participant in a closed room next to a source of radioactive atoms and a sensitive detector, while a second participant sat in a different room. Following Wigner's chain of argument, the first participant should have been able to send messages to the second simply by choosing whether to look at the display monitor attached to the radioactivity detector, since (to Wigner) the act of observing the monitor should have constituted the fundamental quantum act of collapsing their conjoined wavefunction. (In its essentials, Shimony's experiment was not so different from the metaphase typewriter dreamed up by Nick Herbert—minus the LSD and free-for-all party atmosphere in which Herbert and company had sought to use their device to contact the spirit of Harry Houdini.) Suffice it to say, Shimony and his students found no statistically significant results; but they found the topic important enough to warrant publication. Years later, Shimony—an acknowledged leader in the interpretation of quantum theory—reprinted his telepathy null-result article in a volume of his most significant papers.[9]

Gerald Feinberg also lent a hand. An accomplished theoretical physicist in his own right—he studied under a Nobel laureate for his dissertation, and his early prediction of a new type of particle earned three other colleagues their own Nobel prizes when they confirmed its existence— he constantly rubbed shoulders with Nobelists in Columbia's physics department. In 1967 he published a well-known article in the *Physical Review* on particles that might travel faster than light, coining the term "tachyon" in the process.[10] (The article focused on whether such theoretical beasts might be self-consistent within the frameworks of relativity and quantum mechanics; it did not broach potential parapsychological consequences.) A dedicated futurologist, Feinberg penned *The Prometheus Project: Mankind's Search for Long-Range Goals* in 1969, landing him squarely within a network of futurists that also included Ira Einhorn.[11] Perhaps because of these interests, Uri Geller's handlers brought the young Israeli performer to Feinberg at Columbia during Geller's first trip to the United States in autumn 1971, before transferring Geller to

Puthoff and Targ's psi lab for in-depth studies.[12] Young physicists like Elizabeth Rauscher, inspired by Feinberg's work on tachyons, sought him out for advice on their own investigations. Feinberg obliged—he and Rauscher spoke often and at length about her work for the better part of a decade—and he went further, contributing his own paper to a conference on quantum theory and parapsychology. Not long after that, he served as Columbia's physics department chair, becoming the boss of all those Nobel laureates.[13]

David Bohm, whose early model of hidden variables in quantum theory had inspired John Bell to investigate quantum entanglement and nonlocality in the first place, likewise enjoyed close relations with members of the Fundamental Fysiks Group. In addition to conducting his own tests of Uri Geller, Bohm hosted Rauscher, Fred Alan Wolf, and Jack Sarfatti at his home department in London at various times during the 1970s. He shared his emerging ideas about quantum theory and the "implicate order"—worked out as much in conversation with New Age thinkers like Jiddu Krishnamurti as with fellow physicists—with the Fundamental Fysiks Group during a visit in 1978.[14]

The French theoretical physicist Olivier Costa de Beauregard also made several visits to the Fundamental Fysiks Group. By the time he visited the Berkeley group and checked in with Harold Puthoff and Russell Targ at their Stanford Research Institute psi lab in 1975, Costa de Beauregard had published a well-received textbook on quantum mechanics and become a sought-after speaker in physics departments throughout Europe and North America. He also served as director of research for the theoretical physics division of France's prestigious Centre National de la Recherche Scientifique (CNRS).[15] He often dropped hints in his mainstream physics articles of possible links between Bell's theorem, telepathy, and clairvoyance; he made the connections more explicit in his other writings.[16] He joined the Fondation Odier de Psycho-Physique, France's version of the American and British Societies for Psychical Research. As he explained to a journalist in 1981, his physics research dovetailed with his Roman Catholic faith: each had inspired him to become a spiritualist, believing strongly in the notion of mind over matter.[17]

Richard Mattuck showed similar facility with the new material. By the early 1970s, the MIT-trained theorist had published more than twenty major articles on condensed-matter physics in leading journals, using quantum theory to describe bulk properties of materials. Complementing his research articles, he had published a much-beloved textbook, *A Guide to Feynman Diagrams in the Many-Body Problem*, in 1967. And then, for eight years, his name stopped appearing in the mainstream physics journals.[18] By the time the second edition of his textbook appeared, in 1976, Mattuck had turned his attention squarely to the quantum mechanics of psi. He built upon Evan Harris Walker's model of consciousness as a collection of hidden variables, which Nick Herbert and Saul-Paul Sirag had followed closely from their perch in Arthur Young's Institute for the Study of Consciousness in Berkeley. Mattuck drew on many of the calculational tricks he had mastered in his many-body work, summing over the combined effects of many tiny, piecemeal quantum processes, to demonstrate that larger psychokinetic effects would result if consciousness sent out pulses of information, rather than proceeding (as in Walker's original model) via continuous information processing. The summed pulses, Mattuck continued, could account for psi phenomena of the magnitude reported in several laboratory studies, including one of his own. He published several iterations of this work in parapsychology journals and spoke on the same psi lecture circuit as Rauscher, Puthoff, Targ, and Costa de Beauregard. Once he began publishing "regular" physics papers again, they were entirely on hidden variables, Bell's theorem, and the foundations of quantum mechanics—topics he had never broached before his work on psi.[19]

Then there is Brian Josephson. While a graduate student at the University of Cambridge in the early 1960s, Josephson published a short paper on electrical currents that might tunnel between a thin slice of ordinary metal sandwiched between two superconductors. Experimentalists observed the predicted effect within months, and the "Josephson junction" earned Josephson a Nobel Prize in 1973, at the tender age of thirty-three.[20] Today such supersensitive junctions are hardwired into everything from quantum computer prototypes to instruments that

measure neural activity inside the human brain. By the time Josephson accepted his prize in Stockholm, however, his research interests
had turned squarely to Eastern mysticism, the nature of consciousness,
and parapsychology. He traveled to San Francisco late in 1976 to check
out Puthoff and Targ's psi lab and to deliver a talk for the Fundamental
Fysiks Group. Sarfatti's Physics/Consciousness Research Group under-
wrote the expenses for Josephson's two-week trip. A reporter for the *San
Francisco Chronicle* covered Josephson's visit, describing how the young
Nobelist "padded around" Sarfatti's Nob Hill apartment "in maroon sox,"
while the two compared notes on their evolving theories of quantum
entanglement and psi. Josephson continued to speak at conferences
on parapsychology alongside Puthoff and Targ, Rauscher, and others,
even providing the keynote address for the fabled 1977 conference in
Reykjavík, at which Ira Einhorn had mysteriously failed to show.[21] (Fig.
8.1.) When the *New York Review of Books* ran a feature article in 1979 that
was critical of efforts to use quantum theory to explain psi phenomena,
Josephson teamed up with Costa de Beauregard, Mattuck, and Walker to
write a feisty reply.[22] Josephson's passion for the topic has not wavered
to this day. He directs a "mind-matter unification" project at Cambridge
and vigorously defends parapsychology from naysayers.[23]

FIGURE 8.1. Elizabeth Rauscher and
Brian Josephson at a conference on
quantum mechanics and consciousness, held in Spain in the late 1970s.
(Courtesy Elizabeth Rauscher.)

Perhaps the apotheosis of such interminglings—of people and topics that one might have assumed would remain as cleanly separated as oil and water—occurred at a conference in 1986 in honor of Eugene Wigner's ninetieth birthday. Nick Herbert's former roommate from graduate school, Heinz Pagels, helped to organize the meeting under the auspices of the New York Academy of Sciences. (Pagels served as the Academy's executive director.) Olivier Costa de Beauregard also served on the organizing committee. The large meeting, featuring more than fifty invited lectures and twenty poster presentations over four days, was held in New York City's World Trade Center. The conference served as something of a "coming out" party for those researchers who had toiled for years on the foundations of quantum mechanics while most working physicists balked at the topic. In fact, the organizers encouraged those whom they contacted to spread the word to other people who might be interested, since the community was still so diffuse.[24]

To the organizers' delight, the meeting drew several top-notch contributors. In addition to Wigner, three other Nobel laureates or soon-to-be laureates participated, as did heavyweights like John Wheeler. (John Bell contributed a paper, though he was not able to attend in person.) Several core members of the Fundamental Fysiks Group shared the podium with these leading figures. Henry Stapp and John Clauser were each invited to attend, and Nick Herbert contributed a poster presentation.[25] Jack Sarfatti and Fred Alan Wolf were there, too; Sarfatti even caught the ear of a *New York Times* reporter who was covering the conference and made a plug for his idiosyncratic interpretation of Bell's theorem.[26] As one of the plenary sessions was wrapping up, Nick Herbert heard someone calling his name. He looked up and was greeted by Andrija Puharich—Uri Geller's original promoter, and sometime collaborator with Ira Einhorn on paranormal topics. Puharich had come into the city for the conference from his residence-cum-psi-lab in nearby Ossining, New York. Mingling among all those physicists, Puharich appeared decked out in opera cape and cane, accented by a "beautiful blonde babe" hanging on each arm.[27] Such was the clientele at conferences on Bell's theorem and the interpretation of quantum theory.

●

One conspicuous feature of the New York Academy of Sciences conference in 1986 was its heavy emphasis on new experiments. Among the most significant, lauded even to this day, was a variant of Clauser's experiment, performed by the French physicist Alain Aspect. Physicists routinely look back and date the upsurge of interest in Bell's theorem and the foundations of quantum mechanics to Aspect's elegant experiment, the results of which appeared in 1982.[28]

John Clauser and his Berkeley student published the results of their first test of Bell's theorem in 1972. Not long after that, Clauser began to correspond with a French experimental physicist working at the Institut d'Optique in Orsay, just outside Paris. The French physicist was interested in conducting his own version of Clauser's test, and sought Clauser's advice for how best to proceed.[29] A few years later, after Clauser had left the Berkeley laboratory, one of his former postdoctoral advisors shipped several boxes of equipment to the Orsay group, including the sensitive calcite polarizers that had been at the heart of Clauser's jerry-built apparatus.[30] In the meantime, the French experimentalist had been tutored in the niceties of Bell's theorem by Olivier Costa de Beauregard and Bernard d'Espagnat, France's leading experts in the topic.[31]

While Costa de Beauregard and d'Espagnat coached their French colleague, the physicist who would eventually take the lead on the new experiment set off on a separate adventure. After earning his master's degree, Alain Aspect embarked for Cameroon, the small African nation and former French colony. He spent three years there working on various aid projects, the French equivalent of an American Peace Corps mission. To relax in the evenings, he pored over a recent textbook on quantum mechanics by Claude Cohen-Tannoudji, already a major figure in French physics who would go on to receive the Nobel Prize. Inspired by the textbook, Aspect's mind wandered over the classic conundrums of quantum theory. Upon Aspect's return to Paris in 1974, his advisor—the French physicist who had been coached by Clauser, Costa de Beauregard, and d'Espagnat—handed him a stack of papers. On the very top sat Bell's original article on quantum entanglement. And so Aspect dove into the topic for his doctoral research.[32]

In Clauser's original experiment at Berkeley, he and his student had set the angles of polarizers at the two detectors before each round of photons was released. A determined advocate of hidden variables could therefore object that information about the settings at one detector could somehow be communicated to the other detector long before the photons had been emitted. One could imagine that measurements on the pairs of photons could arrange themselves into the startling Bell-like correlations by some as-yet unknown local, causal process. Clauser had recognized the point from his earliest musings on Bell's theorem. In fact, Clauser had mentioned in his first letter to Bell, back in 1969—the very first indication of interest in Bell's work that Bell had received from anyone, anywhere—that one might try to vary the angles of the polarizers at random, while the photons were in flight.[33] Such a feat would realize in practice the basic idea behind John Wheeler's delayed-choice experiment. With helpful guidance from Costa de Beauregard and d'Espagnat, that was what Aspect set out to do.[34]

Before launching into the new experiment, Aspect made a pilgrimage to CERN in Geneva to ask John Bell himself whether such an elaborate experiment would be worthwhile. Bell's immediate response: "Have you a permanent position?" Such was the stigma still attached to the topic. After Aspect assured Bell about his job security—he had obtained a stable teaching position that allowed him to pursue his doctoral work—he and Bell began their discussion in earnest.[35] Aspect's work was further buoyed when d'Espagnat invited him to participate in the summer workshop on Bell's theorem, held in 1976, which d'Espagnat was organizing. The workshop opened more doors; soon one of the participants introduced Aspect to Cohen-Tannoudji, whose textbook had sparked Aspect's interest in foundations of quantum theory during his African sojourn. "That really made a big difference," Aspect recalled recently: only upon seeing Cohen-Tannoudji make frequent visits to Aspect's laboratory did the local physicists show Aspect some grudging respect. Until that time, Aspect says, "most physicists thought I was a crackpot" for wasting so much effort on the interpretation of quantum mechanics.[36]

Aspect's ingenious twist on Clauser's experiment was to use super-fast switches to change the path of a given photon en route from source to detector. Like the switches on a railroad track, Aspect's acoustico-optical switches steered incoming photons toward different destinations. Depending on which orientation the switch happened to be in when a photon arrived, the photon would be directed toward one of two polarizers, set to different angles. The switches changed at lightning speed—about 100 million times per second—so that the orientations of the switches flip-flopped two or three times while the photons were in flight. Now there was no way that a signal traveling at light speed could inform one detector about the settings at the other detector before each side had completed its measurement.[37]

Aspect began making presentations on his planned experiment years before he had collected any data, and thus anticipation began to grow among small circles of self-selected enthusiasts.[38] Elizabeth Rauscher visited Aspect and received a tour of his laboratory late in 1977, on the heels of her presentation to David Bohm's group in London. Through an intermediary, Jack Sarfatti also got in touch with Aspect before the experiment was complete.[39] D'Espagnat likewise became an effective spokesperson. He published a lengthy feature article on Bell's theorem in *Scientific American* in 1979, in which he highlighted Aspect's developing work. D'Espagnat also gave a sneak preview of Aspect's results at the Esalen workshop in February 1982, more than half a year before Aspect submitted his article to the journal.[40] Meanwhile, Aspect reached out directly to John Clauser. Clauser sent detailed comments on an early draft of one of Aspect's articles, and Aspect thanked Clauser for "the tremendous work you have done for editing my paper," clarifying details of the physics as well as smoothing over infelicities of English, which was not Aspect's native language. Aspect also thanked Clauser for his "comments at P.R.L. [*Physical Review Letters*]." It seems that Clauser served as a not-so-blind referee of Aspect's paper as well as its behind-the-scenes editor.[41]

At last Aspect's article appeared in *Physical Review Letters* late in 1982, a decade after Clauser's original experiment had been reported there. Having sent nearly a trillion pairs of photons through his elaborate apparatus, hyperfast switches and all, Aspect found the quantum-mechanical prediction fulfilled to unprecedented accuracy. Aspect had produced the "spooky action at a distance" in his laboratory, demonstrating Bell-styled nonlocality beyond a reasonable doubt. In Aspect's experiment, the hypothesis that the photon pairs were *not* subject to quantum entanglement failed by five standard deviations. In the light of those data, Einstein's (and Bell's) hoped-for notion that the measurement of one photon would have no discernible effect on the second photon slipped to less than a one-in-a-million chance. To those (still in the minority) who were paying attention, Aspect's experiment seemed to seal the deal: the quantum world is inherently nonlocal, after all.[42]

Not long after Aspect's article came out, he spent the summer at an IBM research laboratory near San Francisco. He and Clauser enjoyed sailing together—still one of Clauser's favorite pastimes—while Jack Sarfatti peppered the French physicist with questions about the experiment and haggled over details of their competing interpretations. Aspect also followed (with a hefty dose of skepticism) Nick Herbert's efforts to use an experimental arrangement similar to the one in Orsay to send signals faster than light.[43] Right from the start, Alain Aspect and his justly renowned experiment traveled in overlapping circles with core members of the Fundamental Fysiks Group.

•

Patronage ties also kept members of the Fundamental Fysiks Group entangled with mainstream physicists. By 1975 Werner Erhard had hired Robert W. Fuller to lead the *est* foundation, the philanthropic wing of his fast-growing enterprise. With Fuller's help, Erhard's circle of physicist-interlocutors widened considerably.

Fuller had completed his PhD in physics at Princeton under John Wheeler in the early 1960s before joining the physics faculty at Columbia University. He worked on nuclear physics and gravitation, and coauthored a textbook on mathematical methods for quantum physics based on one of his courses. But as the Vietnam War dragged on, Fuller found himself drifting from physics toward issues of social justice. Following short stints at various research centers, in 1970 he became president of his alma mater, Oberlin College, a prestigious liberal-arts college in Ohio. He was all of thirty-three years old. The experience was intense: sixteen-hour days, seven days a week. Fuller initiated major reforms, including steep increases in the recruitment of minorities among students and faculty, but burned out quickly. He left Oberlin after four years.[44]

While exploring options for what to do next, Fuller made a trip to San Francisco in 1974 and underwent Erhard's *est* training, mostly, he recalls now, to see what all the fuss was about. When Erhard got wind that a former college president was taking the *est* course, he sought Fuller out. Erhard made the pitch: how would Fuller like to take over the *est* foundation and give away a quarter of a million dollars to worthy causes each year? Fuller agreed on one condition: that Erhard abolish the foundation's board, so that only Fuller and Erhard would make decisions regarding the foundation's charitable activities. Erhard agreed. After writing a brief, positive account of the *est* phenomenon, Fuller accepted Erhard's invitation and become director of the *est* foundation in 1975.[45]

Soon after Fuller came on board, Erhard's physics consultants, Jack Sarfatti and Saul-Paul Sirag, met the Esalen Institute's Michael Murphy; that connection, in turn, led to the "Physics month" workshop at Esalen in January 1976. Yet Erhard yearned to make an even deeper impact on physics. Fuller told Erhard about Ernest Solvay, the Belgian industrialist and philanthropist who had sponsored a series of elite conferences in the 1910s and 1920s. Though hardly a household name, the Solvay conferences had long been renowned among physicists. Giants of the field like Albert Einstein, Niels Bohr, Marie Curie, Ernest Rutherford, Erwin Schrödinger, Werner Heisenberg, and their colleagues had gathered regularly, on Solvay's tab, to puzzle through the mysteries of quantum physics. Solvay played host to some of the most weighty debates that marked the birth pangs of quantum mechanics. Like Erhard, Solvay had lacked a university education, and yet his steadfast financial support

had goaded the world's greatest physicists into some of their most important breakthroughs. Erhard listened to Fuller's tales of the legendary
conferences and dreamed of becoming the next Ernest Solvay.[46]

Erhard told Fuller that he wanted to host a new kind of conference,
not just run-of-the-mill meetings: the kind of conference that would
"actually make some difference to the physics community." He asked
Fuller to organize the new conference series; Fuller agreed to organize
the organizers. Together they cast around for a respected physicist who
could encourage the discipline's heavyweights to attend. After confer-
ring with Berkeley's Geoffrey Chew and Caltech's Richard Feynman, they
approached Sidney Coleman.[47] Coleman, then a professor at Harvard,
had done his doctoral work at Caltech in the early 1960s. As one of Coleman's colleagues put it recently, Sidney was "always interested in expanding his horizons," and despite nearly fifteen years on the Harvard faculty
by that time, he still had something of the "California spirit" in him.[48]
A world expert in quantum physics and a celebrated lecturer with a sly
and sarcastic sense of humor, Coleman was intrigued by the offer. He
teamed up with his colleague from MIT, Roman Jackiw, to plan the first
meeting. By July 1976 they had hammered out a topic and guest list.[49]

Coleman's first order of business was to establish some ground rules.
He and his colleagues (including Jackiw) were to have total control over
content, invitation lists, and so on, with no meddling from Erhard or the
est foundation. Erhard readily agreed, asking only that he be allowed to
sit in on the meetings. With that half of the negotiation settled, Coleman
turned to his colleagues, trying to wrangle them to attend the unusual
conference. The invitation letter was vintage Coleman. "The following
information may be of interest to you," he spelled out:

> The *est* Foundation (though a legally independent entity) derives
> its income from Erhard Seminars Training, a San Francisco based
> organization that offers expensive weekend self-improvement
> courses. For what it is worth, my uninformed opinion is that the
> fact that it is possible to make good money this way is yet another
> piece of evidence that we are living in the Golden Age of Silliness.

However, this is irrelevant, because the proposed conference will
be no more devoted to promoting Erhard Seminars Training than
the activities of the Ford Foundation are to pushing Pintos. I have
received explicit agreements to this effect from the responsible
parties, and I promise you that at the slightest sign these agreements are not being kept, I will throw a tantrum and cancel the
conference.[50]

The letter went to a Who's Who of the world's top theoretical physicists.
Of the seventeen physicists whom Coleman and Jackiw invited, six had
already won or would go on to win the Nobel Prize.[51]

Nearly all of the invitees signed up. With Fuller's help and Erhard's
generous backing, Coleman and Jackiw convened their first meeting, on
"Novel configurations in quantum field theory," in late January 1977. By
all accounts it was a lovely affair. The *est* foundation paid all expenses for
the physicists and their spouses to travel to San Francisco. The foundation sent each confirmed participant and guest open round-trip airline
tickets, so they could come and go as they pleased. Lodging was at fine
hotels, with the bill (including all room-service charges) sent directly to
the *est* foundation. Drivers greeted participants at the airport and chauf-
feured them around the city.[52]

An *est* foundation staff person assured physicists in advance of the
meeting that the foundation's intention was to provide a comfortable
place where they could work without distractions.[53] That private space
turned out to be Erhard's San Francisco mansion, Franklin House, which
served as Erhard's personal residence and *est* headquarters. "We took all
the stuff out of my office," Erhard recently explained, "and set it up for the
conference with a blackboard." As it happens, the huge blackboard they
wheeled in was curved in the shape of a parabola, which had an added
bonus: not only could all the participants see each other's scribbles at
the same time, but the curved shape helped project speakers' voices as
well.[54] Erhard and Coleman each aimed to keep things informal, with
long blocks of time reserved for casual conversation rather than clut-
tered with presentations. As Coleman explained to his colleagues, "both

conscious policy and instinctive sloth lead me to keep things as unstructured as possible." Most important, Erhard kept his word: he sat in the
back and kept quiet the whole time. "I couldn't follow ninety percent of
what got said," Erhard recalls, "but I really loved the way they thought
and worked," including "the way they worked with each other. That was
a kind of payback for me." Coleman never had to throw a tantrum.[55]

As Erhard and Coleman had hoped, conversations spilled beyond
the stipulated conference hours. Fuller and Feynman, for example,
discussed quantum physics while out for a morning jog.[56] Conversations continued well into the night as well, though not always about
physics. Feynman, for one, couldn't believe "how well we were taken
care of at the meeting," as he wrote to an *est* foundation staffer. But
he felt sheepish about having Erhard's foundation pick up the tab for
his "binge" at the hotel bar one evening, so he sent in a check to cover
those expenses.[57] Coleman registered his "mild disappointment" that
no one had announced in the mornings that they had stayed up late in
their hotel rooms the night before, calculating some new effect, having been inspired by something they had heard at the previous day's
meeting. "Nobody could say that at our conference," Coleman explained,
"because by the time we got back to our hotel rooms we were all drunk
and stuffed." Even so, Coleman congratulated Fuller on their mutual
success. "Well, they said it was impossible, but we did it: we got both
Feynman and Werner to wear ties."[58]

Three weeks after Coleman and Jackiw's *est*-backed conference,
Erhard considered doling out some additional largesse to his original
physicist-consultant, Jack Sarfatti. The latest idea was for Erhard to
finance a new fellowship for Sarfatti to teach as a visiting lecturer at
the San Francisco Art Institute.[59] Sarfatti was thrilled. Writing to David
Finkelstein—veteran of the Esalen workshops who, unbeknownst to Sarfatti, had also just attended the Coleman-Jackiw meeting at Erhard's
Franklin House—Sarfatti enthused that "I shall be the new Henri Berg-
son of San Francisco. I shall hold an ongoing seminar in *Adventures of
Ideas* discussing Borges, Buddhist logic, QM [quantum-mechanical] logic,
Whitehead, James, Einstein, Bohr, Goethe, Physics as Conceptual Art,

etc." Sarfatti promised "a grand vision to set before the eyes and ears of the San Francisco artistic-literati. I shall sing it and deliver it in poetic cadence. New forms of inquiry, new modalities of thought and expression for the new physics!" In his excitement, he signed his letter, "Professor of Quantum Cabalistic Art."[60] Sarfatti was scheduled to deliver an inaugural lecture entitled "Plato's anticipation of quantum logic" at the Art Institute a few months later. He printed up copies ahead of time and mailed them out to his long list of recipients. John Wheeler thanked Sarfatti for his copy, and recommended further reading: one of Wheeler's favorite studies of the poet Samuel Coleridge.[61]

But it was not to be. Around the time that the visiting lectureship was to begin, Sarfatti began corresponding with MIT's Viki Weisskopf, the senior physicist who had recently coached Fritjof Capra along the road to *The Tao of Physics*. Sarfatti had invited Weisskopf to join an advisory board for the Physics/Consciousness Research Group. Weisskopf declined, as usual in his gentlemanly Austrian manner. "Naturally I am interested in what you are doing and find some of your things reasonable and useful," Weisskopf assured Sarfatti. But two major sticking points remained. "One is your connection to Werner Erhard," about whom Weisskopf held a rather low opinion. "The other is your constant connection to such silly things as ESP, coincidences of events, etc., with quantum mechanics. As you know, it is my strong opinion that they have nothing to do with each other."[62]

One week later, in response to Sarfatti's suggestion that Weisskopf might have misunderstood Erhard and his mission, Weisskopf returned to the matter of Erhard's patronage. Weisskopf, who had fled fascism in Europe as a young physicist, had done some reading about Erhard. By that time Erhard and *est* had begun to receive some negative publicity for purportedly authoritarian tactics.[63] Weisskopf had also spoken with graduates of the *est* training, although he had not received "any feedback, positive or negative, from the physicists" who had attended the recent Coleman-Jackiw conference, including his own department-mate Roman Jackiw.[64] Weisskopf tried to end his letter on a more upbeat note. "I hope you don't interpret this letter as a declaration of war between you and me," Weisskopf closed to Sarfatti. "On the contrary, as I say I am always interested in what people like you are doing and I like to discuss the issues they are interested in with them." But he made clear that he would not participate in any official capacity with the Physics/Consciousness Research Group.[65]

Similar advice came in from Martin Gardner, the *Scientific American* columnist and leading organizer, together with physicist John Wheeler and magician James Randi, of the Committee for the Scientific Investigation of Claims of the Paranormal (CSICOP). "Jack, my friend, take my advice and get out of the psi field," Gardner counseled. "It's sicker than you suspect. Nobody is in the least interested in trying to 'explain' psi by Q[uantum] mechanics, or electromagnetism, or the weak force, or quarks, or tachyons, or anything else. All the funders care about is practical results—i.e., miracles." Gardner hoped Sarfatti could make a clean break. "You're too honest, and know too much science to be wasting your talents trying to get funding for theoretical work on the nature of consciousness" from patrons in the human potential scene. "Do something honest," Gardner suggested, "like, maybe, rob a bank" or "make a porno movie."[66]

Spurred by these correspondents, and deeply insulted at not having been consulted about the Coleman-Jackiw conference, Sarfatti broke with Erhard—one of his principal sources of funds—with gusto. "Until recently I never took a close interest into what Werner and *est* were really about," Sarfatti declared in an open letter that summer. "After all the chap was giving me and my colleagues considerable money, so why be so impolite as to inquire too deeply? After all physicists are notorious prostitutes anyway." The recent *est* foundation physics conference—to which he had not been invited—stung Sarfatti as one insult too many. Sarfatti alleged that Erhard's cronies had reneged on a promise of more cash to come. He further likened Erhard's underwriting of the Coleman-Jackiw conference to "Stalin signing a non-aggression pact with Hitler," it came as such a stab in the back. Perhaps the latest letters from Weisskopf and Gardner provided Sarfatti the face-saving cover he needed to justify his sudden break with Erhard, about whom he had previously expressed only friendship and admiration. Sarfatti announced that "If such distinguished men as Gardner and Weisskopf take their time to keep me honest, the least I can do is to keep Werner Erhard and *est* honest."[67]

Sarfatti then turned his attention to those physicists who had recently begun to enjoy Erhard's patronage. Coleman, Jackiw, and Fuller were already planning a follow-up conference, building on the success of the Coleman-Jackiw meeting of January 1977. For the next annual conference they chose the topic of quantum gravity and again began soliciting participants.[68] Sarfatti got wind of the plans and sent out a blast. Scrawling across the top of his recent letter from Weisskopf—the one that had mentioned no "declaration of war between you and me"—Sarfatti penned a warning and distributed copies far and wide. "Dear colleague," he began: "Werner Erhard/*est* has or may invite you to a 'Quantum Gravity' conference in January, 1978. You are being used as part of a larger plan" to buy "prestige," he alleged. "It is immoral for you to participate once you are aware of the realities behind the appearance." He closed with a pledge. "We will actively picket the gravity conference with media coverage. We urge you to *boycott est*."[69]

Those rumblings quickly made their way back to Erhard and his lawyers. One attorney wrote to Sarfatti with a kind of cease-and-desist order phrased in *est*-speak. The attorney urged Sarfatti to redirect his energies to more positive pursuits. Sarfatti's actions, the lawyer continued, reflected a malice toward Erhard all out of keeping with the generosity that Erhard and his foundation had shared with Sarfatti to date.[70] The lawyer's letter only fanned the flames. Sarfatti shot back with a rambling seven-page letter and then sent out a follow-up memo to physicists who might have been invited to the 1978 Erhard-funded conference on quantum gravity, again warning them against attending.[71]

Erhard, Fuller, and their advisors took Sarfatti's threat seriously enough that they arranged for extra security guards at the January 1978 meeting, expressly to keep Sarfatti out of Franklin House.[72] They needn't have worried. Sarfatti decided to protest by nonviolent means, in ways the San Francisco "artistic-literati" might well have appreciated. He groused about the turn of events with an artist friend and fellow frequenter of the Caffé Trieste in North Beach. Riveted by Sarfatti's moral outrage and bemused by the wider *est* phenomenon—he had long been fascinated by what he called the "daily parade of well-heeled suckers" who flocked to Erhard's seminars—the artist put pen to paper to depict Sarfatti's plight. His first cartoon featured a gaunt, bearded Sarfatti chained to a crumbling "Temple of *est,*" while larger-than-life posters of Erhard's smiling face fluttered in the background. The second image cast Sarfatti as David to Erhard's Goliath, showing Sarfatti struggling mightily against the power of Erhard's microphone. Sarfatti mailed out copies of the pictures along with some of his anti-Erhard letters. The artist, meanwhile, went on to have a successful career, including work on the animated television series *The Simpsons.*[73] (Fig. 8.2.)

Still fuming at what he considered Erhard's betrayal, Sarfatti picked up his own pen as well. He retreated to his spacious apartment on Telegraph Hill—the apartment loaned to him by Esalen's director, Michael Murphy—cranked up Wagner's "Ride of the Valkyries" on the stereo,

Dr. Jack Sarfatti at the Temple of est

FIGURE 8.2. Jack Sarfatti's friend Norman Quebedeau drew whimsical cartoons to depict Sarfatti's dramatic break with Werner Erhard and *est* in the summer of 1977. (Courtesy Norman Quebedeau.)

indulged in some psychedelic mushrooms, and set to work. Within an hour he had banged out a first draft of a radio play he called *Hitler's Last Weapon*. From time to time he emerged from his room "shouting with almost mad glee," as his former roommate recalls, to read a portion of his script while his friends (also tripping on mushroom tea) egged him on.[74] In the play, a scheming physicist helps Hitler accomplish his most audacious feat. While hunkered in the famous bunker in the spring of 1945, they create a perfect quantum clone of Hitler's consciousness, ensuring that the Führer can live on even after the Nazis' defeat. The scientist entangles Hitler's brainwaves with those of a "mule," an unsuspecting Jewish kid from Philadelphia. The child grows up, leaving dead-end jobs as a used-car salesman and encyclopedia peddler, until, some time in the 1970s, the entangled wavefunction collapses, transferring Hitler's most evil characteristics to the new carrier. After infiltrating the California human potential movement, the thinly disguised Erhard character works to seize power as the first psychic dictator of the United States, the "Chancellor of Megalomania." As the play unfolds, hope rests on the shoulders of "Rabbi Sarfatti," lead agent of an undercover unit of the "Higher Intelligence Agency" headquartered in San Francisco's North Beach. (Agent Sarfatti's cover: "the village idiot of Grant Avenue.") Forces of good thus battle evil on the plane of quantum consciousness. Years later producers of the Berkeley-based *Hearts of Space* radio show recorded the play, narrated by the accomplished radio dramatist Erik Bauersfeld. Sarfatti's quantum-psychedelic thriller aired on hundreds of affiliated stations across the country.[75]

Not long after Sarfatti made his flamboyant break with Erhard, Robert Fuller bowed out as well, albeit much more quietly. Fuller had accepted Erhard's invitation to lead the charitable *est* foundation because he thought it could provide a platform for tackling world hunger, a problem whose severity had struck Fuller during an earlier trip to India. Soon after President Jimmy Carter entered office, Fuller worked on his personal contacts, some stretching back to his college president days, to try to encourage Carter to work on the hunger problem. Fuller's contacts helped him get the message high into the new administration, but he was unable to reach

Carter directly until the singer John Denver stepped in. Denver, then at the peak of his fame and a well-placed friend of *est*, paved the way for Fuller to approach Carter directly about a new "Hunger Project." Fuller briefed President Carter about the plan in the White House's Oval Office in June 1977. He also gave Carter's son a copy of a short film, *The Hungry Planet*, by Keith Blume. Carter watched the film that night and soon announced the formation of a Presidential Commission on World Hunger.[76]

As Fuller remembers it, he and Erhard had agreed ahead of time that Erhard would remain a silent partner in the Hunger Project, devoting significant *est* foundation resources to the project behind the scenes but refraining from making any public statements about the effort until Carter had publicly endorsed it. Fuller had feared that the negative publicity beginning to swirl around Erhard and *est* would scuttle any efforts at real progress. All seemed to be going well after Fuller's meeting with Carter until Erhard began talking publicly about the project. As Fuller had feared, a media backlash quickly followed. The San Francisco–based liberal political magazine *Mother Jones* ran a long investigative piece accusing Erhard of manipulating the Hunger Project for "self-aggrandizement." The magazine alleged that the charitable effort was little more than a crass publicity stunt aimed to benefit Erhard's for-profit *est* corporation. (Today the major international Hunger Project charity includes a Nobel-laureate economist and the Queen of Jordan among its honorary members.) Disappointed at what he considered Erhard's broken promise, Fuller resigned from the *est* foundation.[77]

Despite Fuller's departure and Sarfatti's antics, the *est*-backed physics conferences developed into a robust annual tradition. The second meeting, held in January 1978 on quantum gravity, attracted such luminaries as Stephen Hawking.[78] (Fig. 8.3.) Three years later Hawking returned for another *est* foundation conference, at which the earliest battles flared between himself and fellow physicist Leonard Susskind over whether quantum mechanics implied that information could leak out of black holes. Their intense but good-humored debate raged for another two decades before Hawking conceded defeat, acknowledging that information might indeed escape a black hole's cosmic tug.[79] The 1979 meeting

FIGURE 8.3. Participants at the second annual *est* foundation conference on theoretical physics, January 1978. Stephen Hawking sits in the middle of the front row; John Wheeler stands just behind Hawking's left shoulder. (Courtesy Roman Jackiw.)

on phase transitions, focusing on how quantum theory can describe the shifting boundary between order and disorder at the atomic scale, such as ice melting to water, likewise spurred important results. Participants at the meeting thanked Erhard and the *est* foundation for facilitating inspiring discussions in an article published in the prestigious journal *Physical Review Letters*.[80] And so it went: every year for a decade, the *est* foundation's physics conferences attracted star after star of the physics firmament.

Despite the successful series, not everyone was happy with the field's latest patron. One or two of Coleman's and Jackiw's colleagues had turned down their original invitation back in 1976, none too eager to become associated with Erhard or his California concoction.[81] The organizers took the early objections in stride. Coleman even joked about the situation. When planning the fourth annual conference, for example, Coleman addressed his planning memos to Jackiw as "Eminence Grise" and Erhard as "Controversial Public Figure."[82] More serious trouble hit a few years later, during the January 1981 meeting. A reporter for the *San Francisco Chronicle* heard rumors about the upcoming conference, but was frustrated in his efforts to receive a straight answer as to why the world's top physicists were gathering at Werner Erhard's Franklin House residence. An *est* spokesperson didn't help matters when he told the journalist that the meeting "is not secretive, but we don't want to say anything about it." That kind of evasion was just what a reporter in the post–Watergate era needed to hear; the *Chronicle* heralded the story about the "mysterious conference" on its front page.[83] This brand of press attention was more difficult for Coleman to laugh off. He made copies of the *Chronicle* article and circulated a memo to Erhard, Jackiw, and the chairs of previous meetings:

> As nearly as I can determine, this is what happened: There is a physicist manqué and member of the Bay Area physics fringe who views the Franklin House conferences as diabolical; he sent me a letter a few years ago in which he numerologically identified Werner Erhard with both Adolph Hitler and the Beast of Revelations. (This misspelling of Adolf was necessary to make the numerology work.) This person gave a description of the fifth [*est* foundation] conference from his viewpoint to Charles Petit of the *Chronicle*. When Mr. Petit attempted to determine the facts of the matter, the Est functionaries whom he contacted felt themselves constrained by Est's agreement not to publicize the conference. I believe that if they had not felt this way, there would have been no story; "Philanthropic Foundation Sponsors Scholarly Meetings" is no competition for Jean Harris's murder trial on the front page of the *Chronicle*.[84]

The "physicist manqué" whom Coleman mentioned was almost certainly Jack Sarfatti; soon after Coleman sent off his memo the *Chronicle* reported that Sarfatti—"another eccentric genius in North Beach"—was still busy printing up his "jeremiads" against Erhard and *est*.[85] Taking no chances, Coleman typed up a "fact sheet" to be distributed to future conference invitees, emphasizing that neither Erhard nor any *est* personnel participated in sessions or played any substantive role in the conferences' intellectual content. Even so, Coleman acknowledged after the *Chroni-*

cle incident that "there exists a subset of the physics community that is opposed to Est-sponsored conferences in principle."[86]

Coleman and colleagues enjoyed a reprieve the following year. "Nobody refused an invitation because *est* was a sinister force," Coleman noted, and "we were not beseiged by hordes of cranks or journalists."[87] But the calm was not to last. During preparations for the 1984 meeting, one of the invited participants raised a public objection. "My sole reason for declining your invitation," physicist Michael Turner explained, "is the sponsorship of the meeting by the *est* Foundation. . . . It is my belief," he continued, "that *est* sponsors these meetings primarily for the purpose of gaining prestige and legitimacy."[88] As Sarfatti had done years earlier, Turner circulated an open letter to all invited participants, urging them to "think carefully about attending a meeting sponsored by the *est* Foundation."[89] Unlike the year that Sarfatti raised a ruckus, the latest letter came from a rising star of the field who had just been promoted to codirector of the country's first center for particle astrophysics at the Fermi National Accelerator Laboratory near Chicago. The objection touched a nerve. Two weeks later, Jackiw reported to Erhard that "the pesky and persistent anti-*est* allergy" required some attention, since 20 percent of the people whom that year's organizer had invited declined to be involved.[90] Similar problems crept up when planning a later meeting. Coleman advised against holding an *est* workshop on string theory—just then capturing widespread attention as physicists' best hope for unifying gravity with quantum theory—because the leading figure in the field, Ed Witten, refused to participate in an *est*-funded conference.[91]

If the unusual source of patronage rubbed some physicists the wrong way, so too did the curious behavior of the *est* staff and volunteers who helped with local arrangements. Each year Erhard dispatched a number of *est* faithful to help run the meeting, doing everything from chauffeuring physicists around town to waiting on them at meals. By all accounts, the *est* volunteers were attentive, even doting: if there were a chance of rain they were prepared with extra umbrellas. The extreme hospitality was a welcome change for most of the academics, who were still suffering through lean times. But the *est* volunteers seemed to act strangely, in some physicists' reckoning, performing their tasks with great courtesy but subdued affect. Their behavior became so awkward that one of the organizers had to intercede directly with Erhard during the 1984 meeting.[92] (A clinical psychologist who underwent the *est* training in Manhattan wrote of the "zombielike faces of the *est* volunteers who staffed the training room," with their "catatonic stares and rigid posturings.")[93] "The Menace of the Zombie Sycophants is real," Coleman conceded in one of his annual memos. "Many participants feel they are being coddled by individuals whose motives are mysterious, and they find this disturbing." Coleman added a footnote: "On the other hand, some of us can't get enough of it."[94]

During the late 1970s and early 1980s, Erhard's generous funding helped to fill the vacuum still gaping from the collapse of the physicists' Cold War bubble. Securing such funding was no small matter: lack of funds scuttled the plans of Princeton physicist and Nobel laureate Eugene Wigner, who had hoped to organize a conference on foundations of quantum mechanics in the early 1980s.[95] Several participants in the *est*-backed conferences, including Feynman, Jackiw, and Coleman, developed warm personal relationships with Erhard, exchanging frequent letters, holiday cards, and anniversary greetings, stretching over a decade. They occasionally met outside the confines of the physics conferences, enjoying dinners at fine restaurants or taking in short sailing trips. When Erhard's marriage ended—rather publicly—in a disputatious divorce in 1983, some of the physicists received early word from Erhard and replied with personal letters of support. Jackiw wrote to Erhard a few years later to express his appreciation that their relationship extended beyond their official conference duties.[96]

All that came to an end in 1991 when Erhard sold off his *est* assets to a collection of employees.[97] "San Francisco is talking," announced *Newsweek* magazine: the "embattled *est* guru" had "seemingly disappeared under a cloud of bad publicity." The news echoed around the world. A headline in Sydney, Australia, blared, "Lawyers chase seminar guru," the story alleging that Erhard had "pulled up stakes and gone fishing leaving behind an avalanche of litigation." The *Times* of London proclaimed,

"New Age guru goes into hiding." Nonsense, Erhard's attorney clarified: "His whereabouts are definitely known," the lawyer explained to the *Newsweek* journalist.[98] All the same, Erhard had hit a difficult patch. A string of deeply critical media reports had appeared, some laced with lurid allegations (which were later retracted).[99] An assistant wrote to some of Erhard's regular correspondents that he had decided to change gears and close his office.[100] For a time communication flowed with all the clandestine trappings of a John le Carré novel. Correspondents could send letters to a handpicked Erhard confidante, who would read them over the telephone to Erhard and share any pertinent reactions with the original sender. Other times a letter or two would trickle in directly from Erhard, but with no return address.[101]

•

Fundamental Fysiks Group members like Jack Sarfatti only overlapped in their receipt of Erhard funds for a few months with leading lights like Harvard's Sidney Coleman and MIT's Roman Jackiw. The significance of their bond was not the duration of time during which they each benefited from Erhard's generosity, but the fact that physicists from across the discipline's broad spectrum found themselves relying on private patronage from some of the same unusual patrons. Sarfatti's eager embrace of Erhard's funds (at least for a while) proved not to be so different from the creative opportunism with which his more acclaimed colleagues dabbled. Likewise, the great enthusiasm with which Sarfatti and his partners in the Physics/Consciousness Research Group and the Fundamental Fysiks Group threw themselves into New Age implications of Bell's theorem differed in degree, not in kind, from equally earnest investigations by several accomplished physicists, none of whom could be dismissed as less than "real" members of the profession. The multiple entanglements between the Fundamental Fysiks Group and leading physicists of the day strain philosopher Karl Popper's great good hope that clear criteria might demarcate authentic science from pseudoscience. In the face of the Fundamental Fysiks Group's ever-colorful activities, Popper's dream of demarcation seems little more than wishful thinking.

Chapter 9

From FLASH to Quantum Encryption

Nick Herbert's erroneous paper was a spark that generated immense progress.

—**Asher Peres, 2003**

From Nick Herbert's earliest encounters with Bell's theorem and entanglement, something kept nagging at him. If the quantum world really were subject to such "spooky actions at a distance," he wondered, could we harness that fundamental feature and put it to work? In the closing paragraphs of his succinct rederivation of Bell's theorem, published in 1975, he mused about one possible application: "superluminal telegraphy," using entangled quantum particles to send messages from point A to point B faster than light could travel between them. On the face of it, Herbert acknowledged, such faster-than-light signaling appeared inconsistent with Einstein's relativity. "But," he concluded, "the technological advantages of such a rapid communication device seem to make investigations" of such possibilities "of more than philosophical interest."[1]

What would it mean to send signals faster than light? Beyond the apparent violation of Einstein's relativity—that would be bad enough— all manner of strange paradoxes would be unleashed. Seen from the right vantage point, superluminal signals would travel backward in time: a message would be *received* before it was *sent*. No wonder the idea makes the hairs on the backs of physicists' necks stand on end. As one acclaimed textbook author put it recently, physicists are particularly "squeamish about superluminal influences."[2] Such chicanery dredges up all kinds of causal loopholes. You could send a retroactive telegram instructing your grandmother not to marry your grandfather. Or, on a brighter note, you could warn your forebears to divest their stock-market holdings a day before the great crashes of 1929, 2001, or 2008—the ultimate in insider trading. The possibilities would be truly Orwellian: sending messages faster than light could allow us to rewrite history to suit our present-day whims, or, as one wit put it, to "change yesterday today for a better tomorrow." Perhaps, some argued, such signaling was already occurring. After all, what were mental telepathy and precognitive clairvoyance but messages received outside the usual channels?[3]

While his paper on Bell's theorem was in press, Herbert and other members of the Fundamental Fysiks Group continued to brainstorm about the "intrinsically almost obscenely non-local" behavior of entangled particles.[4] In September 1975, Jack Sarfatti gave a presentation to the group on "Bell's theorem and the necessity of superluminal quantum information transfer." A month later, Herbert followed up with his own presentation on "Bell's theorem and superluminal signals."[5] That December, Berkeley physicist and Fundamental Fysiks Group member Henry Stapp also weighed in. As he put it, "the central mystery of quantum theory is 'how does information get around so quick?'" To Stapp, Bell's theorem and the landmark experiment by group member John Clauser led to the "conclusion that superluminal transfer of information is necessary."[6]

And so the agenda was set. The question of superluminal information transfer, and whether it could be controlled to send signals faster than light, would occupy Herbert, Sarfatti, and the others for the better part of a decade. Their efforts instigated major work on Bell's theorem and the foundations of quantum theory. Most important became known as the "no-cloning theorem," at the heart of today's quantum encryption technology. The no-cloning theorem supplies the *oomph* behind quantum encryption, the reason for the technology's supreme, in-principle security. The all-important no-cloning theorem was discovered at least three times, by physicists working independently of each other. But each discovery shared a common cause: one of Nick Herbert's remarkable schemes for a superluminal telegraph. Little could Herbert, Sarfatti, and the others know that their dogged pursuit of faster-than-light communication—and the subtle reasons for its failure—would help launch a billion-dollar industry.

•

Like Nick Herbert, Jack Sarfatti was quick to appreciate some of the practical payoffs that a faster-than-light communication device would bring. In early May 1978, Sarfatti prepared a patent disclosure document on a "Faster-than-light quantum communication system." The document was the first step in a formal patent application. In addition to filing his disclosure with the Commissioner of Patents and Trademarks in Washington, DC, he sent a copy to Ira Einhorn, scrawling across the top: "Ira—please circulate widely!" (This was a year before Einhorn would be arrested for murder; his "Unicorn preprint service" was still in full swing.) Sarfatti's proposal bore several signs of the Fundamental Fysiks Group's discussions. It began by citing Clauser's experimental tests of Bell's theorem, before citing a preprint of Henry Stapp's paper on super-luminal connections, which Sarfatti most likely received directly from Stapp at one of the group's weekly meetings.[7]

Sarfatti's device consisted of a source that emitted pairs of entangled photons, which were directed at two detectors, A and B. The detectors were located far enough apart that no light signal could travel between them before each had completed its measurement on the incoming photons. The experimenter at detector A could choose whether to let the photons pass through a double slit and produce the usual interference pattern on a screen or to insert a slit detector in the photons' path to measure through which slit each had passed. (So far his setup was straight out of John Wheeler's musings on the delayed-choice experiment, which few besides Sarfatti had shown any interest in to date.) Next came the twist: the experimenter at A could change the efficiency of his slit detector. When its efficiency was set to 100 percent, the slit detector would always determine through which slit each photon had passed, and hence there would never be any interference pattern on the screen at A. When the efficiency was set to zero, the interference pattern would always show up. By varying the efficiency of the slit detector on his side,

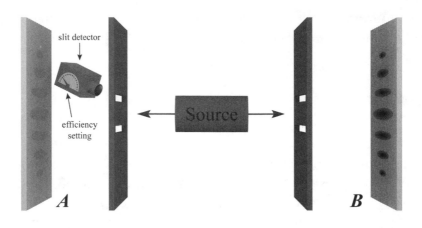

FIGURE 9.1. Jack Sarfatti's design of a faster-than-light communication system, based on his 1978 patent disclosure document. A source shoots out pairs of photons. The experimenter at A can tune the efficiency of a slit detector. When the slit detector operates at maximum efficiency, it always determines through which slit a given photon traveled, washing out the interference pattern on the screen at A. At minimum efficiency, the slit detector fails to determine the photon's path, and the characteristic double-slit interference pattern emerges at A. Because of quantum entanglement, Sarfatti argued, the varying sharpness of the interference pattern at A should be instantly observable in the correlated photons at B. (Illustration by Alex Wellerstein.)

Sarfatti suggested, the experimenter at A could encode a message to the other experimenter, stationed at the faraway detector B. The receiver at B, Sarfatti argued, would see the interference pattern on her end alternate from sharp to washed out over time, as the transmitter at A played with the efficiency of his slit detector—all thanks to the nonlocal correlations of the entangled photons.[8] (Fig. 9.1.)

Potential applications abounded. For one thing, Sarfatti reasoned, such a device could transmit a human voice across vast distances, with no possible eavesdropping. If the slit-detector efficiency at A were controlled by some transducer, such as a microphone, then the pattern of vibrations from the speaker's voice would become encoded in the varying sharpness of the double-slit interference pattern. A loudspeaker on the other end could then retranslate the pattern of interference fringes received at B into sound waves. "The application to deep space communications is obvious," Sarfatti concluded: messages could be relayed instantly across vast, cosmic distances. Benefits would accrue closer to home as well, such as "giving instant communication between an intelligence agent and his headquarters"—that is, espionage. Clearly his prior experiences with Harold Puthoff, Russell Targ, and their remote-viewing experiments at the Stanford Research Institute had left their mark. "In this case," Sarfatti clarified, "we would not use the above system but would use the same principle using e.g. correlated psycho-active molecules, such as LSD, affecting the neurotransmitter chemistry." Presumably the image of CIA agents doped up on LSD, communicating instantly with operatives half a world away via correlated brain impulses, seemed no more far-fetched than the parapsychological effects in which Sarfatti had been immersed for years.[9]

While Sarfatti dreamed of harnessing nonlocality, his proposal began to unravel, thanks entirely to local interactions. Another regular participant in the Fundamental Fysiks Group, physicist Philippe Eberhard, submitted a lengthy article to the Italian journal *Il Nuovo Cimento* the very week that Sarfatti filed his patent disclosure document. Eberhard, who had completed his doctoral training in France, had been working at the Lawrence Berkeley Laboratory as a theoretical physicist since the late 1950s. With Stapp, he began sitting in on Fundamental Fysiks Group discussions from the beginning, delivering several presentations to the group and also participating in the Esalen workshops.[10] From the start, he impressed Nick Herbert as "an energetic, black-bearded, flirtatious Frenchman," who fit right in with the group.[11] Eberhard's article on "Bell's theorem and the different concepts of locality," which he mailed to the journal during the first week of May 1978, clearly owed much to the group's activities. The opening citations were all to works by group members or by John Bell; he closed by thanking group members Henry Stapp, John Clauser, and Nick Herbert for helpful discussions.[12]

Like Stapp, Eberhard emphasized that Bell's theorem and Clauser's experiments decisively demonstrated that the outcomes of measurements at detector B depended on the detector settings at A—just as Sarfatti hoped to exploit with his new device. But, Eberhard went on, the fact that the outcomes depended on the faraway settings did not mean that one could control that dependence to send an intelligible message. The catch was that the telltale correlations predicted by Bell would only be revealed when the two experimenters compared *both* their detector settings for each given run *and* the measured outcomes at each setting. Until they came together to compare notes—a decidedly slower-than-light process—the receiver at B would measure a random-looking series of events.[13]

Consider the original scenario that Bell had analyzed: measuring the spins of entangled electrons, as experimenters varied the orientation of the detectors on each end of the apparatus. Roughly half the time, the electron careening toward detector B would be registered as spin up, and half the time it would be measured as spin down. The individual outcomes at detector B—spin up for electron number 1175, say, and spin down for electron number 1176—might indeed have depended on what settings the experimeter at A had used. But without separate knowledge of what the detector settings at A had been in each case, the receiver at B would have no way of discerning any pattern or message. The measurements from her own detector—strings of spin-up results interspersed with spin-down results—would simply appear to be random: all noise, no signal.[14]

Eberhard's thorough analysis was the first published attempt to reconcile Bell-styled entanglement with the dictates of relativity. It introduced what has become the standard response to the question of whether quantum nonlocality can lead to faster-than-light communication. Eberhard's paper has made a respectable impact in the scientific literature, receiving nearly 120 citations to date.[15] It has also been picked up in more accessible settings. Nick Herbert's friend and former roommate Heinz Pagels, for example, prominently featured Eberhard's argument in his book *The Cosmic Code* (1982). (Pagels, the executive director of the New York Academy of Sciences, later helped to organize the 1986 meeting on the foundations of quantum mechanics in New York City's World Trade Center.) In the light of Eberhard's argument, Pagels lambasted anyone who claimed they could use Bell's theorem to communicate faster than light. Such people had "substituted a wish-fulfilling fantasy for understanding."[16]

Though Eberhard's article has usually been invoked as the death knell for superluminal telegraphy, Eberhard in fact ended on an upbeat note. As he made clear, his no-go theorem rested on several assumptions, any one of which might break down. Perhaps quantum mechanics was not the final word on the behavior of the microworld; perhaps the principles of Einstein's relativity required fine-tuning; perhaps physicists' intuitive model of causality was the weak point. Eberhard mentioned that members of the Fundamental Fysiks Group were at present scrutinizing each of those options. "Consequently, any attempt to discourage the work that is being performed" would be "either futile or counterproductive."[17]

On that final point, Eberhard and Sarfatti certainly agreed. Sarfatti worked tirelessly to find some loophole by which his superluminal signaling scheme might survive Eberhard's test. Henry Stapp got into the mix, too, having quickly appreciated the significance of Eberhard's demonstration. Back and forth the arguments flew. Perhaps, Sarfatti suggested, the Eberhard-Stapp argument had neglected some means by which the various experimental arrangements in Sarfatti's scheme—slit detector in the photons' path or not—could be distinguished by an observer at the faraway detector.[18] Other times, Sarfatti leapt on Eberhard's parting observation that quantum mechanics itself might be surpassed by some more general theory, in which controllable superluminal signaling might survive. After all, Sarfatti reminded his interlocutors, "superluminal precognitions"—psi visions of the future—"exist as facts in abundance in my own laboratory of the mind. Am I to ignore facts simply because old men are afraid to experience them?"[19]

A Bay Area newspaper reporter captured some flavor of the intense discussions at the time. In his article about "maverick physicist Jack Sarfatti" and his quest for faster-than-light communication, the reporter noted that few physicists outside Sarfatti's immediate circle paid much attention to his latest schemes. Perhaps that was because Sarfatti was leading "an admitted 'bohemian' existence in San Francisco's North

Beach, hanging out in espresso coffee houses and working on his theories or talking with a small group of followers." Henry Stapp offered a different explanation: Sarfatti was "slow to admit his mistakes," as the reporter put it. Stapp and Sarfatti had "argued for a year before Sarfatti admitted" that his original scheme would not work.[20]

In the midst of those heated arguments, Sarfatti floated a new corporate vision. Having been cut off from Werner Erhard's *est* funds for nearly two years, Sarfatti's financial situation had become dire. Even bohemians needed to eat; it was time to drum up some new patrons. In place of the Physics/Consciousness Research Group Sarfatti created "i^2 Associates, a Meta-Corporation of the Emerging Post-Industrial Order." Its goal: "To initiate and catalyze faster-than-light quantum communications technology based on the development of quantum correlation physics of the Einstein-Podolsky-Rosen effect."[21]

Where once Sarfatti had looked to human-potential gurus like Erhard and Esalen's Michael Murphy for underwriting, this time he trained his eye on the defense establishment. The principal objective of i^2 Associates, Sarfatti proposed, would be the "transformation of the entire United States national defense communications network to untappable, unjammable, zero-time delay, quantum correlation systems." By relying on Bell-styled quantum entanglement rather than ordinary signals, such as radio waves or light, the superluminal communication system would be impervious to the weaknesses that hampered present-day communications, be they poor weather or security breaches. In a footnote, Sarfatti acknowledged the problem of whether the superluminal correlations could be controlled or modulated, so as to produce intelligible messages. Not to worry: "Dr. Sarfatti believes he may have found a fruitful approach to the modulation problem," and i^2 Associates would make that its first priority.[22] After all, even Eberhard had called for further research.

Sarfatti promised a slew of benefits once the hurdle of controlling the superluminal correlations could be cleared. Alongside improved "control of nuclear missiles in flight" and "more reliable communications for Polaris-type submarines under the sea," his efforts would also lead to the "development of more accurate remote viewing," beyond the results that Harold Puthoff and Russell Targ had logged in their psi lab at the Stanford Research Institute. There was even the prospect of developing "a proper quantum psychopharmacology" that could deliver "highly specific drugs to induce sharply selective experiential effects," thereby "amplifying desired behavioral capabilities."[23]

Desperate for sponsors, Sarfatti circulated his i^2 Associates proposal far and wide.[24] Among the recipients was the Central Intelligence Agency. A CIA analyst prepared a formal review of Sarfatti's proposal. Suddenly, thanks to Sarfatti's prodding, the CIA had to go on record about how best to interpret quantum mechanics. The analyst cited several recent expositions of Bell's theorem—from sources such as *Scientific American*—to concede that "non-local reality implies an interconnectedness which appears to allow instantaneous signalling, i.e., superluminal signals, giving credence to Dr. Sarfatti's claims." However, the analyst noted that no consensus had emerged among experts about just how to interpret quantum entanglement and its relation to relativity. "It would seem there are interpretations of advanced physical theory, i.e., quantum theory and relativity, which are consistent with the possibility of superluminal communications." But Sarfatti's leap to controllable signaling piled "speculation on top of speculation." Given the theoretical uncertainty, the CIA analyst cautioned, "the only convincing experiment would be the demonstration of macroscopic subliminal signals." (Sarfatti relished the analyst's typographical error—or Freudian slip?—that had turned "superluminal" into "subliminal.") In the end, the analyst adopted a wait-and-see attitude. "These concepts are genuine basic research and should be evaluated and funded by those in the federal government charged with that responsibility." Funding "should not come from Intelligence Community research programs."[25] Close, but no cigar.

•

While Sarfatti tried to interest new patrons, Nick Herbert entered the superluminal fray with a design of his own. Herbert aimed to exploit differences between various polarizations of light. Classically, polarization refers to the direction in which a light wave's electric field varies.

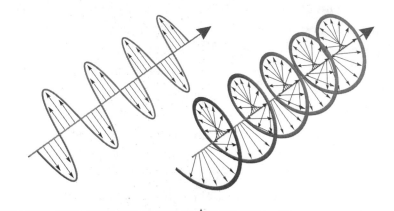

FIGURE 9.2. The electric field associated with a light wave may vary in many different patterns. *Left*: In linearly polarized light, the electric field (small black arrows) oscillates up and down, perpendicular to the direction of the light wave's travel (large gray arrow). *Right*: In circularly polarized light, the electric field traces out a helical shape. (Illustrations by Alex Wellerstein.)

The field may vary along a straight line ("linear polarization"), oscillating up and down and up and down as the wave speeds on its way. Or the electric field may spin around in a circle ("circular polarization"), tracing out a helical shape as the light wave travels.[26] (Fig. 9.2.) For each type of polarization, light waves can come in one of two varieties. Linearly polarized light, for example, can be polarized either horizontally—its electric field waving back and forth in the horizontal direction—or vertically. The corkscrew pattern of circularly polarized light, meanwhile, can be either right-handed or left-handed. (Hold out both of your hands, stick out your thumbs, and curl your other fingers toward your palms. Your thumbs point in the direction that the light wave travels. The other fingers on your right hand curl in the direction that the electric field varies in right-circularly polarized light; likewise for the fingers on your left hand and left-circularly polarized light.)

A polarizing filter, such as the plastic often used in sunglasses, acts like a picket fence. Linearly polarized light, whose polarization axis lines up with the orientation of the filter, passes through unhindered; light waves whose polarization axis differs from the filter's orientation by 90 degrees are blocked. That's why polarized sunglasses allow you to avoid squinting at ocean waves while relaxing at the beach: most of the light reflected off the water is horizontally polarized, so the vertically polarized filters cut down on most of the glare. Circularly polarized light, meanwhile, can be broken down into equal parts horizontal and vertical (linear) polarization. When circularly polarized light shines on a polarizing filter, half of the light makes it through.

At the quantum level, polarization behaves much like electrons' spin. Just as electrons can only exist in one of two spin states—either spin up or spin down along a given direction—photons, the tiny quanta that make up light waves, come in various polarization states. Physicists often abbreviate the states by their initials: H for horizontal linear polarization; V for vertical linear polarization; R for right-handed circular polarization; and L for left-handed circular polarization. Pairs of photons that emerge from a common source in a state of zero total angular momentum, as in Clauser's experiments on Bell's theorem, show perfect correlations. If one member of the pair is measured to be linearly polarized in the H state, then its companion must be linearly polarized in the V state (when measured along the same direction); if one photon is measured to be circularly polarized in the R state, then its twin must be circularly polarized in the L state.

All that was old hat by the time Nick Herbert began brainstorming about superluminal signaling. The basic ideas about polarized light had been around since the early nineteenth century—right from the dawn of the wave theory of light—and the application to individual photons had been postulated during the 1920s.[27] What Herbert wanted was some way to exploit observable differences between individual H, V, R, and L photons; some property that could be used to encode messages between distant experimenters.

As luck would have it, Herbert stumbled across a handy volume of reprints not long after finishing his PhD at Stanford. His dissertation had been on nuclear physics, and by the time he wrapped up that work he had had his fill of the subject. He cast about for something new to think about and quantum optics caught his eye.[28] Just a few years earlier, the American Association of Physics Teachers had bundled together some of the most important articles on the subject and republished them as *Quantum and Statistical Aspects of Light*. Included was an article dating all the way back to 1936 by a physicist working at Princeton who had managed to measure the angular momentum of circularly polarized light. That physicist, Richard Beth, had received some pretty impressive help: Beth thanked both Albert Einstein and Boris Podolsky for extensive discussions, just months after they had published their famous EPR article.[29]

In Beth's clever arrangement, he used a special device called a "half-wave plate" suspended from a thin quartz fiber. When right-circularly polarized light shone on the half-wave plate, it set the plate spinning in one direction; left-circularly polarized light spun the half-wave plate in the opposite direction. Moreover, the half-wave plate flipped the light's polarization: incoming light that had been right-circularly polarized emerged as left-circularly polarized, and vice versa. (A diagram of Beth's novel half-wave plate graced the cover of the 1963 reprint edition in which Nick Herbert first discovered Beth's article.) As Einstein helped confirm, the amount of rotation that Beth measured in his device was consistent with the notion that individual photons in each state of circular polarization carried equal and opposite units of angular momentum.[30]

Beth's early device had measured the angular momentum for light waves, that is, huge collections of photons all acting together. Herbert imagined a similar device, appropriately sensitive, that could measure the angular momentum of individual photons. He reasoned that R and L photons would each impart a fixed amount of rotation to the half-wave plate (in opposite directions), whereas H and V photons—because they were not in states of definite angular momentum—would pass through the hypothetical device unaffected: their polarization would remain unchanged, and the half-wave plate would not rotate.

Here was the distinction between photon states that Herbert had been looking for. In the spring of 1979, he wrote a preprint detailing his design and sent it on its way. Herbert, who has a knack for puns, limericks, and the like, called his paper "QUICK," an acronym so clever even he can't remember what it stood for anymore. He sent a copy to Ira Einhorn, who put Herbert's paper into circulation just before being arrested for the murder of Holly Maddux; Herbert's paper was likely in one of the last Unicorn-network mailings that Bell Telephone would ever send out. Herbert also mailed out copies on his own, to a mailing list he had been cultivating for a few years as part of his "C-Life Institute." Like Sarfatti's i^2 Associates, Herbert's C-Life Institute never amounted to more than a fancy name for his post office box. He chose the "C" to signal his interest in the physics of consciousness.[31]

In Herbert's QUICK scheme, an experimenter at detector A measures the polarization of the photon headed his way. At detector B there is a half-wave plate inserted in the photon's path. By choosing whether to measure linear or circular polarization on his end, the experimeter at A could control whether the plate at B would rotate. Whenever the experimenter at A chose to measure linear polarization, he would find H half the time and V half the time. He wouldn't be able to control which state of linear polarization resulted on any given run—the data would be a random series of H's and V's, averaging out to fifty-fifty each over the long haul—but he could guarantee that when measuring linear polarization, he would always get either H or V. Upon measuring linear polarization at A, the twin photon heading toward B would instantly enter into the complementary state of linear polarization. If experimenter A's measurement result were H, then the photon heading toward B would be V, and vice versa. Whether photon B entered the H state or the V state on a given run didn't matter—neither, according to Herbert, would set the half-wave plate in motion. On the other hand, if experimenter A chose to measure circular polarization, then the twin photon heading toward B would instantly enter into a state of circular polarization. Whether it entered state R or L, the photon heading toward B would make the half-wave plate rotate. Thus the experimenter at A (the transmitter) could dispatch a message to B instantaneously, simply by choosing whether to measure linear or circular polarization.[32]

Herbert's paper made waves among the remnants of the Fundamental Fysiks Group. Although the group's weekly meetings had wound down a few months earlier, former members, including Herbert, Jack Sarfatti, Saul-Paul Sirag, Henry Stapp, and Philippe Eberhard, hashed out Herbert's design in June 1979. Stapp immediately challenged the idea, building on Eberhard's argument that statistical averages should wash out any superluminal effects. Sarfatti countered that Herbert's QUICK scheme evaded that problem. Herbert's imagined device seemed to promise an immediately distinguishable effect for each individual photon: the half-wave plate at B either rotated or it didn't.[33]

Thanks to Einhorn, others began to grapple with Herbert's QUICK paper, far from the California crew. The paper made its way to GianCarlo Ghirardi, an Italian physicist working at the International Centre for Theoretical Physics—the selfsame center at which Sarfatti had spent much of his sabbatical in 1973–74. Ghirardi had been interested in the interpretation of quantum theory for some time, dating all the way back to his graduate work in the late 1950s. But like so many of his generation (John Bell included), Ghirardi had learned to keep those interests on the sideline. He worked on "more fashionable" topics, "fields in which you might hope to get a permanent position," as he put it recently. The strategy worked. By the mid-1970s he had secured a double position in Trieste, both at the International Centre for Theoretical Physics and as physics department chair at the neighboring university. At last he felt safe turning his attention squarely to foundational matters.[34] It didn't take long before his plate was full. Around the time he received Herbert's paper, Ghirardi attended a meeting at nearby Udine, in northeastern Italy, and heard a presentation by his Italian colleague Franco Selleri.[35] Selleri and his group in Bari, Italy, had been among the earliest and most active researchers on Bell's theorem anywhere in the world. (Selleri, a contemporary of Ghirardi's, had likewise turned to the topic relatively late in his career. He later explained that the university at Bari was new when he was hired and had no other theoretical physicists on staff, so Selleri had more leeway to pursue idiosyncratic or unpopular interests.)[36] Selleri and company had hatched a scheme for superluminal signaling remarkably similar to Herbert's. With a Trieste collaborator, Ghirardi dug into Herbert's and Selleri's proposals and isolated a fatal flaw.[37]

The main weakness, Ghirardi realized, was that Herbert had worked in a kind of semiclassical approximation. He had tacitly assumed that quantum mechanics applied only to the pair of photons, and not to the apparatus itself. Any half-wave plate that functioned as Herbert's scheme required—flipping individual R photons into L photons and L photons into R photons, while letting H and V photons pass through unaffected— would run afoul of a fundamental quantum limit, akin to Heisenberg's uncertainty principle. The original half-wave plate, devised back in 1936, had worked because the experimenter sent zillions of photons at the half-wave plate at a time. To get the same results at the single-photon level, Herbert's half-wave plate would need to be infinitely massive. But then it would be too heavy to rotate whenever an R or L photon zoomed past.[38]

Ghirardi and his colleague wrote up their analysis in November 1979, and followed up with a more general demonstration a month later. Together with Eberhard's paper from 1978, the papers by Ghirardi have been hailed as the earliest rigorous proofs that Bell-styled nonlocality could peacefully coexist with Einstein's relativity.[39] Though they reached similar results, however, Ghirardi and his coauthors struck a rather different tone in their conclusions than Eberhard had done in his earlier article. Where Eberhard had argued that it would be "futile or counterproductive" to discourage further work on the topic, Ghirardi saw things differently: "To conclude, we have considered [it] worthwhile to illustrate explicitly the general proof of the impossibility of superluminal transmission, even though it is quite elementary, to stop useless debates on this subject."[40]

On this last point, Ghirardi fell somewhat short of his goal. In fact, his demonstration of the weakness of Herbert's device only spurred on the quest. From the objections by Stapp and Eberhard to Jack Sarfatti's original proposal, Nick Herbert realized the importance of exploiting individual quantum events, rather than statistical averages. From Ghirardi's intervention, Herbert came to appreciate the importance of amplifying the tiny distinctions between various quantum states, to evade fundamental limits on signaling. The rules of the game were set. Herbert got back to work.

Within a year he had devised an alternate design. Like his QUICK proposal, his new scheme relied on distinguishing H, V, R, and L photons. Again the experimenter at detector A could encode messages to the faraway experimenter at B by choosing to measure linear or circular polarization. But this time, the signaling did not rely on experimenter B manipulating finicky, single photons. Rather, the photon en route to B first passed through a laser gain tube—the amplifying mechanism at the heart of real-world lasers, which had been around for nearly twenty years by that time. "Laser," after all, is an acronym for "light amplification by stimulated emission of radiation." In Herbert's scheme, the incoming photon would do the stimulating; the laser gain tube would take care of the amplification. He cooked up a new acronym for his latest design: "FLASH," for "First Laser-Amplified Superluminal Hookup."[41]

Herbert latched on to the idea of using a laser because laser light is special. The light that comes out of a laser gain tube is not just amplified—lasers are not just bigger, brighter lightbulbs—but *coherent*. That is, in principle the light that exits the laser is perfectly in phase with the incoming radiation: its electric field oscillates in the same way, at the same rate, and in the same direction. That is just another way of saying that the polarization of the amplified, outbound signal is exactly the same as the polarization of the incoming light. Send in a weak signal of horizontally polarized light, and get out an intense beam of horizontally polarized light. A leading textbook at the time emphasized the phase coherence of laser output as the laser's defining quality.[42]

Armed with this new gadget, Herbert laid out his FLASH design. Experimenter A would make his measurement first; that would force the twin photon, headed toward experimenter B, into a state of either linear or circular polarization. Then photon B would enter the amplifying tube. Out would come a burst of identical copies: a beam of laser light consisting of photons all in the same, still-unmeasured state of polarization as photon B. (Herbert, who had logged all those hours working at the photocopy-machine company, called this the "perfect photon xeroxing provided by the laser effect.")[43] Next the experimenter at B could make some quick measurements on the light beam, no more intrusive than the kinds of manipulations that physicists had managed to perform since the 1930s. In particular, the experimenter at B could send the light through a beam splitter, such as a half-silvered mirror. That would bounce half of the incoming light toward one set of detectors (call them station 1) while allowing the rest to pass through toward a second set of detectors (station 2). Station 1 was equipped to measure linear polarization; station 2 measured circular polarization. In an instant experimenter B would know, from the output of these two sets of detectors, whether experimenter A had set his own device to measure linear or circular polarization. For example, if experimenter A chose to measure linear polarization and got the result H, then the laser gain tube on the other side would spit out a burst of photons all in the state V. Assume for simplicity that the tube released 100 photons, each with V polarization (although real lasers by that time could amplify incoming signals by a factor of several million). Thanks to the beam splitter, 50 of those photons would get diverted toward station 1 (set to measure linear polarization), while 50 sailed toward station 2 (set to measure circular polarization). At station 1, Herbert wrote, the results would be clear: 0 photons in state H, and 50 photons in state V. Station 2 would show 25 photons in state R and 25 photons in state L.[44] (Fig. 9.3.)

Excited about the latest design, Herbert wrote up a new paper in January 1981. He submitted a copy to the journal *Foundations of Physics*, the relatively new journal that welcomed speculative papers on philosophical topics; Herbert had published a generalization of Bell's theorem in the same journal a few years earlier.[45] He also prepared a preprint version to circulate on his own. As it happened, Herbert finished his FLASH paper the very month that Ira Einhorn hopped bail and fled the country, just days before his murder trial was set to begin. With Einhorn out of the picture, Herbert had to rely on his own informal network to spread the word. Back in 1979, he had distributed the QUICK paper under the aegis of his C-Life Institute. This time Herbert circulated the FLASH paper as a preprint of the "Notional Science Foundation," another play on his deep interest in the intersections of physics and consciousness.[46] (Both the "Institute" and the "Foundation," of course, shared the same address: Herbert's post office box in Boulder Creek, California.)

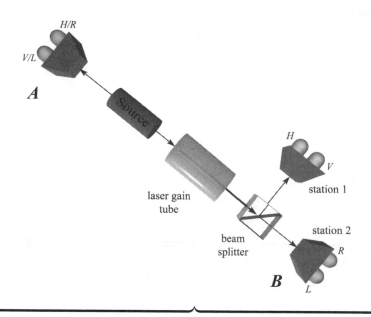

FIGURE 9.3. Nick Herbert's FLASH design. A source emitted entangled pairs of photons. The experimenter at A measured either linear or circular polarization, after which the photon heading toward experimenter B entered a laser gain tube. The burst of identical copies would then be split by the beam splitter, half going toward detectors to measure linear polarization and half to be measured for circular polarization. (Illustration by Alex Wellerstein, based on Herbert [1982], 1174.)

Had he done it? Could FLASH really communicate messages faster than light? Right after mailing out his preprints, Herbert convened another of his Esalen workshops on Bell's theorem and the nature of reality. Most of the familiar faces from the Fundamental Fysiks Group were there, including Henry Stapp, John Clauser, Saul-Paul Sirag, and Elizabeth Rauscher. Herbert presented his latest scheme. As his annual report to his Esalen sponsors indicated, his proposal "was described, discussed heatedly, but not refuted."[47] Soon after that Berkeley's Philippe Eberhard sent Herbert detailed comments. Eberhard thought he had isolated a flaw, but Herbert countered that challenge, too. "Does this mean," Herbert wrote back—you can almost see the schoolboy grin on his face— "that I can now count on your support for raising investment money for commercial exploitation of FTL [faster-than-light] communication?"[48] After several years of intense effort, it looked like the quest for what Jack Sarfatti repeatedly called "the Holy Grail" had come to fruition.[49]

•

The gears of academic peer review grind slowly, but grind they do. The same week that Herbert responded to Eberhard's critique, in March 1981, the editor of *Foundations of Physics* sent FLASH out to referees.[50] One reviewer later recalled that he knew the instant he received Herbert's FLASH paper that it must be incorrect, because of its incompatibility with relativity. On the other hand, he couldn't find any error. Whatever the flaw might be, the reviewer reasoned, it must be a nontrivial one, some juicy nugget whose careful elucidation would nudge forward the community's understanding. Still convinced that the paper was "obviously wrong," the reviewer wrote back to the journal's editor recommending publication: the provocative paper was sure to spur further development of the field.[51]

The journal editor had also asked GianCarlo Ghirardi to review Herbert's paper. Ghirardi was an obvious choice, especially after his careful debunking of Herbert's QUICK proposal just a few years earlier. Like the first reviewer, Ghirardi struggled with Herbert's latest paper—"I must confess that it took some time for me to spot the mistake," he explained recently. But within a few weeks he had found what he was looking for, and he wrote up a brief report for the editor. Unlike the first reviewer, Ghirardi recommended a flat-out rejection.[52]

The problem, Ghirardi explained, was that no laser gain tube—not even an ideal one, the kind imagined in thought experiments—could function the way Herbert's plan required. The flaw didn't come from any limiting feature of a given laser design; it came from quantum theory itself. Quantum mechanics is what physicists call a "linear" theory. That means that when you add together two different solutions to the governing equations, the result is also a solution. Indeed, linearity is what lies behind many of the conceptual sticking points in quantum theory, such as Schrödinger's cat. In one solution, the cat lies dead in its box. In another equally valid solution, the cat purrs merrily, awaiting its release.

And in a third solution—made possible thanks to the linearity of quantum theory—the cat is caught in some suspended state, neither dead nor alive. Likewise with the famed double-slit experiment. When both slits are open the interference pattern emerges because the photon's wavefunction is the sum of two possibilities: photon traveling through top slit plus photon traveling through bottom slit.

That same linearity applied to the photon polarization states in Herbert's FLASH proposal. The states R and L could be broken down into linear combinations of the states H and V, and vice versa. For Herbert's scheme to work, the laser gain tube needed to behave like a "photon xeroxer," as Herbert had put it, making duplicate copies of whichever polarization state entered the tube. Suppose the initial photon were in the state R. Even before the photon entered the laser tube, it would exist as a combination of equal parts H and V. If the experimenter had chosen to measure the photon's linear polarization at that point—before it entered the laser gain tube—she would have had an equal probability of finding it in state H or in state V, that is, all in one state or all in the other, with equal odds. A laser tube that really could amplify any of the polarization states equally would produce 100 duplicate copies of that same combination: a state that had a fifty-fifty chance of containing all 100 photons in the H state, or all 100 photons in the V state. It would never produce 50 H photons and 50 V photons at the same time, as Herbert had assumed. "In other words," Ghirardi explained in his brief report, "it is impossible that for all 4 states of polarization . . . the laser gain tube acts simply as a duplicator producing N photons of the same type."[53]

Without the ability to make perfect copies of an arbitrary polarization state, Herbert's FLASH scheme fell apart. The experimenter at B would always see random patterns coming from the detectors at station 1 and station 2: on any given run, each station would find all of its signal in just one state (all H or all V at station 1, and all R or all L at station 2). Any intended message from experimenter A would be hidden within the random strings, just as in Eberhard's original critique. To Ghirardi, therefore, the decision seemed obvious. "The basic device of the suggested experiment violates the linear nature of quantum mechanics and therefore all the proposal is incorrect. For these reasons I consider that the above paper does not deserve publication in *Foundations of Physics*."[54]

Ghirardi thought he had sealed the deal, and in less than a single page, no less. The journal editor, however, received two reports from leading experts that came to opposite conclusions: the first reviewer recommended immediate acceptance, while Ghirardi urged summary rejection. Faced with the conflicting recommendations, the editor did what most journal editors would do: he asked Herbert to revise his paper in light of the reviews and resubmit it for further consideration. (A kind of editorial version of Schrödinger's cat. Editing an academic journal appears to be a linear process, too.) Herbert went back to his draft and beefed up his discussion of how the sought-for signal might be gleaned from real-life noise, perhaps by sending an initial triggering pulse and then scrutinizing various coincidences between detectors.[55] He seems not to have caught the real thrust of Ghirardi's critique at this time—not terribly surprising, given how terse Ghirardi's report had been. Nonetheless, Herbert hammered out a new draft and dutifully mailed it back to *Foundations of Physics*. It arrived on January 15, 1982, almost exactly a year after he had submitted his original version. He mailed out preprints of the new version, too, including a copy sent directly to John Bell.[56]

Imagine Ghirardi's surprise when his Trieste colleague, Tullio Weber, received a request to referee the new version of Herbert's paper. (Weber had coauthored with Ghirardi the earlier critical analyses of Herbert's QUICK scheme.) Weber made quick work of the task, essentially repeating the argument from Ghirardi's previous report, this time condensing the entire analysis to a single paragraph. Despite the brevity, Weber tried to make clear the essential point. "In my opinion," he closed, the design of a thought experiment "must not present intrinsic contradiction with the theory it pretends to test."[57]

Independently of Ghirardi and Weber, Henry Stapp had also caught a glimpse of how the linearity of quantum theory might threaten Herbert's FLASH scheme. Just as Weber was sending in his referee report, Stapp argued vehemently with Herbert at their annual Esalen workshop. But, Herbert shot back, Stapp's argument seemed too strong: it also appeared to rule out the ordinary operation of lasers, which amplify incoming signals all the time. They tussled round and round in Esalen's "big house" without reaching any definite conclusion.[58]

•

As Ghirardi and Weber tried to clarify their objections to the journal editor, and as Stapp argued with Herbert near Esalen's hot spring baths, parallel stories began playing out elsewhere. Copies of Herbert's paper made their way to other little groups of physicists here and there, and a few began to take notice. One copy landed in the hands of Wojciech Zurek and Bill Wootters, two recent PhDs who worked closely with John Wheeler at the University of Texas at Austin. Around the time they finished their PhDs, Wheeler contrasted "the outgoing, 'take-on-anything' spirit of Zurek" with the "quieter and more reflective" style of Wootters.[59] Together they made a powerful team.

Zurek and Wootters had been office-mates as graduate students at Austin. They quickly fell under Wheeler's spell, showing greater-than-average interest in foundational matters.[60] After a particularly inspiring visit to Wheeler's summer cottage in Maine, Zurek enthused, "Every day I come back to the discussions on physics and philosophy we had (always in rooms overlooking the sea!). I often wonder how would contemporary physics look if it were done in the rooms overlooking [the] ocean, and not in [a] 'publish or perish' jungle."[61]

Wheeler's reputation helped to shield the young students from other professors' disdain for such philosophical patter. Aside from Wheeler's strong backing, Zurek recalls the general attitude in the hallways at that time: graduate students like himself and Wootters had received "a very obvious and loud message that thinking seriously about foundations was a waste of time and a detriment to one's career."[62] Nonetheless, Wheeler and his small circle of students soldiered on. Wheeler brought in a steady stream of visiting scholars to keep discussions fresh. He also organized a brand-new seminar on quantum measurement, the mysterious process by which arrays of quantum probabilities get reduced to single, measured results: an entire semester spent puzzling over paradoxes like Schrödinger's cat.[63] Zurek and Wootters polished up a term paper for that class—a fine-grained analysis of the double-slit experiment, calculating precisely the limits and trade-offs imposed by wave-particle duality—and published it in 1979, as one of their earliest forays into the foundations of quantum theory.[64] The following year, PhD in hand, Zurek helped Wheeler coteach the seminar. They went on to publish a major collection of articles together, culled from the readings for the class. The mammoth volume, entitled *Quantum Theory and Measurement*, filled over 800 pages, and quickly became a standard reference. As Wheeler explained soon after the book's appearance, "Zurek was the brain and dynamo of this enterprise."[65]

How exactly Nick Herbert's FLASH paper made it onto Zurek's and Wootters's desk remains unclear. Certainly Wheeler had been an early and frequent recipient of Ira Einhorn's mailings, and both Jack Sarfatti and Nick Herbert had long been in the habit of mailing their latest notes and papers directly to Wheeler themselves.[66] But Zurek recalls Wheeler keeping "the 'Californians'" at arm's length, at least where his students were concerned. The attitude at Austin seemed to have been that it was one thing for an acclaimed physicist like Wheeler to interact with Sarfatti and crew, but quite another for graduate students who were interested in the interpretation of quantum theory to do so. The students already faced an uphill battle among skeptical faculty, and meddling with curious characters outside the mainstream might only exacerbate the situation. Zurek, at any rate, has no recollection of Wheeler passing along papers or preprints by Sarfatti or Herbert to his students.[67]

By the late 1970s and early 1980s, however, several other vectors linked Sarfatti, Herbert, and company to Wheeler's circle. One likely conduit was Larry Bartell, a physical chemist based at the University of Michigan whose scientific interests ranged far and wide. (As Bartell put it recently, a physical chemist is someone who studies "anything that is interesting.") Much of his research had focused on electron diffraction, in effect conducting real-life double-slit experiments by shooting electrons at crystals and measuring the resulting interference patterns. That work, in turn, had sparked his interest in the foundations of quantum theory. He spent a sabbatical with Wheeler's group at Austin in the spring of 1978, where he befriended Zurek and Wootters. A few months later, Bartell submitted an article building on Wootters's and Zurek's early double-slit paper.[68]

At that same time, Bartell also began corresponding with Sarfatti and Herbert; they traded papers and commented on each other's drafts.[69] In the spring of 1980, following another brief visit with Wheeler's group at Austin, Bartell wrote a longer paper on "concrete new tests" of Bell-styled entanglement. Thanking Wheeler and Wootters for helpful discussions, he closed his article by challenging the superluminal communication schemes of Sarfatti's 1978 patent disclosure document and Herbert's 1979 QUICK paper, both of which he had received straight from their authors.[70] Bartell remained in touch with Sarfatti and Herbert throughout the early 1980s, and Herbert and Saul-Paul Sirag, in turn, invited him to participate in their February 1983 workshop at Esalen. Bartell still reminisces about his encounters at that "touchy feely place in Big Sur with mixed-sex nude natural baths and natural food." ("You can imagine it took little persuasion to get people to accept such an invitation to such a pleasant place," he notes.)[71] Meanwhile, Zurek and Wheeler selected one of Bartell's 1980 articles for inclusion in their *Quantum Theory and Measurement* volume, which appeared just as Bartell was taking in the delights that Esalen had to offer.[72] Still other links—only slightly more attenuated—connected Zurek and Wootters with the remnants of the Fundamental Fysiks Group.[73]

Zurek and Wootters shared a hotel room during a small, informal workshop in San Antonio in March 1982 that John Wheeler organized. As Zurek recalls, the informal discussion at the meeting meandered around to a topic that had been on Wheeler's mind for some time. Wheeler had been seeking some way to dramatize the consequences of Bell-styled entanglement. He was after some way to amplify the effects of nonlocality to a macroscopic scale: to do for Bell's theorem what his delayed-choice thought experiment had done for the question of quantum measurement. Zurek, meanwhile, had also been ruminating on what might happen if one could amplify an entangled state.[74] In his notes from that meeting, he scribbled some questions to himself about whether one could rotate the polarization of individual photons: the selfsame process that had been at the heart of Herbert's QUICK scheme, as Bartell and Wootters had already discussed in some detail. Not quite two weeks later, Zurek began playing around with the idea of using a laser to amplify an individual photon, such as one member of an entangled pair—the same basic mechanism as in Herbert's FLASH proposal.[75]

With those ideas swirling in his head, Zurek set off for yet another conference. This time it was to Perugia, Italy, for a conference celebrating the ninetieth birthday of Louis de Broglie. De Broglie, a famous architect of quantum theory who had first proposed that wave-particle duality extended to matter as well as light, had long been an outspoken icono-clast when it came to how quantum theory should be interpreted. The birthday meeting in April 1982 featured similar iconoclasts, including Franco Selleri, the Italian physicist who had proposed a superluminal scheme quite similar to Herbert's QUICK idea back in 1979.[76] Since that time, Selleri and Herbert had been corresponding; Selleri even invited Herbert to come visit the group in Bari, though Herbert's lack of funds scuttled those plans. But Herbert could still afford postage. He had sent Selleri a copy of his FLASH paper, and the two had traded extensive cor-respondence about it by the time of the de Broglie meeting in Perugia.[77] Selleri was besotted with Herbert's latest proposal, and featured it in his lecture at Perugia. More important, he pushed copies of Herbert's preprint on several other colleagues, and helped convince an experi-mental physicist from Pisa to mount a real test of Herbert's design. The Pisa physicist, in turn, dutifully reported on the early steps of his experi-ment at the Perugia meeting.[78] A graduate student who participated in the meeting later recalled the scene: "People around me were all talk-ing about a 'Flash communication' scheme, faster than light, based on entanglement."[79] Herbert's preprint had become the talk of the confer-ence. Thanks largely to Selleri's efforts, references to Herbert's FLASH paper recur throughout the published proceedings of the meeting.[80]

Wojciech Zurek's research notebook records his impressions of the lectures at the Perugia conference in which Herbert's scheme had been laid out in detail.[81] Thus by April 1982—if not before—Zurek had collided head-on with Herbert's FLASH design. Immediately upon his return from Italy, Zurek's thoughts returned to the discussion he and Wootters had shared in their San Antonio hotel room the previous month. He convinced himself that linearity was the key: the linear nature of quantum theory would place the ultimate limit on superluminal signaling, by making it impossible to duplicate arbitrary quantum states. He wrote up a few pages, extending the argument to the general case, abstracting away from details of Herbert's R, L, H, and V polarization states, and sent them to Wootters to review. Together they clarified that certain quantum states could indeed be amplified, even in the light of linearity. One could design a device that would make perfect copies of a *known* incoming state or of a state orthogonal to it—replicating states R or L, for example, but then not replicating H and V, and vice versa. What could never be built, at least according to the laws of quantum mechanics, was a device that could make perfect copies of an unknown or arbitrary incoming state. After some minor edits they submitted their paper, under some ponderous (and now forgotten) title, to the *American Journal of Physics*, a perfectly respectable if less-than-flashy journal.[82] (The journal, then as now, focuses on pedagogical matters; it is the forum in which several physicists hashed out their competing schemes for enrolling Fritjof Capra's *The Tao of Physics* in the classroom.)

Luckily for his subsequent career, Zurek also sent a copy of the paper to his former mentor, John Wheeler. Wheeler pressed his young colleagues to aim higher. In particular, he convinced them to take the unusual step of withdrawing their paper from consideration at *American Journal of Physics*—most likely after it had already been sent out for reviews—and submitting it instead to *Nature*, one of the world's leading and most influential scientific journals. More than that, Wheeler lent his famous knack for catchy titles; a decade earlier, for example, Wheeler had coined the term "black hole." The essence of Wootters's and Zurek's argument, Wheeler gleaned, was that "a single quantum cannot be cloned." With Wheeler's blessing, Zurek and Wootters chose that exact phrase for the new title of their paper.[83] In their now-famous article, Herbert's FLASH preprint assumes pride of place. Indeed, the very first sentence of Wootters's and Zurek's article reads: "Note that if photons could be cloned, a plausible argument could be made for the possibility of faster-than-light communication," followed by a citation to Herbert's paper. Wootters and Zurek mailed their paper to *Nature* in early August 1982, where it sailed through the reviewing process. It appeared in print in late October.[84]

Across the Atlantic, while Zurek and Wootters were piecing together their version of the "no-cloning theorem," Dennis Dieks was following the same series of steps. He had gotten his hands on Herbert's FLASH paper from a fellow member of a loose, informal discussion group in Amsterdam. Members of the group, who had taken to calling themselves "The Quantum Club," ranged far and wide. The group consisted of physicists and philosophers, and included everyone from American expatriates who had fled the Vietnam-era draft to eager young Dutch students just finding their scholarly legs. Like Wheeler's circle in Austin, Texas, the Amsterdam group had been connected to members of the Fundamental Fysiks Group in recent years by sinews like Ira Einhorn's network and by the Swiss-based newsletter *Epistemological Letters*.[85]

Dieks, a young physicist in Utrecht at the time, recalls that most of his discussion-mates in the Quantum Club were quite taken with Herbert's argument. Yet something didn't seem quite right to him. After the discussion had broken up, he continued to puzzle through the argument, and within a few weeks he, too, had isolated essentially the same flaw as Ghirardi, Weber, Wootters, and Zurek had done. Dieks wrote up a brief article laying out the basic critique, emphasizing up front "the crucial role of the linearity of the quantum mechanical evolution laws in preventing causal anomalies."[86]

In Dieks's paper, Herbert's FLASH paper assumed an even larger role than in the Wootters-Zurek no-cloning article. Indeed, Dieks's entire paper revolved around Herbert's preprint. Other than a citation of Eberhard's and Ghirardi's earlier no-go theorems against superluminal signaling, Herbert's was the only paper that Dieks even mentioned. The citation, meanwhile, betrayed some of the boundedness of Herbert's circulation scheme. Copies of his papers could travel far and wide, but their meanings or associations did not always stay firmly attached. Dieks cited Herbert's paper as a preprint from the "National" Science Foundation, missing Herbert's "Notional" joke entirely.[87]

Unbeknownst to all at the time, Dieks submitted his paper to the prestigious *Physics Letters* journal (based in the Netherlands) exactly one week after Wootters and Zurek sent their paper to *Nature*, in August 1982. His articulation of the no-cloning theorem encountered a few more bumps en route to publication than Wootters's and Zurek's did. Soon after sending his paper off, Dieks learned to his dismay that his enthusiastic pals in the Quantum Club weren't the only ones who bought Herbert's argument: so did the referee to whom *Physics Letters* had sent Dieks's article. The referee, a fellow Dutchman, went so far as to call Dieks on the phone to explain to him why, based on Dieks's own description, Herbert's FLASH scheme would work! Under pressure, Dieks made a few cosmetic changes to his brief paper, and it appeared in *Physics Letters* in late November 1982, three weeks after the Wootters-Zurek paper had been published in *Nature*.[88]

•

If GianCarlo Ghirardi had been surprised when his Trieste colleague Tullio Weber had been asked to review Herbert's revised paper—thinking that he had already killed off the submission a year earlier—a bigger shock was in store for him. One day in April 1983, while flipping through some back issues of journals in the library, he happened to come across Wootters's and Zurek's no-cloning article in *Nature*. (Part of the reason he stopped on that page, Ghirardi later recalled, was the paper's unusually appealing title.) Skimming the first few paragraphs, Ghirardi knew immediately that the article, which had already been in print for more than six months, contained the same basic argument that Ghirardi had composed for his referee report on Herbert's FLASH paper two years earlier. Worse still, he noticed from the way that Wootters and Zurek cited Herbert's paper that "FLASH" had been accepted by *Foundations of Physics*, and would soon be published![89] (In fact, Herbert's revised version of "FLASH" had already appeared in the December 1982 issue of *Foundations of Physics*, though Ghirardi had not caught up on his reading yet.)

Together with Weber, Ghirardi composed a hasty letter to the editor of *Foundations of Physics*. "With this letter we want to call your attention to a very peculiar situation," they began. They reminded the editor of the basic timeline of events, concluding with the latest shockers (publication of Wootters-Zurek in *Nature* and Herbert in *Foundations*). "You will understand that we are really upset by this situation," they continued. The reason for their displeasure is most telling. They were upset, they explained to the editor, because other experts in the field knew that Ghirardi and Weber often acted as referees for *Foundations of Physics*. Given their earlier published critiques of Herbert's QUICK scheme, their colleagues would easily surmise that they had been referees of Herbert's FLASH submission, too. And so Ghirardi and Weber feared that their own reputations were now in danger: others might think that they had failed to find the flaw in Herbert's latest paper. That would be one insult too many, on top of having been scooped by the Wootters-Zurek paper. Like Wootters and Zurek, who had first submitted their paper to *American Journal of Physics*, neither Ghirardi nor Weber had any inkling at this early stage that the no-cloning theorem might have any real-world significance. In closing their letter, Ghirardi and Weber shifted to damage control. "We are then asking you to let us know who is the referee responsible for the acceptance of Herbert's paper and on what basis the journal has decided to follow his judgment in place of ours. If this point would not be clarified we are not willing any more to act as referees for your journal."[90]

Faced with such a blast, the journal editor took a most unusual step. "In my memory of the 13 years during which I have been associated with *Found. Phys.*," the editor began, "this is the first time I have been taken to task by a referee." The occasion seemed so significant, in fact, that the editor wrote out a lengthy justification of his decision to accept Herbert's paper. He waved off Ghirardi's and Weber's demand to know who the other referees had been—anonymity of referees is considered sacrosanct in academic publishing—but he did share some of the backstory with them. For one thing, he had received written responses on Herbert's FLASH submission from half a dozen other referees beyond Ghirardi and Weber.

To give a flavor of the variety of responses, the editor quoted from several reports, referring to them by number to conceal the reviewers' identities:

> 1) "This is an important result. The article is beautifully written and I recommend its publication."
>
> 2) "We have not been able to identify any fundamental flaws with the proposed experiment that reveal the origin of the paradox." (The "we" here are two authors.)
>
> 3) "I spoke to several people in Europe, and everybody believes that it (Herbert's FLASH Experiment) should work."[91]

Though the editor couldn't know it at the time, his referees were in good company. Even Richard Feynman had been stumped by the FLASH setup, unable to poke any holes in Herbert's design until Wojciech Zurek walked him through the no-cloning argument at a blackboard. ("I think his reaction was short of full satisfaction," Zurek dryly recalled.)[92]

Of course, the *Foundations* editor had received some reports that were critical of Herbert's paper, too. But, the editor explained to Ghirardi and Weber, no consensus had emerged among those reports as to just what Herbert's error had been. Rather than decide the matter in secret by himself, the editor had decided to publish Herbert's paper. In that way, "*Found. Phys.* could be instrumental in stimulating the controversy surrounding Herbert's paper and hastening its resolution in a public forum." And, indeed, subsequent events, such as the publication of the Wootters-Zurek article in *Nature* and the Dieks article in *Physics Letters*, had shown the editor's instincts to be "dead right." His only regret was that the rebuttals by Wootters, Zurek, and Dieks had "bypassed" *Foundations of Physics*, appearing in those other (vastly more prestigious) journals.[93] In sum, the editor assured Ghirardi and Weber, "I took the action I did for the noblest of reasons. . . . All the same, I wish I could make amends for the mental agony you have suffered." Perhaps the Trieste physicists would like to submit their own article on the matter to *Foundations of Physics*?[94]

Ghirardi and Weber didn't have to wait for the editor's suggestion. They had already written up a lengthy article of their own, building on

Ghirardi's notes from his first inspection of Herbert's paper back in April 1981. They mailed the latest paper off to the journal *Il Nuovo Cimento* (in which their earlier critiques of superluminal communication schemes had appeared) the same week that they sent off their spirited letter of complaint to the *Foundations* editor. (They were too fed up with *Foundations of Physics* at the time to consider submitting anything there.)[95] At least some cognoscenti appreciated their detailed analysis of the way that quantum theory's linearity scuttled any design for superluminal telegraphy. An Italian colleague showed their preprint to Rudolf Peierls, a grand old man of the field and John Bell's former advisor, who offered his congratulations. Not only had Ghirardi and Weber arrived at "the right answer"; Peierls was "glad that the reason why Herbert's proposed experiment is nonsense, has been presented so clearly."[96] Their paper came out in November 1983, a year after the articles by Wooters, Zurek, and Dieks had been published.[97]

•

While Jack Sarfatti, Nick Herbert, and their brilliant critics—Stapp and Eberhard, Ghirardi and Weber, Wootters and Zurek, Dieks, and others—wrestled with superluminal communication, a parallel set of efforts began to coalesce. The activities were never truly independent—the hiccups of history had already entangled several key players—but by the mid-1980s, the two lines had become thoroughly enmeshed. The result: the first proposals for quantum encryption, a whole new way to protect information.

In each of its guises, the no-cloning theorem seemed to be about limitations: things that quantum theory will not allow. In short order, that fundamental limitation had been transformed into an asset, or, as computer programmers might say, the "bug" had been turned into a "feature." Faster-than-light communication? Probably not. But how about slower-than-light communication that could be perfectly secure: an encryption system for sending secret messages that could never be hacked, stolen, altered, or imitated?

Encryption of one kind or another has been around for millennia.

Indeed, for as long as people have needed to send secret messages—strategies for battle, missives between star-crossed lovers—they have sought some way to keep those messages out of the wrong hands. Some have looked to physical mechanisms, such as invisible ink. Others have invented secret codes or ciphers, scrambling the original message so that the intended meaning could only be unpacked by someone in possession of the encryption key. The stakes can be huge: think of the heroic effort undertaken by British mathematicians during World War II to crack the Nazis' "Enigma" code, with which the Germans had been coordinating deadly U-boat and *Luftwaffe* attacks. The 1970s saw another uptick in efforts to design reliable encryption devices for both government and commercial purposes, as electronic computers expanded ever more quickly into the daily routines of business and corporate life. Those years saw the birth of public-key encryption, such as the RSA algorithm, named for its MIT inventors, Ronald Rivest, Adi Shamir, and Leonard Adleman.[98]

Public-key encryption involves a carefully choreographed dance among two or more parties. They exchange some information in public—even in the midst of potential eavesdroppers—while keeping other information tightly guarded. The public exchange of information allows them to devise a secret key without taking the trouble of meeting somewhere in person, a handy feature when trying to execute secret plans at a distance. For example, the two parties can publicly exchange a series of large, arbitrary numbers, dozens of digits long. They also select a distinct set of large numbers, but keep these secret, even from each other. After they conduct a few ordinary mathematical operations on their publicly shared and privately held numbers—nothing more complicated than multiplication and division—they trade answers in public. By broadcasting some numbers and guarding others, they may divine each other's hidden numbers and use them as the basis for an encryption key.[99]

Encryption algorithms, such as RSA, rely on the practical difficulty of breaking down very large numbers into their smallest (prime-number) constituents. That is, they employ mathematical operations like multiplication and division that are in principle reversible. It just so happens, as the inventors of RSA and others were able to show, that beginning with a big number and trying to isolate its prime factors takes a long time using realistic, buildable electronic computers—in some cases, a *really* long time, several times the age of the universe.[100] These types of encryption thus offer de facto security: banks and governments can trust that their messages will remain secure for all practical purposes. Today we all trust these systems, whether we realize it or not, whenever we send an email or make a purchase on the Internet. But what if the security of those messages could be protected de jure—not by this or that government regulation, but by a law of nature?

The path toward quantum encryption began with some creative brainstorming by a young physics graduate student named Stephen Wiesner. His father, Jerome Wiesner, had worked on radar as an electrical engineer at the wartime Radiation Laboratory at MIT. Wiesner *père* made his career at MIT, rising through the administrative ranks and serving as president of the Institute from 1971 through 1980. He had also been a highly placed science advisor in the Kennedy and Johnson administrations.[101] Stephen Wiesner grew up reading about quantum mechanics, information theory, and electronic communication, often borrowing books from his father's shelves. He enrolled as an undergraduate at Caltech in 1960, where his lab partner for freshman physics turned out to be John Clauser (who would go on to conduct the first experimental test of Bell's theorem and join the Fundamental Fysiks Group). In addition to talking about physics, Clauser and Wiesner sprung together to buy a used car; Wiesner can't remember whether it cost a total of $15 or $16 (roughly $100 in 2010 dollars). Aside from his friendship with Clauser, things did not go well for Wiesner at Caltech. He flunked out and transferred to Brandeis University, in the Boston area near where his family lived.[102] There he befriended a fellow undergraduate, Charles Bennett.[103] Though neither knew it at the time, John Bell visited Brandeis and wrapped up his famous article on Bell's theorem during their senior year, in 1964.

Wiesner entered graduate school a few years later at Columbia University, not realizing that Clauser was also a graduate student there. (By that time, Clauser was working on astrophysics, so his office was in a building different from the main physics department.)[104] Wiesner came to Columbia at a propitious moment: Columbia saw some of the worst rioting of any American campus during the spring of 1968, as a wave of unrest over the Vietnam War crested across the nation's universities. Wiesner had planned to study high-energy physics, but suddenly the laboratory was closed and classes canceled. "This gave me the chance to forget about what I was supposed to be doing and reflect on what seemed really important to me," Wiesner put it recently.[105] As chaos reigned all around him, he wondered whether quantum theory might enable some foolproof means of securing order. Could one make "quantum money," for example, money that could never be counterfeited, even in principle?

Wiesner walked through the argument. Perhaps each dollar bill could carry a unique serial number—as they do now—as well as a set of trapped photons hidden inside special boxes. The issuing bank would insert the photons in definite states of polarization (box 1, state R; box 2, state H; and so on), and keep a sealed record in its archives of the arrays of polarizations that went with each serial number. Anyone who wanted to make a copy of the bill would need to open up the boxes and make measurements of each photon's polarization. But how could they know whether a given photon was in a state of linear or circular polarization? If the photon in box 1 really had been set up in state R (that is, a definite state of circular polarization), but the counterfeiter happened to choose to measure linear polarization instead, he would have a fifty-fifty chance of finding H or V; he would never find R. And so on down the list: the counterfeiter would need to know, ahead of time, whether each photon was in a linear or circular polarization state before even attempting to make a measurement or produce a copy. The bank, meanwhile, could easily check any bill against its own records to detect fakes. For the whole scheme to work, Wiesner had to assume that the photons in the original dollar bill could not be duplicated without disturbing their original polarizations—an assumption he did, in fact, make, though without providing any justification.[106]

Wiesner wrote up his brief paper while the real-world events on campus continued to swirl. He passed it along to a department secretary, who agreed to type up a clean preprint copy since all ordinary business had ground to a halt. Looking back, Wiesner emphasizes that neither he nor the secretary received "any permission from the higher-ups. Discipline had broken down. The paper couldn't have been produced a year earlier or a year later." Indeed, once order was restored and classes had resumed, none of Wiesner's professors at Columbia showed any interest in his odd little paper—no more than John Clauser's Columbia advisors appreciated his budding interest in Bell's theorem. None of the referees from the various journals to which Wiesner submitted his paper knew what to make of it, either.[107] And so the paper languished in a kind of *samizdat* gray zone, much as Nick Herbert's QUICK paper would do, circulating here and there in crude photocopied form.[108]

A few years later Wiesner ran into his old lab partner, John Clauser, in Berkeley. Wiesner had set his quantum money paper aside, completed his dissertation on mainstream particle physics, and hit the road. Living as a self-styled hippie, Wiesner caught up with Clauser and got a tour of his in-progress experiments on Bell's theorem. Neither remembers talking about the quantum money proposal during that visit.[109] Nick Herbert, whose QUICK and FLASH schemes share so many features with Wiesner's idea, seems not to have met Wiesner at that point; he first learned about quantum money many years later.[110] Their collective wavefunction, still so full of possibilities, had yet to collapse.

Back on the east coast, however, Wiesner's idiosyncratic ideas did start to percolate. One of the few people who took notice was Wiesner's old friend from Brandeis, Charles Bennett. Bennett was just finishing up his PhD at Harvard, where he worked on computer models of molecular behavior. From Harvard, Bennett moved to the IBM research laboratory in Yorktown Heights, New York. He and Wiesner had stayed in touch since their undergraduate days. Meanwhile, at IBM, Bennett's interests shifted more and more from computer simulations of physical systems to the nature of computation and information in their own right. How should scientists conceive of information, computation, and communication in the light of quantum theory? And did those topics offer any insights, in turn, into the nature of quantum mechanics? As he explored the new terrain, he returned often to Wiesner's unpublished thought-piece.[111] A few years later Bennett and Wiesner teamed up to combine the insights about quantum money with various encryption methods. Rather than lock away all the information about each dollar bill, they realized, perhaps they could reveal *some* of the relevant information publicly while keeping the rest hidden. Maybe that could become the basis for some new form of public-key encryption, a quantum version of the RSA algorithm.[112]

By that time, Bennett's work had caught John Wheeler's eye. Wheeler invited Bennett to a tiny workshop at Austin on the foundations of quantum theory early in 1984.[113] There Bennett first met Bill Wootters; he crossed paths with Wojciech Zurek at a different conference around the same time. From them he learned about their recent work on the no-cloning theorem.[114] Now armed with a solid proof that arbitrary quantum states, such as a photon's polarization, cannot be duplicated, Bennett had the final piece of the puzzle. That December, he and his Montreal colleague Gilles Brassard presented a paper at a computer science conference. Ever since, their paper has been known by the simple abbreviation "BB84," for Bennett, Brassard, 1984. It offered the first blueprint of a provably secure encryption system.[115]

Like Herbert's FLASH scheme, the BB84 protocol relies on encoding messages using the polarization states of photons. There is one key difference. After experimenters A and B conclude their measurements on the photons, they open up a conventional communication channel—telephone, email, carrier pigeon, you name it. They compare notes on what their detectors had been set to, but *not* what each measurement outcome had been at those settings. Then they can see on which runs their detector settings happened to have matched. Zeroing in on the subset of runs with matching detector settings, they would then know what results their partners should have measured each time, based on the perfect correlation of entangled quantum systems. If both A and B happened to have set their detectors to measure circular polarization for the photons of run 1139, for example, and if experimenter B checked her log and saw that her detector measured L on that run, then she would know that experimenter A must have registered R on that same run. No one else could know that: even with the public chatter about which set of photons they were considering, and what their detectors had been set to measure for that round, no one could know what actual results each had found. The results would be pure quantum randomness: perfectly correlated between A and B, but unpredictable in advance.

To test for any tampering, experimenters A and B could use the open, public channel to compare notes on a few selected measurements. "For photon 1157, when we both happened to have our detectors set to measure linear polarization, did you get *H*?" experimenter B might ask. If A responds yes—and if that matches B's expectation, given her own log of results—then they can be certain that no one had intervened with the photons of that run. Having publicly exposed the detector settings and measurement outcomes of that run, experimenters A and B would toss those results out of their sample. They would thus sacrifice a small handful of results to ensure that the rest were accurate, their security beyond question. Using the results that were left over, experimenters A and B would have a shared stock of provably secret data—in essence, a string of ones and zeros—with which to encrypt their messages.

The only reason to consider such a jerry-built scheme was the no-cloning theorem. Without that key result—and thus without Nick Herbert's FLASH provocation—the carefully choreographed exchanges of the BB84 protocol would be for naught. Bennett and Brassard explained right up front: usually one assumed that digital communications could always be "passively monitored or copied, even by someone ignorant of their meaning." Not so with their quantum system, which no eavesdropper could possibly access, even in part, without destroying the sought-after signal and announcing her presence. Any effort to intercept the photons en route to make clandestine measurements would irreversibly disturb their quantum state. Thanks to Wootters's and Zurek's result, moreover, no eavesdropper could make clones of the source photons, retaining some for nefarious purposes while sending perfect copies on their way toward the unsuspecting experimenters at A and B.[116]

A follow-up proposal, published a few years later, demonstrated an even more efficient way to test for eavesdropping, by making more direct use of Bell's theorem.[117] Since then, experiments have roared ahead. In recent years, quantum encryption has moved beyond the laboratory to several real-world demonstrations, such as the 2004 bank transfer in Vienna and the 2007 electronic voting in Geneva. Both of those demonstrations used entangled photons shot through fiber-optic cables. Other groups have demonstrated the ability to send robust quantum-encrypted signals even further, down fiber-optic cables as long as 115 miles.[118] Another group successfully broadcast quantum-encrypted signals nearly 100 miles through open air: far enough to demonstrate that quantum encryption could be used to bounce signals from an earthbound station to an orbiting satellite and back, opening up the possibility of creating a worldwide network of quantum-secure communications.[119]

Given the obvious potential for government and military applications, the Defense Advanced Research Projects Agency (DARPA) has lavished millions of dollars in funding. National laboratories such as Los Alamos and the National Institute of Standards and Technology (formerly the U.S. Bureau of Standards) maintain active groups in quantum cryptography, as do similar government organizations throughout Europe and Japan.[120] The private sector has shown comparable interest. In addition to several start-up firms specializing in quantum cryptography, most of the major electronics corporations now sport their own internal divisions dedicated to the topic, including IBM, Hewlett-Packard, Toshiba, Mitsubishi, and NEC.[121] The recent flurry of activity has attracted feature articles not just in the places one might expect—*Scientific American*, *Physics World*, *New Scientist*, *Wired Magazine*—but also in *BusinessWeek*, *The Wall Street Journal*, and more.[122] Thanks to breakthroughs like the no-cloning theorem and the BB84 protocol, the foundations of quantum mechanics have made their way onto the business pages.

•

How far we have come from Jack Sarfatti's vision of drugged CIA agents instantly receiving brain-wave communiqués from headquarters; or, indeed, from Nick Herbert's metaphase typewriter, with which he had hoped to contact the spirit of Harry Houdini, and which Sarfatti credited as a major motivation behind Herbert's FLASH scheme.[123] Scratching just below the surface, however, some startling continuities begin to emerge.

Sarfatti was not so far off the mark in trying to interest the defense establishment with "untappable" quantum-communication systems. Herbert's FLASH design, while clearly unworkable, elicited a world of cutting-edge physics in its wake.

Back in 1983, the editor of *Foundations of Physics* had mentioned in his letter to the Trieste physicists GianCarlo Ghirardi and Tullio Weber that a few different critiques of Herbert's FLASH proposal had come in. The no-cloning variety was one. Another focused on how lasers actually work at the single-photon level—a question that had never been broached until Herbert's paper forced the issue onto the table. By the time he wrote his letter, the editor could point to some publications along these lines.[124] The author of one of those articles heard about Herbert's FLASH proposal at the Perugia conference in honor of Louis de Broglie in April 1982. Another likewise reported that he undertook his investigation in direct response to Herbert's paper. Even Harvard's Roy Glauber, who would later earn the Nobel Prize for his contributions to quantum optics, got in on the act, deriving his own, neat argument against Herbert's proposal.[125]

They each discovered a second reason why FLASH would fail. When stimulated by just a single photon, as in Herbert's plan, lasers would spontaneously emit "noise" photons in random, uncorrelated states of polarization, at comparable levels to the stimulated, coherent radiation. In ordinary operation, lasers are stimulated by billions of photons at a time, and the "noise" photons don't compete with the main output signal. But at the ultimate quantum limit, those unavoidable, extra photons would muddle the statistics at experimenter B's detector and wash out Herbert's predicted signal, even without taking into account the no-cloning theorem. (Herbert followed these developments closely, featuring them in his presentations at Esalen during the mid-1980s.)[126] Alongside the no-cloning theorem, Herbert's FLASH paper thus prompted a second major development: the first proof that no perfect amplifier could ever be built.

Despite all these developments, Charles Bennett recently dismissed efforts like Sarfatti's and Herbert's to design superluminal telegraphy. To Bennett, their determined quest for faster-than-light communication was no different from the perennial hunt for perpetual motion machines.[127] It was clearly not meant to be a flattering comparison. Physicists often invoke perpetual motion machines as the ultimate hokum, the obsession of confused hacks, scheming charlatans, or both.

The comparison bears further consideration. Perpetual motion schemes only seem tainted with the scent of carnival from our vantage point today. Back in the late nineteenth century, however, careful scrutiny of perpetual motion proposals helped to elicit and clarify some of the crowning achievements in the study of heat, energy, and molecular motion. Only after a small band of experts had struggled through such profound conceptual knots as the conservation of energy and the meaning of entropy could scientists begin to dismiss perpetual motion machines in an intellectually legitimate way.[128] Chasing perpetual motion machines today—as rogue inventors around the world continue to do—may rightly be dismissed as folly. Pursuing them a century or more ago proved to be remarkably productive, spurring major advances.

So, too, with the Fundamental Fysiks Group's obsession with superluminal telegraphy. No one had produced a single principled argument to reject faster-than-light communication, in the light of Bell's theorem and quantum entanglement, until Sarfatti, Herbert, and others forced the issue. Only by puzzling through their detailed proposals, step by step, did the community discover deep, first-principles reasons why such schemes must fail. On our side of that dividing line—nearly three decades after developments like the no-cloning theorem—it is easy to dismiss their schemes as so much sophistry or self-delusion. Roy Glauber at Harvard put it best when he tackled Herbert's proposal. "The same infernal ingenuity that once went into devising perpetual motion machines is now suggesting means for communicating faster than light," he explained. "Some of these are interesting schemes," Glauber made clear: "they too might just be capable of teaching us something."[129] Indeed, many textbooks on quantum information science have elevated the various critiques of superluminal communication to a founding principle known as the "no-signaling theorem." The no-signaling theorem posits that no operations using entangled states can allow faster-than-light communication. Herbert's FLASH scheme, in other words, has been elevated to a litmus test: if two quantum states appear, on paper, to be distinguishable, but if that distinction could be exploited to send signals faster than light, then the two states must not be distinguishable after all. Q.E.D.[130]

And yet, even as the latest textbooks tout the no-cloning theorem and its no-signaling spin-off, the fracas that started it all has faded from view. To date, the Wootters-Zurek no-cloning article in *Nature* has garnered more than 1100 citations in the scientific literature, reflecting its crowning importance and its high journal visibility. Dennis Dieks's piece in *Physics Letters* has received roughly a third as many citations, still placing it in the upper tiers of influential physics papers. And what of Nick Herbert's paper, which triggered it all? Herbert's FLASH article has been cited just seventy-two times since its publication, most often in out-of-the-way philosophy journals.[131] Herbert's efforts certainly do not deserve equal credit to those of Eberhard and Stapp, Ghirardi and Weber, Wootters and Zurek, Dieks, or the others, but they clearly deserve *some*. Jack Sarfatti's and Nick Herbert's tireless pushing on the matter of Bell's theorem and the ultimate implications of entanglement forced others to take those questions seriously; they put the matter onto other physicists' agenda. Theirs was a mistake, but a wonderfully productive mistake.

Twenty-five years after Herbert's FLASH paper appeared in print, the scheme received a proper laboratory test. In 2007, a team of Italian physicists who had been in contact with GianCarlo Ghirardi and Wojciech Zurek built a real-life version of Herbert's imagined device. They fired it up, scrutinized their data, and drew their conclusion: alas, it hadn't worked.[132] Or had it? Herbert had anticipated just such a turn of events. Notes from the Esalen workshop of 1985 explain:

> NICK HERBERT, after ten years of trying to signal faster than the speed of light, finally succeeded; unfortunately, there was no physical evidence of this historic precedent as, at the moment of his triumph, he instantaneously popped into the past where he was still trying to prove you can signal faster than the speed of light.[133]

Ain't it always that way?

Chapter 10

The Roads from Berkeley

A few years ago Elizabeth [Rauscher] and I were talking about how important this group had really been in my own genesis of discovering ideas and hearing ideas that were really important to me, and getting feedback on my own ideas, and having a support group of people. It had been a pretty lonely road up until then. Having a support group that was really a critical mass made this a very important group, and I am glad that we are here together to honor it.

— **George Weissmann, 2000**

After meeting every week for nearly four years, the Fundamental Fysiks Group disbanded early in 1979. The proximate causes seemed clear at the time. Elizabeth Rauscher and George Weissmann, the group's founders, had each completed their dissertations and were no longer available to manage the group's logistics. Weissmann had secured a postdoctoral appointment in Europe, and Rauscher had become busy teaching as an adjunct assistant professor at Berkeley. Henry Stapp, the senior staff scientist at Lawrence Berkeley Laboratory who had been involved with the Fundamental Fysiks Group from the beginning, tried to keep the group going for a while but without much success. His other responsibilities at the lab probably left little time for running the weekly discussion group. Some in the group also suspected that the lab's administration had grown wary of allowing so many outsiders onto the site, which, after all, was a major government laboratory. Having started as a collection of about ten down-on-their-luck physicists, the Fundamental Fysiks Group had grown to attract forty or fifty participants—few of them affiliated with the laboratory—to several of its weekly sessions.[1]

Other factors likely played a role. Tensions and jealousies had emerged, stoked by the first glimmers of fame that books like Fritjof Capra's *The Tao of Physics* and Gary Zukav's *The Dancing Wu Li Masters* had attracted. Not long after Zukav's book appeared, Sarfatti wrote a tart note to his onetime friend and collaborator, Saul-Paul Sirag. Sarfatti warned Sirag to stay clear of the "human potential narcissism" that he felt had begun to infect the others. "I am not looking for 'followers' but for tough minded professionals," Sarfatti continued. "We do have a mission. If you do not feel that way then the idiots and frauds of the Esalen clique have destroyed your critical judgment." (By this time Sarfatti had fallen out of favor with Esalen director Michael Murphy; Saul-Paul Sirag had taken over the reins with Nick Herbert for the annual Esalen workshops on Bell's theorem.) "You are still a sorcerer's apprentice," Sarfatti closed. "Fame is spoiling you."[2]

The feeling was mutual, at least for a while. When Sirag and Herbert founded their own spin-off group, the Consciousness Theory Group, they adopted one clear policy. The eclectic group ranged far and wide in its efforts to understand consciousness, experimenting with psychedelic drugs and consorting with psychics, shamans, and "sex magicians." But, recalls Herbert, "we had our limits." The one thing they dared not do: they refused to tell Jack Sarfatti where the group was meeting. After their years of experience with the Fundamental Fysiks Group, the Physics/Consciousness Research Group, and the Esalen workshops, they had become fed up with Sarfatti's tendency to "monopolize the meeting with his own Obsession Du Jour."[3] Sarfatti, meanwhile, took out a personal ad in the Berkeley student newspaper, the *Daily Cal*, challenging Fritjof Capra to a "duel of wits." Claims and counterclaims began to swirl about who had stolen whose ideas, as major royalties began to accrue for some group members' best-selling books.[4]

And so the once close-knit Fundamental Fysiks Group fell apart. The paths that core members have followed since that time are just as diverse and unexpected as the chance conjunctions that brought them together.

Their individual journeys map out a range of possibilities, illustrating different ways one could carve out a career on the margins of modern-day physics. Some continued to rely on private patronage, of the sort they had earlier enjoyed from Werner Erhard, Michael Murphy, and George Koopman. Others made the leap to full-time authors, following the path that books like *The Tao of Physics* and *The Dancing Wu Li Masters* had opened up. Still others became self-supporting entrepreneurs. And a few managed to remain close to mainstream physics, even as the occasional episode continued to remind them of their curious place in the disciplinary terrain.

•

Following Jack Sarfatti's dramatic falling-out with Werner Erhard, his financial situation became desperate. His luxurious Nob Hill apartments were gone, replaced by a dingy little apartment "where two steps take you from the door to the smeary window that looks down on a rooftop and the bathroom is down the hall," as a journalist described it in a *San Francisco Chronicle* profile. Sarfatti picked up menial jobs around San Francisco to try to make ends meet, but rarely stayed in them for long. He lost his job as a hotel porter, for example, because he couldn't or wouldn't learn how to shine shoes properly.[5] He kept up his photocopy-and-postage activities (modeled on Ira Einhorn's famous network) as long as he could, at one point complaining to physicist John Wheeler that he was "starv[ing] with no food, spending my last penny for xerox and mail."[6] All the while Sarfatti hoped that his latest corporate visions—still based on his faster-than-light communication schemes—might capture the right person's imagination.

Sarfatti began to pull out of his downward spiral in the early 1980s. Perched at his regular location (Caffe Trieste, North Beach, San Francisco), he had fallen in with a curious crowd: politically conservative thinkers who were drawn to certain New Age ideas. Chief among them was A. Lawrence ("Lawry") Chickering. A graduate of Yale Law School, Chickering worked for the conservative magazine *National Review* before returning to his native California in the early 1970s to direct the statewide Office of Economic Opportunity under Governor Ronald Reagan. Near the end of Reagan's term, Chickering founded a new political think tank in San Francisco, the Institute for Contemporary Studies, and convinced such leading conservatives as Edwin Meese and Caspar Weinberger to join the Institute's board. Chickering quickly became known as the intellectual leader of the "New Age Right." Where others had seen only left-leaning collectivist ideas on display at Esalen or in the Eastern mysticism craze, Chickering discerned a strong element of "personal responsibility." Borrowing from *est* and the human potential movement, Chickering tried to hone a new "therapeutic vocabulary," as he explained to a journalist: some new means of discussing contentious political issues in a way that emphasized each faction's common ground. When Reagan was elected president in 1980, and Meese and Weinberger joined the new cabinet, Chickering suddenly had the ear of the White House. Sarfatti, in turn, had the ear of Chickering.[7]

Chickering sent memos to highly placed bureaucrats in Reagan's Defense Department touting Sarfatti's work and lobbying for funds to support further research. At a March 1982 dinner in Washington, DC, hosted by Secretary of Defense Weinberger—until recently a board member of Chickering's think tank—Chickering struck up a conversation with the undersecretary of defense for research and engineering. He followed up with a long letter a week later, to describe in more detail "the work of a physicist friend of mine which just might have profound implications for certain aspects of the technology of warfare."[8]

Chickering mentioned the CIA memorandum from 1979 that had expressed some interest in Sarfatti's ideas, and then made his pitch. "Jack says that if in fact we can control the faster-than-light nonlocal effect," then one could make "an untappable and unjammable command-control-communication system at very high bit-rates for use in the submarine fleet. The important point here is that since there is no ordinary electromagnetic or acoustic signal linking the encoder with the decoder in such a hypothetical system, there is nothing for the enemy to tap or jam." "I know this sounds like science fiction" or even "occult 'sympathetic magic,'" Chickering admitted, "but no one honestly knows for sure at this point." Wouldn't it be in the nation's interest to invest a little of the Pentagon's discretionary funding to test Sarfatti's hypothesis, rather than ignoring the idea until some rival country ran with it instead?[9] Sarfatti had already drawn up a proposal to establish a think tank of his own, "PSI: Physical Sciences Institute" (an all-too-obvious reference to "psi" or parapsychology), to advise the Reagan administration on "potential defense applications of the 'new physics' as they emerge." Total cost: a meager $250,000 per year for five years (nearly $600,000 per year in 2010 dollars), mere chump change on the scale of the Pentagon's budget.[10] Chickering could see the merit of the proposal, as he wrote to the undersecretary of defense: "God knows what other clever ideas he [Sarfatti] and his crew of eccentric geniuses might come up with if they were properly supported."[11]

Chickering's memo did not generate funds for Sarfatti. The undersecretary suggested that Sarfatti confer with the JASON group, an elite corps of civilian physicists whom the Pentagon consulted on defense matters, while an Air Force colonel held out for a "summarization of Dr. Sarfatti's latest findings" before promising any funds, and a different assistant secretary of defense expressed only preliminary interest.[12] But the connection with Chickering introduced Sarfatti to a whole new network of people. Around the time of his memo to the Pentagon, for example, Chickering and a friend (the wife of the Reagan administration's new ambassador to France) met in Paris with physicist Alain Aspect, right in the midst of Aspect's groundbreaking experiments on Bell's theorem, to convey messages from Sarfatti.[13] When an editor of the journal *Foundations of Physics* compared Sarfatti's unusual position to that of another "rogue" physicist who also sought to challenge physics orthodoxy without a stable institutional position, Sarfatti was quick to draw a distinction. "The difference is that I am now getting a sympathetic hearing at the highest levels of President Reagan's Administration—I mean the *highest.*"[14]

Newly immersed in Chickering's circle, Sarfatti's political leanings swung solidly to the right. He began to write with characteristic ire about the leftist excesses of people and groups with whom he had enjoyed close relations only a few years earlier. A typical rant dismissed "charlatans and

'New Age' anti-rationalists of the drug-crazed and meditation-glazed 'counter-culture,'" with their "pop-Eastern mysticism."[15] His ideas about harnessing quantum entanglement likewise began to reflect the latest political hues. For example, Sarfatti imagined fulfilling Reagan's famous call to render nuclear weapons "impotent and obsolete"—the phrase Reagan used in March 1983 when announcing his new Strategic Defense Initiative, or Star Wars program—by shooting entangled quantum particles at enemy missiles from space-based battle stations. The particles would induce harmless nuclear reactions inside the warheads, rendering the fissionable material inert. Unlike many of his other brainstorms about Bell's theorem, this one made it into print, appearing in the journal *Defense Analysis* in the mid-1980s.[16]

As Sarfatti began to explore new ideas and social circles, he emerged as something of a Renaissance courtier. He flitted from wealthy patron to wealthy patron, surviving—at times flourishing—on generous stipends rather than regular employment. One of his first benefactors following Erhard was a Berkeley-based architect. In the 1990s his major sponsor was a Silicon Valley executive and UFO enthusiast, so Sarfatti's attention turned squarely to exotic theories of relativity and gravitation in an effort to explain how aliens might have achieved efficient interstellar travel.[17] (Patrons such as the crown prince of Lichtenstein have lavished funds on other researchers with the same goal in mind.) After another dramatic falling-out with the Silicon Valley sponsor, Sarfatti found a new benefactor. Though Sarfatti remains cagey about revealing his present patron, suffice it to say that he drives around San Francisco in a mint-condition Jaguar (full leather interior) and keeps in constant contact with a diverse network of followers with his iPhone.[18] These days Sarfatti bristles at the term "hippie." As he put it recently, "I am a counter cultural radical conservative who hob nobs with Reaganites [and] billionaires," even if at one time he "went slumming perhaps with hippies in hot tubs."[19]

As the Fundamental Fysiks Group was breaking up in 1979, meanwhile, Saul-Paul Sirag worked hard to keep some sort of group together. He continued to organize (with Nick Herbert) the Consciousness Theory Group, which had begun in Arthur Young's Institute for the Study of

Consciousness in Berkeley and then moved over to the San Francisco offices of toy manufacturer and parapsychology enthusiast Henry Dakin. Sirag also became a regular participant in a related group known as the Parapsychology Research Group, founded by remote-viewing expert and former Stanford Research Institute physicist Russell Targ. Sirag served as president of the group in 1988–89, hosting the monthly meetings. All the while he organized the annual Esalen workshops with Nick Herbert.[20]

For the first few years Sirag survived on small consulting fees associated with this work. Dakin hired him as a personal advisor to help sort bogus claims about the paranormal from more promising leads. Sirag also tutored Esalen director Michael Murphy and other fixtures of the human potential movement on modern physics, squirreling away some cash whenever he could.[21] To supplement these modest fees, Sirag took a job as a night watchman at a high-end apartment building in San Francisco. The job proved to be a perfect match for the diligent and hardworking Sirag. Left alone for long stretches of time each night, he pored over physics textbooks and wrote some research articles. The self-study paid off: Sirag, a college dropout, managed to publish not one but two original physics articles in the world-leading journal *Nature*. (Most PhD physicists who pursue ordinary academic careers never manage to get even one of their research articles through the rigorous peer review at *Nature*.) He followed those up with a few papers presented at national meetings of the American Physical Society in the 1980s, taking the occasional break from his nightly patrols.[22]

Sirag stayed in his night-watchman job for a decade. By scrimping and saving and living quite frugally, he and his girlfriend (now wife) managed to buy a house in the suburbs north of Berkeley. That was long before housing prices in the area began to skyrocket. Once the value of his house ballooned in the late 1980s, Sirag and his wife sold the house and moved to Oregon. A few years later, he picked up some more consulting fees from Jack Sarfatti's new patron (the Silicon Valley UFO buff). After their bitter squabbles in the early 1980s, Sarfatti and Sirag managed to patch things up. Sirag stayed with Sarfatti in San Francisco whenever he was in town, and he contributed a lengthy chapter to Sarfatti's memoirs.[23]

And so both Sarfatti and Sirag managed to land on their feet, buoyed by some well-timed interventions from "angel" donors.

•

Other members of the Fundamental Fysiks Group followed a different path. They morphed from PhD scientists with some university teaching experience to full-time writers. Fritjof Capra proved most successful at making the transition, building on the surprise success of his first book, *The Tao of Physics*. A later reviewer marveled that Capra's *Tao of Physics* "inspired so many imitators that he could be said to have started a genre." In short order, a dozen editions of the book appeared worldwide: four English-language editions published in the United States and United Kingdom, plus eight editions in various European and Asian languages. The number has continued to grow, up to forty-three editions in twenty-three languages at latest count.[24]

By the time he began shopping his next book project, *The Turning Point*, major presses lined up for a chance to publish his work. "The difference between *Tao of Physics* and *Turning Point* was amazing," Capra observed recently. *Turning Point* appeared in 1982 from the mega publisher Simon and Schuster, with a paperback edition from Bantam and another bevy of international and foreign-language editions following close behind. In that work, Capra returned to themes he had explored in *Tao*—the emergence of holism in recent science (such as quantum physics) and the breakdown of reductionist thinking—but moved the discussion more squarely into other fields, such as economics and ecology.[25]

If *The Tao of Physics* had introduced Capra to a generation of readers hungry to ponder links between modern physics and age-old questions of meaning and metaphysics, *The Turning Point* solidified Capra's position as a major New Age thinker, a "devotee of that grab bag of countercultural magic that has come to be called new-age philosophy," as one reviewer put it. The book courted a wide range of admirers. Hugo Chávez, the flamboyant and outspoken president of Venezuela, hailed Capra's *Turning Point* as a must-read as recently as 2006.[26] Capra wrote a screenplay based on the book with his brother, a filmmaker. Released in 1992, the movie *Mindwalk* featured three fictional characters: a quantum physicist, a politician, and a poet. An original score by the minimalist composer Philip Glass punctuated the characters' ruminations on science, holism, and a looming ecological crisis. "Though not exactly spontaneous and seldom witty," the *New York Times* film critic noted, "it is good serious talk, a sort of feature-length op-ed piece."[27]

With his reputation (and finances) secure, Capra moved more squarely into environmental activism. He followed *Turning Point* with a book on the burgeoning Green movement in Western Europe, *Green Politics* (1984); a combination memoir and intellectual autobiography, *Uncommon Wisdom: Conversations with Remarkable People* (1988); and a series of books on ecology, systems thinking, complexity theory, and sustainability—at least five books since 1990. Between 1984 and 1994 Capra directed a think tank devoted to ecological issues, the Elmwood Institute, based in Berkeley. In the mid-1990s, Capra and colleagues transformed the Elmwood Institute into the Center for Ecoliteracy, also in Berkeley, which encourages primary and secondary schools to incorporate ecology and sustainability in their curricula.[28] In short, Capra performed a rather remarkable transition from unemployed physicist to celebrity author and environmental activist in the space of just one decade.

Other members of the Fundamental Fysiks Group meandered their way toward careers as writers as well, though they never matched Capra's commercial success. Fred Alan Wolf found that path quite by accident. His cash flow had been interrupted when he resigned from his teaching post at San Diego State; the pension he had accrued of a few thousand dollars didn't last long. The consulting fees that Werner Erhard and George Koopman funneled through the Physics/Consciousness Research Group helped for a while, but after Sarfatti had his blowout with Erhard that source dried up as well. Wolf left the Bay Area to return to La Jolla. To save on rent, he moved into a commune that had been set up in a private house in a suburban neighborhood. A modest inheritance when his mother passed away helped for a while, but soon his financial situation became bleak.[29]

Wolf, who had left a tenured full-professor position, desperately tried to think of ways he could combine his passions for quantum physics and teaching, and make some money at the same time. He transformed himself into a kind of motivational speaker. He incorporated some of the things he'd learned from Erhard—"that 'positive-thinking' bullshit," as he put it recently—and crafted a live act that was part magic show (he had long been an enthusiastic amateur magician), part performance art, part quantum physics lecture, and part human potential slogans. He developed larger and larger tricks, such as flashing a strobe light on a bouncing ball on stage, to give his audiences some idea about quantum randomness and probabilities. He even gave himself a stage name: "Captain Quantum." In July 1979, Wolf (as Captain Quantum) landed a gig as the opening act for some sold-out lectures by psychedelics guru Timothy Leary, held in a 2000-seat auditorium in Los Angeles. A few months later a niche magazine devoted to occult sciences and science fiction, *Future Life*, ran a feature article about Wolf's act, including a cartoon of Wolf as Captain Quantum.[30]

Around that time Wolf got the idea to transform his act into a popular book. He had gotten a taste of the popular-book business a few years earlier, from his collaboration with Jack Sarfatti and Bob Toben on *Space-Time and Beyond* (1974). Wolf found, to his relief, that *Space-Time and Beyond* had sold well enough that he could attract his own literary agent and get a publisher's contract for the new project fairly quickly. (Ira Einhorn, the agent for the earlier book, was no longer available on account of the pending murder trial.) And so Wolf's first solo venture, *Taking the Quantum Leap: The New Physics for Nonscientists*, appeared in 1981.[31] Plugging the book on a local Seattle television morning talk show, Wolf brought along some dice and performed sleight-of-hand tricks. He emphasized for viewers that quantum probabilities, as random and mixed up as the throws of the dice, gave each person license to craft his or her own destiny. Just as the electron's path was not predetermined, so too could people make their own way in their "youniverse." (The lessons from Erhard seem clear.) Indeed, the show's host enthused that thanks to Wolf's book, "we can understand ourselves and our place in the universe, we can understand perhaps psychic phenomena better and perhaps even a concept of God more clearly with the help of the scientist."[32] The era of quantum therapeutics had arrived. As novelist and social critic Tom Wolfe had foreseen, the holistic and communal language running through so much New Age material could just as easily support a "me first" ideology.[33] Wolf's quirky book, full of cartoons and pithy aphorisms, connected with readers, generating some desperately needed royalty payments as well as critical acclaim. *Taking the Quantum Leap* received the American Book Award for nonfiction, beating out physicist Freeman Dyson's celebrated book *Disturbing the Universe*, which had been nominated in the same category that year.[34]

Since then Wolf has made his living from book deals and royalties. He published five more popular books about physics in the 1980s, writing about staple themes from his Fundamental Fysiks Group days, such as quantum physics and consciousness.[35] His 1991 book, *Eagle's Quest*, documented his effort to absorb the lessons of Native American shamanism, to complement his earlier excursions into Eastern mysticism. Since then he has published several more books exploring relations between quantum physics and spirituality.[36] Most recently, Wolf consulted on and appeared in the controversial 2004 film *What the BLEEP Do We Know?* Like many of Wolf's books, the movie built on fundamental mysteries of quantum physics to ask larger questions about free will, consciousness, and our perceptions of reality. Critics scoffed; one reviewer asked, in desperation, "do we have to indulge in bad physics to feel good?" Major film festivals refused to show it. And yet the quasi-documentary became an underground hit, earning more than $11 million in its first two years from late-night screenings and DVD sales. Though Wolf received no payment for his contributions to *What the BLEEP*, the film's success redounded to him as well. He found he could charge larger speaker's fees on his lecture circuit; sales of his latest books showed new signs of life; and his name (and face) recognition improved enormously, thanks to the brief animated cartoons that run throughout the film and the film's sequel starring Wolf as "Dr. Quantum." The cartoons were also made available to schools and posted on YouTube; at the time of writing, some have been viewed more than 1 million times.[37] Like Fritjof Capra, then, Fred Alan Wolf has introduced much larger audiences to the questions that first brought him to the Fundamental Fysiks Group more than three decades ago.

Of the Fundamental Fysiks Group members who turned themselves into writers, Nick Herbert was slowest to make the transition. Herbert had left his job at Smith-Corona Marchant in the mid-1970s to spend more time with his newborn son. He sank his life's savings into a down payment on a house in Boulder Creek, California, not too far from the Esalen Institute in Big Sur. He and his wife started a home school for their son and some of his friends, using facilities at a nearby YMCA. They bartered; they picked up odd jobs; Nick worked for a while as a dishwasher. But those jobs failed to cover the mortgage, so Herbert went on public assistance: a "Family Development Grant" in the parlance of the day, that is, welfare. In the early 1980s his friend from graduate-school days, Heinz Pagels—by then executive director of the New York Academy of Sciences and an accomplished popular-book author—introduced Herbert to a high-powered literary agent in New York. Herbert's first book, *Quantum Reality: Beyond the New Physics*, appeared in 1985.[38]

Quantum Reality was a distillation of Herbert's years of discussions and presentations in the Fundamental Fysiks Group and the Esalen workshops, pitched for readers with little mathematical background. He paid homage to Capra's *Tao of Physics* and Wolf's *Taking the Quantum Leap*, then focused even more squarely than the others on Bell's theorem and entanglement. His book contained detailed discussions of work by Fundamental Fysiks Group participants Henry Stapp and Philippe Eberhard, as well as by Esalen regular David Finkelstein. John Clauser's experiments on Bell's theorem—which Herbert had witnessed firsthand and discussed often with Clauser in the Fundamental Fysiks Group and the Esalen workshops—featured prominently, as did the more recent experiments by Alain Aspect in Paris. Herbert even included the sheet music for an original song, the "Bell's theorem blues." ("Doctor Bell say we connected. He call me on the phone. But if we really together baby, how come I feel so all alone?") But overall the book played it fairly straight, hewing close to quantum theory and rarely veering off to more mystical or paranormal associations.[39]

One reviewer crowed that Herbert's *Quantum Reality* had gone "a heroic distance in making quantum reality comprehensible. His book is filled with exciting moments when even a neophyte gets the feeling he is 'almosting it,' as Stephen Daedalus [*sic*] puts it in *Ulysses*."[40] The book's appeal was not limited to neophytes or James Joyce fans. A physicist and dean of science at a small college marveled that "Herbert's clever analogies and lucid explanations" were "outstanding, and should be borrowed by teachers of quantum physics." He urged his colleagues to adopt Herbert's "ambitious" book for classroom use, echoing advice from earlier physicist-reviewers of Capra's *Tao of Physics*. Indeed, when Herbert's book appeared, a decade after Capra's *Tao*, there still did not exist a single textbook on quantum theory that treated Bell's theorem. As late as the mid-1990s, Herbert's *Quantum Reality* still appeared on syllabi for undergraduate physics courses.[41]

Herbert's book clearly filled a niche. It sold well—more than 100,000 copies, not including sales of German and Japanese translations—and it remains in print a quarter century later.[42] With his agent's help, Herbert quickly followed up on the success of *Quantum Reality* with *Faster than Light: Superluminal Loopholes in Physics*, which came out in 1988. A Japanese translation was published the following year. In that book, Herbert marched through many of his own hard-fought efforts to harness quantum entanglement to send superluminal signals. He presented the arguments from his QUICK and FLASH papers in accessible form and provided clear explanations of the many important results those papers had instigated, such as the no-cloning theorem.[43] A third book followed soon after that, entitled *Elemental Mind: Human Consciousness and the New Physics*. There Herbert tackled most directly the questions of physics and consciousness that he had been exploring since the earliest days of the Fundamental Fysiks Group, including possible relations between Bell's theorem and parapsychology. He described in some detail his playful metaphase typewriter—the device with which he had hoped to contact the spirit of Harry Houdini back in 1974—as well as Evan Harris Walker's ideas about consciousness as a quantum hidden variable that had inspired Herbert's giddy experimentation. Though the book attracted some fans, it never matched the commercial success of Herbert's earlier books.[44]

Buoyed by the royalties and various odd jobs, Herbert was able to return his attention to faster-than-light schemes. His latest preprint ("ETCALLHOME: Entanglement Telegraphed Communication Avoiding Light-speed Limitation by Hong Ou Mandel Effect") circulated via the Internet in December 2007. An email announcing his latest effort teased, "O my! Another faster than light signaling scheme that's just begging for refutation." This time a physicist pointed out a hole in the argument within hours rather than months, but that only seemed to amuse Herbert more.[45] In between his active blogging he self-publishes witty physics limericks and racy "quantum tantra" poems about encountering bare Nature. Every now and then he makes the trip over to nearby University of California at Santa Cruz to catch the latest physics department colloquium.[46]

•

Both of the Fundamental Fysiks Group's founders, Elizabeth Rauscher and George Weissmann, followed a third path after the group disbanded. They became self-employed entrepreneurs. Rauscher didn't adopt that role right away. After completing her dissertation in 1979, she was hired on at the Lawrence Berkeley Laboratory as an adjunct assistant professor. She left after one year: she had grown tired of being the only woman on the physics staff at the laboratory and having to fight discrimination every day. Life had been difficult enough as a female graduate student at the lab; now the atmosphere seemed downright hostile.[47]

Rauscher picked up some adjunct teaching at nearby John F. Kennedy University, a small school in the Bay Area that offered extension-school courses for working adults.[48] She also founded her own consulting company, Tecnic Research Laboratories. She had worked on some Navy contracts as a graduate student at Berkeley, and capitalized on the experience to land some new contracts of her own. In short order, her consulting company was working on contracts from the Naval Surface Weapons Center and the Naval Ocean Systems Command to study ionospheric effects on signal propagation. A separate three-year contract came in from the major aerospace manufacturer Martin Marietta, as part of its production work on NASA's space shuttle program. Rauscher's tiny laboratory won a contract from Martin Marietta to investigate ways to increase the strength of joints between metallic parts made by plasma arc welding. Other companies hired her firm to consult on semiconductors and related electronics projects. She later picked up some teaching on the side as an adjunct at the University of Nevada, Reno, where she mentored several graduate students and published articles on atomic physics, including an item in the prestigious journal *Physical Review Letters*.[49]

Rauscher also expanded to biomedical topics. From her earliest work as a consultant to the Stanford Research Institute psi lab, she had been interested in phenomena like biorhythms and homeopathic (or "alternative") medicine. During the mid-1980s, she combined her work in sensitive electronics with these long-standing interests. She received three patents in the United States and one more in Europe for devices designed to use ultra-low-intensity electric and magnetic fields as a noninvasive pacemaker for keeping a patient's heart rate regular. The gentle vibrations were also designed to act on the brain to reduce pain.[50] While pursuing those products, Rauscher volunteered for several years as president of the Parapsychology Research Group (founded by remote-viewing physicist Russell Targ), and kept up a steady stream of writing about quantum physics and parapsychology.[51] More recently she has joined forces with the California-based nonprofit Institute of HeartMath to design what they call a "Global Coherence Monitoring System." Hearkening back to Rauscher's earlier Navy contract work, the group aims to track minute fluctuations in the earth's magnetic field, both on the ground and in the ionosphere. The goal is to determine "how the earth's field affects human heart-rhythm patterns or brain activity, and more importantly how human stress and emotions are influenced by fluctuations in the earth's field." The group aims to understand how "effects of collective emotion-based human energetics" can increase an individual's stress level; and to monitor whether shifts in the planet's "collective human emotionality" might be correlated with—and hence used to predict—earthquakes, volcanic eruptions, or "similar planetary scale events." Her interest in entanglement has truly gone global.[52]

The Fundamental Fysiks Group's other founder, George Weissmann, also caught the entrepreneurial spirit. After completing his PhD in Berkeley he returned to his native Switzerland for a postdoctoral fellowship at the Swiss Federal Institute of Technology, known as the Eidgenössische Technische Hochschule or ETH—Einstein's alma mater. The fellowship was in theoretical particle physics, but Weissmann spent most of his time trying to replicate psychokinetic experiments that he had first learned about prior to leaving Berkeley. While in Zurich he realized that his heart just wasn't in mainstream particle physics anymore. He heard about some traditional herbal remedies from Central Asia that a Tibetan family had begun to import into Switzerland. The herbal concoctions seemed to be wonder drugs, and Weissmann, who had long been interested in Tibetan Buddhism, was naturally curious. He returned to Berkeley and established the Padma Marketing Corporation to import and sell the product under the name "Padma 28 Tibetan Herbal Food Supplement."[53]

Setting up shop in 1981, Weissmann jumped into the nutrition supplements business early, just as it was ramping up into a multi-billion-dollar industry. While many of the food-supplement players at the time remained scattered, small scale, and quasi-underground, Weissmann built Padma into a national brand. Marketing materials at the time proclaimed that Padma 28 could treat atherosclerosis and improve blood circulation, lower cholesterol, and improve mental functioning such as memory and alertness. Later claims touted the product's efficacy at reducing asthma, skin allergies, hemorrhoids, depression, and more. Business was booming, and Weissmann soon introduced new products beyond the original twenty-two-herb recipe. Then the U.S. Food and Drug Administration stepped in. To officials at the FDA, the claims made on behalf of Padma 28 went well beyond the domain of health-food supplement; they qualified Padma 28 as a new drug, subject to strict regulations governing clinical testing, labeling, and marketing. Consumers Union, the nonprofit organization famous for publishing the magazine *Con-*

sumer Reports, also went after the product. They singled out Padma 28 as the single most dangerous product on the food-supplements market—not necessarily because the herbal remedy was harmful in itself (although they did raise concerns about one toxic ingredient), but because its marketing encouraged users to forgo professional medical treatment, even for life-threatening chest pain that could indicate a heart attack.[54]

In April 1986, the Food and Drug Administration ordered all shipments of Padma 28 to be destroyed. That made little impact on the herbal remedy's circulation, so the FDA took a bolder step the next year, issuing a permanent injunction against the product. Such injunctions were quite rare: the FDA issued only fourteen injunctions that year, out of thousands of investigations. But in the case of Padma 28, the feds noted that "the defendants were aware that their activities violated the law; and the government believed that, unless restrained by the court, the defendants would continue such violations."[55] Weissmann still believed in the remedy's benefits—indeed, biomedical researchers in Europe continue to investigate whether the compound might be effective against leukemia, hepatitis B, and other illnesses—and two years after the FDA injunction, advertisements for the product continued to circulate under the new name "Adaptrin." But the years of regulatory injunctions and court battles had taken their toll. Disheartened, Weissmann dissolved his company in 1989.[56]

The entrepreneurial bug proved hard to shake, and a few years later Weissmann started a new business with his son. Again Weissmann's interest turned to matters of health and nutrition, inflected with a certain New Age flavor. His son was a vegan and Weissmann a vegetarian, so they started a business to make realistic-tasting meat substitutes: "Veat," as in "vegetarian meat." Beginning in the early 1990s, Weissmann again built up a national brand. Soon the company had overseas production facilities, and its soy-based chicken substitutes were winning taste-test competitions at industry conventions.[57] Yet the day-to-day operations of managing a large company never held the same appeal for Weissmann as the excitement of the start-up phase. Feeling financially secure, he sold his interest in Veat around 2000 so he could work full-time on a book project he had been nursing since his days in the Fundamental Fysiks Group. The book project, which Weissmann has titled the "Quantum Paradigm," represents his decades-long effort to devise a unified theory of consciousness, parapsychology, and quantum theory.[58]

•

Two core members of the Fundamental Fysiks Group remained more closely associated with mainstream physics, though each has endured reminders of their unusual (at times marginal) status. Henry Stapp, one of the few frequent participants in the group who had a regular position as a physicist at the time, has continued to work on quantum mechanics and consciousness from his post as senior staff scientist at the Lawrence Berkeley Laboratory.[59] After years of publishing articles on Bell's theorem and the deep meaning of nonlocality, he was approached at a conference in the early 1990s by a physicist-turned-parapsychologist. The researcher wanted to know why Stapp had never cited any of the parapsychologist's experiments on extrasensory perception or psychokinesis, phenomena (so the researcher continued) that seemed manifestly similar to—perhaps even the result of—quantum entanglement. Stapp replied with some frankness that he had never examined the parapsychologist's experiments very closely and was inclined to assume that any statistically significant results claimed on behalf of psi effects arose from errors of experimental design if not "outright fraud." But it was one thing, Stapp continued, to harbor doubts about a set of experiments and quite another to invest the time and effort to investigate or replicate the experiments himself. Thus Stapp had elected to say nothing at all in public about the experiments.[60]

Sensing, perhaps, that Stapp's reticence stemmed more from pragmatism than hard-set ideology, the parapsychologist proposed a simple, new test that the two could conduct together. Stapp would never have to leave his Berkeley office; he would never have to interact directly with any human subjects (clairvoyant or otherwise); and he could design procedures and protocols to satisfy himself that no chicanery was at play. Stapp's interest was piqued. He had sat in the midst of the Bay Area's psi efflorescence for decades; and here was a parapsychologist who seemed open to a fair, transparent test. Stapp agreed to the partnership.[61]

They hatched an elaborate plan whereby the parapsychologist, working in Europe, would mail batches of cardboard sheets to Stapp. The sheets had rows upon rows of numbers covered by thick black tape. Upon receiving a package, Stapp was to wait for several days, look up the weather report from the *New York Times* for the intervening period, and extract from the newspaper—by a procedure known only to him, never communicated to his faraway lab partner—a pair of randomly selected numbers. After that (in what came to resemble a cross between Rube Goldberg contraption and John le Carré spycraft), Stapp fed those weather-related numbers into a computer program of his own design, which in turn spit out a new series of data: random strings of plus signs and minus signs. For each plus sign in a series, Stapp removed the black tape from a number in one column of the sheets sent from the parapsychologist; for each minus sign, Stapp removed the tape from a number in the other column. Further machinations ensued, until Stapp had reduced the jumble of numbers into a series of signs, plus and minus. If the signs were truly random, then their sum should average to zero: like flips of a coin, one should find as many heads as tails on average. Stapp's job was to perform standard statistical tests on his rows of plus signs and minus signs to check whether their sum showed any bias, deviating systematically from zero.[62]

And so it went for months at a time. The parapsychologist, for his part, recruited people whom he considered especially sensitive or gifted with parapsychological powers—he tended to use children and teenagers who studied martial arts—and asked them to focus all their attention on nudging the still-unexamined numbers under the black tape in Stapp's office toward a positive bias. That is, their task was to apply psychokinesis backward in time, to change the already-printed-but-not-yet-observed numbers that lay under the thick black tape in Berkeley, such that Stapp's numerical recipe would produce more plus signs than minus signs.

To Stapp's great surprise the sum of pluses and minuses did indeed trend positive. Analyzing the first half of the data—he and the parapsychologist had agreed ahead of time how many sheets they would run through over the course of the experiment—Stapp's series of signs deviated from zero by two standard deviations. The likelihood that such a bias would occur merely by chance was less than one out of twenty. As it happens, I was an undergraduate intern at the Berkeley laboratory around this time, and I fondly remember discussing the experiment with Stapp in his office. When he came to the statistical results his eyes lit up, like a seasoned storyteller spinning a ghost story around a campfire. Such results, if they continued to hold up, would surely require some sort of explanation.[63]

Intrigued, Stapp thought hard about how one might accommodate such a "causal anomaly" within established principles of physics. He turned his attention to a recent suggestion from a Nobel laureate about modifying the basic equations of quantum mechanics, which had been put forward for quite a different reason. Stapp realized that the modified equations could account for effects like those in the recent parapsychology experiment while still reproducing the usual, well-tested behavior of atoms predicted by ordinary quantum theory. With his new theoretical model worked out, Stapp wrote an article on his findings. The article began by describing the experiment and its empirical results before introducing the new theoretical ideas that were meant to explain them, following what Stapp considered to be the appropriate format for any theoretical article that aimed to account for experimental findings. He mailed off his paper to the *Physical Review* in March 1993, as he had done scores of times during his long career, and waited to hear back from the referees. That's when the trouble started.[64]

The referees and editor accepted Stapp's paper. After a little while, however, Stapp received a second letter from the journal editor. Upon further consideration, the letter explained, the editor suggested that Stapp rearrange the paper, relegating details of the experiment to an appendix at the end. The goal was to shift the emphasis from parapsychology to theoretical physics. Bemused by the request—wasn't the point of theoretical physics to explain empirical results? Stapp asked himself—Stapp nonetheless complied, and submitted a revised version of the paper that September. Still feeling uncomfortable, the journal editor sent the revised paper back to a referee, a most unusual step to take after already accepting an article. The referee, too, was stymied. "I have had to think rather long about this one," the new report began. It continued:

> I as a referee am faced with the following problem. Here is a theory of ESP, which the author claims is consistent with an amended quantum theory. It seems to me he has every right to both create and publish such a theory. This part of the paper is for me not in question, either as to its appropriateness or result. But to present the included experiment as though it was a physics experiment is misleading, and will provoke an unpleasant response from the physics community. I thought putting it in an appendix would help, but now I am not sure. . . . I think this paper may well bring forth a flood of crackpot contributions from others of clearly dubious objectivity, who do not have the training of this author.

The referee apologized for "backtracking," but qualified that "it's no favor to him [Stapp] to allow him to publish something that will needlessly arouse the ire of everybody, without first trying to make it as palatable as possible."[65]

The editor agreed with the referee's advice, and offered Stapp a deal: if Stapp would remove *all* details of the experiment and cite a separate published account by the parapsychologist, then he could publish his theoretical model all on its own. Alternately amused and aggravated by the exchange, Stapp accepted the editor's offer. Nearly a year and a half after its initial submission, Stapp's heavily revised article appeared in the *Physical Review*.[66]

Getting the paper into print proved to be only half the battle. Six months after his paper appeared, the editor-in-chief for all divisions of the *Physical Review*—the senior editor for the entire American Physical Society, who had not been involved with Stapp's submission—sent Stapp a long and agitated letter chastising Stapp's work and regretting that the paper had ever been accepted. He granted that Stapp and others were "legitimately interested in such matters as human intervention in experiment, or, even, of the nature of thought and its relation to physical 'reality.'" But, the editor continued, "at the present time such ideas belong to the world of philosophy, not to the world of physics." Stapp's gravest offense, as the editor saw it, was lending credibility to parapsychologists' claims.[67]

The discussion spilled beyond private correspondence. *Physics Today*, a trade magazine mailed to all professional physicists in the United States, ran not one but two batches of overheated letters that had been sparked by publication of Stapp's article. Some held up Stapp's paper as a sign of all that was rotten in the state of physics. Others bristled that changing editorial and refereeing policies at the *Physical Review*—as the editor-in-chief had promised to do, in direct response to the Stapp episode—amounted to ideological censorship.[68]

And what of the empirical results, the mysterious bias in the long strings of plus signs and minus signs that had captured Stapp's attention? By the time the entire experiment had run its course, and Stapp had processed all of the agreed-upon sheets from the parapsychologist, the statistical effect had entirely washed out. The average of all those plus signs and minus signs averaged to zero, just as Stapp had expected in the first place. Try as they might, the young karate kids in Europe had not been able to nudge Stapp's sums after all.[69]

Like Henry Stapp, John Clauser managed to craft a career as a working physicist, though he has had to wander, institutionally, quite a bit further than Stapp. Even after publishing the first-ever experimental test of Bell's theorem during his postdoctoral fellowship at Berkeley in the early 1970s, Clauser could not secure an academic physics job. During the months and years after his major article appeared, supporters feverishly wrote letters to befuddled physics department chairs across the country, trying to land Clauser a tenure-track position. The department chairs needed reassurance that Clauser's work, or, indeed, the topic of Bell's theorem and entanglement, constituted legitimate research. "I believe he shows promise of becoming one of the most important experimentalists of the next decade," wrote one supporter, "in spite of the fact that Clauser's results spell trouble for my own pet theory." But it was to no avail. Clauser never did secure an academic position; nor did the Lawrence Berkeley Laboratory make much effort to keep him after his postdoctoral appointment expired.[70]

Throughout the travails, Clauser found some much-needed support in the Fundamental Fysiks Group. "No subject was off limits," he recalled recently, and "no subject was summarily dismissed as pure quackery." That attitude contrasted sharply with the other messages he received around the laboratory. "I was given the distinct impression that, in the opinion of the faculty, my work did not amount to any real new physics, was well outside of acceptable 'mainstream' inquiry, . . . and probably amounted to a major waste of time and resources," he explained. What a relief when the Fundamental Fysiks Group drummed up another "very lively discussion" each Friday afternoon. "We definitely had a lot of fun. We asked some fundamental questions, and I think we got some reasonable answers, too."[71]

In the mid-1970s, with his appointment at Berkeley winding down and no nibbles of interest from the academic job market, Clauser moved to nearby Lawrence Livermore Laboratory, the defense laboratory dedicated mostly to nuclear weapons research at which Elizabeth Rauscher and Fred Alan Wolf had previously worked. He joined a group working on nuclear fusion for (civilian) energy production, a long-sought means of safely reproducing the nuclear reactions that power the stars, to generate energy here on Earth. Through that work, Clauser cultivated skills with a new set of experimental techniques. Whereas he had earlier worked on optical experiments, he now shifted to manipulating X-rays.[72]

About a decade later he tried to set off on his own, intending to follow an entrepreneurial path as Elizabeth Rauscher and George Weissmann had done. Clauser realized that he could tweak some techniques for high-precision X-ray imaging to design a tool that could identify underground deposits of oil and natural gas. His clever idea suggested that one could survey huge fields at one go, tens of kilometers on a side. He filed for patents and briefed executives at big oil companies. But nothing came of his efforts. In hindsight, he suffered from poor timing: the oil crisis of the early 1970s had faded from most executives' view, and they felt flush with the oil deposits that had already been identified. No one at the time saw much need to invest in unproven techniques to hunt for unknown oil reserves. Once again Clauser had to scramble to reinvent his career.[73]

His next step was to move into medical imaging. Still fascinated by X-rays, Clauser tweaked yet another experimental technique he had mastered at Berkeley and Livermore. He designed a new method for high-contrast imaging. By exploiting the precision interference techniques he had honed in other contexts, Clauser could produce vast improvements in sensitivity for radiography, especially in soft tissues. He converted his garage at home to a makeshift laboratory and began to apply for new patents. He also applied for research grants from the National Cancer Institute, part of the National Institutes of Health. This time his luck began to improve. He secured a string of research grants to develop noninvasive imaging tests for breast cancer. On his own, in his garage, scrounging for spare parts, Clauser designed a "passive biopsy": a means of identifying various masses or growths inside the body without having to intervene surgically.[74]

In his roundabout way, Clauser thus fashioned a research career in medical physics. He kept up his interest in Bell's theorem and quantum entanglement, and spoke often at conferences during the 1980s and (especially) 1990s, after the topic moved squarely into the mainstream. Four decades after he first latched onto Bell's theorem and began doggedly pursuing experimental tests, Clauser's efforts received commensurate recognition when Clauser and two other physicists won the 2010 Wolf Prize in Physics. Many physicists consider the Wolf Prize, awarded each year by a private foundation in Israel, to be second in prestige only to the Nobel Prize. Clauser shared the $100,000 prize with two other entanglement pioneers: Alain Aspect, whose experiments in Paris during the early 1980s had further captivated members of the Fundamental Fysiks Group; and Anton Zeilinger, the Vienna-based physicist who masterminded the 2004 quantum-encrypted bank transfer (among other things). At long last Clauser's many years of hiding in the metaphorical closet, trading ideas about Bell's theorem with other members of the Fundamental Fysiks Group, paid off.[75]

•

In the late 1990s, decades after the Fundamental Fysiks Group had broken up, George Weissmann and Elizabeth Rauscher struck up a conversation. Feeling nostalgic for the old days, each shared with the other how important the group had been in their lives, an importance they had only come to appreciate with the passage of time. They decided to organize a reunion. They would bring the old group back together to reminisce, to trade stories of what they had done since those days, and to share their latest ideas about physics and consciousness. Equally important, Weissmann and Rauscher realized, the reunion should commemorate what their ragtag group had stood for, what it had achieved.[76]

Just like in the old days, they set to work. Henry Dakin, the toy manufacturer and onetime patron of the spin-off Consciousness Theory Group, offered to host the reunion in his office space in San Francisco. Rauscher and Weissmann tracked down mailing addresses, telephone numbers, email addresses, and fax numbers. They invited their past friends and colleagues to reconvene for a twenty-fifth anniversary meeting of the Fundamental Fysiks Group, to be held on November 18, 2000.[77]

In their invitation, Rauscher and Weissmann wrote poignantly about "those heady days," back when the group had been active. "At a time when we had been very lonely in our individual investigations," the Fundamental Fysiks Group had provided "a community of kindred spirits which acted as a resonance board and source of both constructive critique and encouragement for our ideas." Even more: the group had "provided the first peer group in which we were free to truly pursue our ideas beyond conventional limits, in which we could dare to express our deepest and boldest thoughts."[78]

Their peers responded in kind. Nearly all the core members from the old days attended. In addition to Rauscher and Weissmann, John Clauser and Nick Herbert reconnected with Jack Sarfatti, Fred Alan Wolf, and Saul-Paul Sirag; Fritjof Capra traded stories with Russell Targ. Each person spoke for fifteen minutes on life and work since the Fundamental Fysiks Group, followed by a roundtable and free-form discussion among group members. An audience of a few dozen more observed the proceedings, some of whom had been occasional participants in the Fundamental Fysiks Group in the past, others of whom had begun to follow individual members' work since that time.[79]

Elizabeth Rauscher opened her presentation by projecting a series of photographs of group members from the 1970s. The photos elicited just the mix of nostalgia and wonder—"how could we ever have looked so young?"—that one might expect. But she concluded on a more serious note. Standing there before her former friends and colleagues, Rauscher intoned her articles of faith:

> What I conclude about life so far is this: The search for truth is the fundamental driving force behind me and what I do. The telling of truth has gotten me in a lot of trouble. I believe in nonlocality; I believe nonlocality is real. Quantum mechanics is probably very fuzzy stuff. Reality is better described by more than four dimensions. Most of everything, I think, is spirit, and a little is condensed out as matter. I believe in remote viewing, precognition, psychokinesis—because I did it—remote healing effects on at least bacteria systems, electromagnetic effects on biological systems. UFOs are a question mark. Life and death exist. And interestingly enough, it only takes one experiment to find out that you believe in life, because it only takes the birth of one child. And I have to tell you the truth: ghosts are real. I and another person— my business partner, colleague, co-patenter, and husband—also saw the same ghost, and described it the same way. We can have peace, love, and joy on this planet, instead of war, crime, and violence. Instead of warships, we need peaceship.[80]

A fitting credo for the Fundamental Fysiks Group.

Ideas and Institutions in the Quantum Revival

If a culture cannot afford an area in itself where pure nonsense happens, and where it is not practical, it has no objectives, it was for no reason whatsoever ... then this culture is dead.

—**Alan Watts, 1967**

We need some periods of anarchy when new or irreverent thoughts and changes can come forward.

—**Stephen Wiesner, 2009**

B rimming with ideas, the friends gathered for another session of their informal discussion group. Ever frugal—none could find regular employment as a physicist—they sat around a sparse table in their favorite café for another evening of spirited discussion. Driven by broad philosophical questions and desperate to make sense of the equations of modern physics, they read more widely than their teachers or textbooks had encouraged. Sometimes they switched gears, shifting from highbrow philosophical prattle to the nuts and bolts of machines, both real and imagined. Unhindered by academic pretense, they asked open-ended, even childlike questions. Sometimes their answers seemed downright silly, even to themselves. But still they kept at it, pounding relentlessly on the accepted interpretations, always searching for the weak spots in the edifice of modern physics.

Just another session of the Fundamental Fysiks Group? No. This group had convened seventy years earlier in Bern, Switzerland, under an equally grandiose name: the "Olympia Academy." A young Albert Einstein had founded the group. Physics degree in hand, Einstein had been unable to land an academic position. Perched outside the mainstream,

Einstein and his discussion-mates kept up their intellectual assault on electrodynamics and early atomic theories, honing a vision that would help launch the twin revolutions of relativity and quantum theory.[1]

Is that the answer? Did the Fundamental Fysiks Group succeed in transforming our ideas about quantum theory and communication because they arranged their group in a fashion similar to Einstein and company's? Alas, no. Physicists have come to know the "Einstein syndrome" all too well. Despite earnest protestations from unsolicited correspondents, not every person with a background in physics and a penchant for asking bizarre-sounding questions is the next Einstein.[2]

I got a taste of the phenomenon as an undergraduate. A curious individual rode into town on his bicycle, took a job as a dishwasher at the local inn, and began holding court on the campus green. He had convinced himself that the entire universe was one big plutonium atom, Earth just a tiny electron orbiting around the cosmic nucleus. He became so enamored of his theory that he legally changed his name to Ludwig Plutonium. Like members of the Fundamental Fysiks Group, he had been trained in physics and mathematics, and had even spent some time as a teacher. Like them, he turned to some unconventional means of spreading the word, largely by taking out full-page advertisements in the student newspaper, densely filled with equations.[3]

We are back to philosopher Karl Popper's dilemma: how to separate legitimate science from the many pretenders. Popper and most philosophers looked to the world of ideas, hoping to isolate some special intellectual approach or scientific method that might plainly distinguish authentic science from the chaff. Historians and sociologists largely agree that the philosophers' quest came up empty. Nor have alternative schemes, sought in the realms of institutions or social relations, fared any better.[4] The three examples above—Einstein and the Olympia Academy, members of the Fundamental Fysiks Group, and Ludwig Plutonium—make clear that marginalization bears no necessary relationship to Popper's long-sought demarcation criteria. Geniuses like Einstein have been shoved to the margins; but so too have other eccentric characters, whose ideas have rightfully been ignored.

Instances of marginalization will not solve Popper's demarcation riddle. But examples like the Fundamental Fysiks Group can teach us other important lessons about how science works. The hippie physicists' exploits illuminate the relationship between science and counterculture, the intertwining of ideas and institutions, and the ultimate roots for the revival of interest among today's leading quantum physicists in questions that had once been derided as "mere philosophy."

●

Neither the next Einsteins nor crackpots who could just be ignored, members of the Fundamental Fysiks Group toiled on the margins of physics during an unusually turbulent time. Amid the worst budget crunch and jobs crisis the field had seen for decades, the physics discipline's boundaries became especially plastic. The Cold War template had been challenged. Ways of doing physics and models for being a physicist were up for grabs again. Several group members grew their hair long; they explored the psychedelic drug scene; they daydreamed about quantum physics while reclining, naked, in Esalen's hot spring baths. Yet still they remained part of "real" physics. The Fundamental Fysiks Group's stubborn tenacity—our inability to dismiss the group from the court of science, even as members stretched the typical role of the physicist—provides insights into the wider phenomenon of science and the counterculture.

The American counterculture attracted its first chronicler from the academy rather early in its course. Princeton-trained historian Theodore Roszak tried to capture some of the ferment in his 1969 study, *The Making of a Counter Culture*. He grabbed a term ("counterculture") that had bounced around among academic sociologists and used it to brand the still-inchoate youthful rebellion. By that time journalists had already devoted dozens of breathless stories to stunts like the January 1967 Be-In in San Francisco's Golden Gate Park, or the carnivalesque antiwar march that summer led by Yippie protesters Abbie Hoffman and Jerry Rubin, in which tens of thousands of dazed hippies tried to exorcise evil spirits from the Pentagon by using their communal energy and hip vibe to levitate the massive building, hoping to hasten the end of the Vietnam War.[5]

On first blush, Roszak seemed to blame technocracy for the rise of the counterculture. At the time he was writing, technocracy—the unquestioned rule of elite experts toiling away at their jobs, machinelike, while modern society careened further from its values or moorings—seemed a dystopia made real. How else could one characterize Defense Secretary Robert McNamara's by-the-numbers prosecution of the war in Vietnam, with spreadsheets of abstract kill ratios camouflaging the bloody realities on the ground? But Roszak pressed further. On his reading, the counterculture emerged not just as a rejection of technocracy, but of the entire "scientific world-view of the Western tradition." The flower children aimed not just to stamp out the rule of experts, Roszak concluded. Their goal was nothing less than the "subversion of the scientific world view" itself. To Roszak, the disaffected youth—wrapping themselves in their beads, buckskin, and bell-bottoms, swooning from one Eastern or occult enthusiasm to the next—fled from science "as if from a place inhabited by plague."[6]

As the 1960s and 1970s have receded to the domain of history, a new generation of historians has tried to make sense of the counterculture and New Age movements. By and large they have echoed Roszak's analysis. Whatever these movements might have sprung from, these historians agree, the counterculture and New Age revival represented a flat rejection of modern science. The hippies and flower children craved authenticity, spontaneity, and experience, we are told: the enlightenment of Eastern mystics rather than the Enlightenment of Western science.[7]

Despite this fast-hardening conventional view, however, we now know that leading lights of the counterculture and New Age movements were anything but antiscience. In fact, quantum theory and Bell's theorem served as intellectual anchors for many New Age speculations. Icons of the counterculture—from *est* inventor Werner Erhard to Esalen founder Michael Murphy, psychedelic proselytizers Ira Einhorn, Timothy Leary, and more—proved to be eager patrons of the new physics. Rather than reject modern science they actively sought it out, paying handsomely for the privilege to learn about the latest developments.

The Fundamental Fysiks Group certainly benefited from this countercultural appetite for modern physics. Others did as well. During that turbulent time, similar groups sprang up, as loose and informal as the Fundamental Fysiks Group, whose members likewise chased the deep, metaphysical questions that their mainstream scientific colleagues had overlooked or ignored. Consider, for example, the "Dynamical Systems Collective" (also known as the "chaos cabal"), a bunch of physics graduate students at the University of California at Santa Cruz. Toiling in the lean years of the 1970s, well after the physicists' Cold War bubble had burst, they strove to understand the onset of chaos, the knife-edge division between order and disorder. Like the Fundamental Fysiks Group's obsession with Bell's theorem and quantum entanglement, chaos theory offered a juicy set of questions that could be pursued on the cheap. No need for megabuck particle accelerators; a leaky kitchen faucet would do, its maddeningly irregular *drip, drip, drip* inspiring improvisational inquiries into mathematical patterns behind seemingly random phenomena. In their own way, the Dynamical Systems Collective—never more than a clubhouse name for the musty old beach house, littered with beanbag chairs and secondhand furniture, in which the graduate students gathered to brainstorm—sought the same kinds of hidden connections and holistic convergences on which the Fundamental Fysiks Group had fastened. They, too, let their imaginations run wild, leaping from the mathematics of chaos to questions about determinism, free will, and the nature of thought and intelligence. No surprise, then, that the two groups found each other. A representative from the Santa Cruz collective told the Fundamental Fysiks Group all about their research into chaos at the first Esalen workshop on physics and consciousness in 1976.[8]

The Santa Cruz graduate students were able to pursue their passion with a new tool. They had latched onto some of the earliest programmable personal computers, bulky machines much closer to do-it-yourself hobby kits than the sleek consumer products so familiar today. Yet even those boxy computers embodied a countercultural ethos. They were machinic ideals, the first computers designed for personal use. Another set of Bay Area dreamers, inspired by the flowering counterculture and setting up shop in Silicon Valley, wondered what might happen if electronic computers could be something other than the room-sized mainframes that powered the military-industrial complex. What if everyday people could access limitless information, harnessing computers for personal exploration and play? "Small is beautiful" became their rallying cry; "information wants to be free" their new mantra. These counterculture visionaries—people like the *Whole Earth Catalog*'s Stewart Brand and the founders of Apple Computer, Steve Jobs and Steve Wozniak—were hardly antiscience technophobes.[9] Nor were their kin among the "appropriate technology" enthusiasts, Bay Area biotechnology pioneers, or environmental activists.[10] Few if any among these groups fit the neo-Luddite profile whom Roszak and others assumed were filling out the counterculture ranks. Alongside the Fundamental Fysiks Group, cutting-edge science and technology found an easy place within the burgeoning counterculture.

•

The Fundamental Fysiks Group and similar offshoots from mainstream science flourished during the 1970s. It was an unusual time. The Cold War–era institutions, in which most group members had expected to make their careers, sustained their first serious challenge since the end of World War II. More than just budget lines were at stake; the mammoth symbols of science-based modernity had lost their sheen. Roszak was certainly correct that most young hippies looked with disdain upon the once-mighty signifiers of big science, the nuclear reactors, particle accelerators, and hulking computer mainframes. Few among them supported the massive federal outlays that had paid for the huge machines, or paid to train generations of scientists and engineers to use them.[11] Yet the momentary pause in the Cold War drive for bigger, better, faster— the 1970s hiccup in the progress narrative of modernity—was hardly a dead time for science. Surprising flowers sprang from cracks in the infrastructure. They were small in scale, the kind of thing that could be financed by private patrons rather than the Atomic Energy Commission or Department of Defense. But they were long in legacy, rescuing crucial insights like Bell's theorem and quantum entanglement from near-total obscurity.

The Fundamental Fysiks Group championed Bell's theorem and the "spooky actions at a distance" inherent in quantum theory at a time when most physicists still ignored the interpretation of quantum mechanics altogether. As we now know, their intuitions about the deep significance of Bell's theorem paid off. And yet a topic whose importance strikes so many physicists today as self-evident had been shunned for a full decade after Bell's original publication, castigated as less than "real" physics. Bucking the trend, members of the Fundamental Fysiks Group sought out Bell's theorem and wrestled with it, quickly coming to dominate world-wide publications on the topic. Indeed, an early review article on Bell's theorem by group member John Clauser has become a classic, cited nearly 800 times in the scientific literature.[12] Clauser's detailed article grew directly out of his years-long engagements in Fundamental Fysiks Group discussions and the Esalen workshops.

Members of the group worked relentlessly for years on end to make sense of Bell-styled nonlocality, to sit with the quantum weirdness and follow where it might lead. Spurred by the group's creative and spirited pursuit, physicists began to unpack exactly how the beguiling quantum connections could be squared with Einstein's relativity. The delicate interplay between quantum entanglement and relativity inspired one physicist—a close friend and collaborator of group member John Clauser's—to coin the phrase "passion at a distance," a kind of Abelard and Heloise of the microworld.[13] Getting all that straight—"passion" but not "action"—sprang directly from the Fundamental Fysiks Group's efforts and instigations. In fact, *every single one* of the now-standard responses for how to accommodate Bell-styled nonlocality with Einstein's relativity came either from participants in the Fundamental Fysiks Group or from other physicists' concerted efforts to comprehend or critique their ideas. Recent breakthroughs like quantum encryption, meanwhile, rest on the bedrock of the no-cloning theorem, itself a direct offshoot of the group's brainstorming sessions.

The physicists in the Fundamental Fysiks Group were not alone in their quest. They hashed out many results in dialogue with a handful of others who had likewise ignored prevailing fashions and delved deeply into the interpretation of quantum theory: physicists like John Wheeler, Abner Shimony, Eugene Wigner, Gerald Feinberg, Bernard d'Espagnat, and Heinz Pagels. The group's missives ricocheted off other physicists as well—younger physicists like Alain Aspect, Wojciech Zurek, Bill Wootters, Dennis Dieks, GianCarlo Ghirardi, and Franco Selleri—often with astounding results. Several popular books written by members of the Fundamental Fysiks Group, meanwhile, became publishing sensations, bringing news of Bell's theorem and entanglement to far larger audiences.

In the years since the Fundamental Fysiks Group was active, topics like Bell's theorem and quantum entanglement have moved squarely to the center of legitimate physics. With good reason, Louisa Gilder's recent book about Bell's theorem, *The Age of Entanglement,* carries the telling subtitle *When Quantum Physics Was Reborn.*[14] A host of breakthroughs followed in the wake of Bell's theorem: not just quantum encryption, but quantum computing, quantum teleportation, indeed the entire kit-and-caboodle of quantum information science. Today it is no longer uncommon for Nobel laureates to debate the interpretation of quantum theory. The latest textbooks foreground topics like Bell's theorem and quantum entanglement; most now include whole chapters on the still-evolving contests over how best to interpret the quantum formalism. Physicists continue to revise curricula to strengthen students' conceptual grasp of quantum weirdness, including nonlocality and its uneasy coexistence with Einstein's relativity.[15] Where once influential leaders of the field had castigated philosophy as a waste of time—even when it came to plumbing the deep mysteries of quantum mechanics—the latest journals, conferences, and books on quantum information science feature contributions from card-carrying philosophers alongside those from professional physicists. Even the journal *Foundations of Physics,* once a quirky and fledgling venture that published items like Nick Herbert's FLASH scheme and Abner Shimony's null-result on quantum telepathy, is currently edited by a Nobel laureate.[16]

Given the astounding importance of Bell's theorem and quantum entanglement today, why has the Fundamental Fysiks Group been written out of physicists' history? When experts in quantum information science mention "LSD" these days, they don't mean acid trips at Esalen. Invariably they have in mind the Lloyd-Shor-Devetak theorem, which stipulates an upper limit to the rate at which one person could send quantum information to a receiver using a quantum channel.[17] More generally, when physicists look back on the sea change of the 1970s and 1980s—the era when working on the interpretation of quantum theory slowly gained legitimacy again, a toehold on the terrain of "real" physics—most credit the spate of new experiments rather than any hippie enclave. It was one thing to twist one's mind up in knots over quantum weirdness, so this line of thinking goes, but quite another to confront honest-to-goodness data in the laboratory.[18]

That line of reasoning makes sense; physicists often argue that theirs is an empirical science, driven first and foremost by experiments. Yet if experiments drove the revival of work on interpreting quantum mechanics, then some portion of the Fundamental Fysiks Group's story should be familiar today. After all, group member John Clauser sweated through the world's first laboratory test of Bell's theorem back in 1972. He went on to assist several other experimentalists who sought to conduct their own tests. Likewise, Alain Aspect, whose brilliant experiments in 1982 with time-varying switches managed to close a potential loophole in Clauser's original tests, interacted with several members of the Fundamental Fysiks Group. Not only did he inherit critical pieces of equipment from Clauser's Berkeley lab, but he was also coached on the topic by physicists like Olivier Costa de Beauregard and Bernard d'Espagnat, in between their visits with the Fundamental Fysiks Group and participation in Esalen workshops. If experiments played a decisive role in legitimating interpretive work, then surely the Fundamental Fysiks Group deserves some modest portion of the credit.

But the argument that physics is a science driven by experiment is not without controversy. Many of the bitter debates over string theory in recent years, for example, have turned on the contested role of experiments (or the lack thereof) in moving the field forward.[19] One might also wonder why, if physics is an experimental science, editors at the *Physical Review* forced Fundamental Fysiks Group member Henry Stapp to rewrite his paper in the early 1990s, removing all reference to the experiment that had gotten him thinking about modifying the equations of quantum theory.

More to the point: as John Clauser, Alain Aspect, and their colleagues learned the hard way, experiments can't force the community to pay attention to them.[20] No one welcomed Clauser's pathbreaking experiment on Bell's theorem back in 1972—virtually no one outside the Fundamental Fysiks Group, at any rate. When Clauser's colleagues tried to replicate the experiment a few years later at a different university, they found their requests for funding denied.[21] For years after his momentous experiment (and several clever follow-ups), Clauser received little credit for his efforts. Consistently passed over for academic jobs, Clauser never caught a break in the physics job market. Prominent colleagues lobbied hard to convince department chairs across the country that Bell's theorem constituted legitimate physics, and that Clauser was at the top of the game when it came to laboratory skill and creativity, but to no avail.

Even the breathtaking experiment that Aspect completed in Paris a decade later failed to do the trick. Like Clauser, Aspect found few physicists who showed much interest at the time. In fact, he stopped working on topics like Bell's theorem soon after completing his now-famous experiment. He struck up an entirely new line of experiments—on laser cooling of atoms—in what he figured would be a more fashionable field. For years after he had completed what is now considered the most important experimental demonstration of quantum nonlocality, Aspect mostly found himself discussing his work with paranormal enthusiasts (for whom he had little patience) rather than fellow physicists. At one point he attended a private meeting in Paris with someone who claimed to enjoy special powers such as extrasensory perception. Aspect recalls being astonished by what he considered the naïveté of his fellow observers, who numbered among Paris's elite bankers and other professionals. Years later, upon receiving the gold medal from France's CNRS in 2005, Aspect was invited to bring one guest to the award ceremony. He chose Gérard Majax, a French magician who has adopted the same mantle as James "The Amazing" Randi: self-appointed debunker of paranormal claims.[22]

Other physicists of Clauser and Aspect's generation experienced the same frustrations. Anton Zeilinger, the Viennese physicist who shared the 2010 Wolf Prize with Clauser and Aspect, and whose extraordinary experiments include the 2004 quantum-encrypted bank transfer, recalled similar hostility from his colleagues when he began his experiments on Bell's theorem and quantum foundations. "There was a widespread negative attitude of the scientific community towards foundational work of that kind," he explained. John Bell himself had tried to warn the young Zeilinger, much as he had cautioned Aspect. "In my early talks about fundamental physics experiments I could sometimes feel the dislike of such work by the older members of the audience sitting in the front row," Zeilinger wrote. The disrespect shown toward serious efforts to interpret quantum theory—even when driven by ingenious and original experiments in the laboratory—made life difficult for Zeilinger early in his career, much as it had done for Clauser and Aspect.[23]

Beyond Clauser's, Aspect's, and Zeilinger's individual experiences, publication patterns reveal larger trends within the field. Citations to Bell's theorem in the literature remained flat following publication of Aspect's experiments. As late as 1990, contributions on the subject were still dominated by the small circle of physicists who had already begun working on Bell's theorem before Aspect completed his experiments. No stampede of physicists rushed to work on the interpretation of quantum theory in the wake of the new experiments.[24] Only many years later, after interpretive work had returned to the fold, did experiments like Clauser's, Aspect's, and Zeilinger's assume a retrospective importance in physicists' reconstructions—an importance I believe they richly deserved, but which none of these experimentalists enjoyed at the time.

●

The return of interpretive work was no simple response to new data or experiments. Institutions, not experiments, proved to be the most critical factor in driving the change. It was a slow-grinding cultural shift—a subtle change of values and styles, a reorientation of research and pedagogical priorities set in motion by the sudden collapse of the Cold War bubble. Like most cultural shifts, the changes took hold gradually, difficult to grasp in real time and with few signposts to which one could point. Members of the Fundamental Fysiks Group had a hand in those developments. Looking back, their activities illuminate larger trends.

My argument is certainly not that everything the Fundamental Fysiks Group touched turned to gold. Group members were often wrong in the particulars, and many of their enthusiasms have failed to pan out. (There are very good reasons to doubt that extrasensory perception—whether or not it is real—has much to do with quantum theory, for example.)[25] But members of the group were right in what mattered most. They were among the first to ask the big questions again, to return to a spirit of doing physics that had animated Einstein, Bohr, and their generation fifty years earlier. That approach to physics had fallen out of the mainstream during the decades of runaway growth that the physics profession enjoyed after World War II. The Fundamental Fysiks Group's open-ended, small-scale, informal, discussion-based bull sessions were far better suited to the discipline's stark new realities after the Cold War bubble had burst.

The Fundamental Fysiks Group's legacy thus extends even beyond the particulars of Bell's theorem. They took on the prevailing Cold War model for pursuing physics. Their critique did not center on military patronage per se. Several group members had worked at major defense laboratories like the Lawrence Livermore Laboratory and retained consulting ties with the defense-contractor site at the Stanford Research Institute. Although Jack Sarfatti would occasionally fulminate against the physicists of his teachers' generation—the " 'scientific laborers' who made the first atomic bombs"—for having "sold out to the Defense Department," he, too, sought funds from the Central Intelligence Agency and the Pentagon for his latest ideas about how to harness Bell's theorem and entanglement for long-distance communication with the submarine fleet or to disable nuclear-tipped ballistic missiles in flight.[26] Some group members remained convinced that George Koopman, the former (or "former"?) military intelligence analyst, was funneling money to their Physics/Consciousness Research Group from deep-cover U.S. Air Force or Defense Intelligence Agency funds—and they were glad nonetheless to have Koopman's cash in hand.[27] Their critique of the Cold War routines and of the military's influence on science remained distinct from the more familiar us-versus-them, New Left-against-the-Establishment pattern at that time.[28]

Their concern, instead, was to broaden the physicists' range of approaches or methods beyond the hyperpragmatism that had marked the earlier Cold War years. They strove to expand the physics profession's collective mental space, to push beyond what they considered a narrowness of vision that had hardened after a quarter century of instrumentalist thinking. They laced their investigations with more of the Dionysian spirit than the strictly Apollonian; as Sarfatti put it in 1976, physicists needed more "Mythos" to leaven the "Logos." At one point Sarfatti quoted Werner Erhard's words back to him, in a follow-up grant proposal to Erhard's charitable foundation. "Werner makes a distinction between 'junior' and 'senior' scientists," noted Sarfatti. The distinction turned on neither age nor rank, but on the researchers' spirit of inquiry. Junior scientists remained trapped in a literalist mode, confusing their equations for the stuff of the world; they "confuse the symbol for the experience." Stymied by whether an electron could be both a particle and a wave, they stop trying to find meaning in their equations altogether. But that practical, pragmatic mindset had blinded nearly the entire discipline to major breakthroughs like Bell's theorem.[29]

With this broader critique, members of the Fundamental Fysiks Group were among the first to refashion the daily practice of physics, along lines that some of their elders came to advocate as well. The severity with which physicists' Cold War bubble burst in the early 1970s occasioned sustained soul-searching by many members of the profession. The National Academy of Sciences convened a blue-ribbon panel, the Physics Survey Committee, to assess the damage and plot a new course forward. MIT's Victor Weisskopf and sixteen equally prominent colleagues from across the country organized the massive study. The group's recommendations filled several thick volumes, totaling nearly 3000 pages. Like members of the Fundamental Fysiks Group, the committee was particularly concerned about the direction that physics had taken during the boom years of the 1950s and 1960s. In their estimation, the massive buildup of physicists' infrastructure had come at the cost of too close an association between physics and technological applications (military or otherwise). "Since the pursuit of physics always has been related to technological development, it could be assumed that the main purpose of physics education is to further our control over nature for our own immediate benefit. Such a view is both narrow and false." Just like Sarfatti and pals in the Fundamental Fysiks Group, the Physics Survey Committee concluded that the boom years had exaggerated one among many legitimate styles in physics, and that the pragmatism required for "technological development" had crowded out other important approaches. Now that society (including its military patrons) no longer seemed willing to pay for physicists' "technological development" at anywhere near previous levels of support, the time had come to reenvision the discipline. Physicists needed to resurrect "the quality of a quest," and to make physics come alive as a "humanistic subject" once more.[30]

The Physics Survey Committee argued that the best way to accomplish such a top-to-bottom refashioning was through education. Curricula for graduate students needed to be revamped, they argued, opening up space once again for the types of students against whom the 1960s model had "discriminated." They had in mind, for example, "the thoughtful theorist with a philosophical bent but a distaste for the routine of problem solving. Einstein's claim concerning the deadening effect of our conventional formal education should serve as a constant reminder of the damage that might be unwittingly inflicted." Only by critically reexamining past pedagogical practices could the discipine accomplish its much-needed "self-renewal."[31]

The new classroom conditions facilitated the kinds of reforms for which the Physics Survey Committee had called. As graduate-student enrollments plummeted—down by a third in just four years, sliding to one-half of the post-Sputnik peak by the end of the 1970s—faculty across the country found that they could afford to incorporate a wider range of pedagogical techniques. Essay questions that required students to articulate their ideas with words rather than just grind through algebra began to appear again on doctoral students' comprehensive exams. Such problems had been routine in departments across the country during the 1930s and 1940s, only to disappear when the student population exploded in the 1950s and 1960s. A new spate of textbooks on quantum mechanics also began to appear in the mid-1970s, having been conceived and written after the dramatic falloff in enrollments. Unlike books from the boom years, the new graduate-level textbooks now bulged with essay assignments and discussion questions: more than 40 percent of the homework problems now called on students to describe, in words, their evolving understanding of how to interpret the quantum formalism. Undergraduate textbooks on quantum theory followed the same trend. One prominent book from 1978 even labeled certain homework problems as "speculative question" and "interpretive question."[32]

Beyond the formal curricula, leading physics departments from Stanford to Harvard began to offer informal seminars for graduate students with titles like "Speculations in physics" during the mid-1970s, after enrollments had plummeted and budgets taken a nosedive.[33] Slowly, grudgingly, physicists began to make space again for the kind of free-wheeling, philosophically attuned discussions that had animated the founders of quantum mechanics back in the 1920s, and that Elizabeth Rauscher and George Weissmann had craved to re-create with the Fundamental Fysiks Group.

As pedagogical space began to open up once again for a kind of interpretive or philosophical mode, the problem became what to fill it with. Textbooks remained slow to incorporate cutting-edge developments like Bell's theorem. Materials from the Fundamental Fysiks Group thus began to fill in the gaps. Some of the earliest lesson plans on Bell's theorem, published in the *American Journal of Physics*, emerged directly from interactions with members of the Fundamental Fysiks Group. Fritjof Capra's

The Tao of Physics kept its place on physicists' syllabi for fifteen years.[34] A few years after Nick Herbert's *Quantum Reality* appeared, the author of a textbook on quantum mechanics quoted liberally from the book and recommended it to students for further reading. Nearly twenty years after that—as recently as 2008—one of the latest major textbooks on quantum mechanics, written by a Stanford professor, likewise directed students to Herbert's book as one of only four "readable accounts" on quantum theory worthy of students' attention.[35] Hatched in the hot tubs of Esalen, Herbert's lucid book continues to educate physicists young and old about Bell's theorem and other essential quantum mysteries.

Twenty-five years after John Clauser completed his first experimental test of Bell's theorem, meanwhile, I had the opportunity to redo the experiment in one of my physics courses. As a PhD student in theoretical physics, I was required to complete at least one undergraduate-level course in experimental physics. The main lesson I learned throughout the semester was that I have absolutely no aptitude for working with experimental apparatus. (That should have been clear enough already.) Nonetheless, using up-to-date electronics and computers to collect the data, even an all-thumbs student like me could reproduce Clauser's stunning results about Bell's theorem and "spooky actions at a distance" in just a few afternoons.[36] Without mentioning anyone by name, the lab manual from which my fellow students and I worked even pushed us to consider the kinds of faster-than-light schemes that Jack Sarfatti and Nick Herbert had labored over decades earlier. "The experiment suggests a number of philosophical questions," the manual prodded. Might my faraway lab partner be able to manipulate detectors on his side of the apparatus so as to modulate the outcomes of measurements on my end, thereby sending signals faster than light? "Should you quit school and invest your last dollar in his fledgling company? Why or why not?"[37] By the mid-1990s, not only had Clauser's farsighted experiment at last entered physicists' pedagogical canon but so too had the broader style of engaging the deep, foundational issues that the Fundamental Fysiks Group had worked so hard to foster.

Patterns of activity like these suggest a gradual shift in attitudes and assumptions, a measured reappraisal of the boundaries of legitimate physics after the Cold War template had faltered.[38] The exuberant efforts of the Fundamental Fysiks Group heralded broader changes to come. They were the boisterous leading edge, forging a new vision of what physics could be.

•

The young hippie physicists of the Fundamental Fysiks Group had to blaze a new trail themselves, carving out their own patrons, forums, and communication outlets during the lean years of the 1970s. Just as their group was winding down, the fortunes of the physics profession began to rebound. The Reagan administration launched a second boom in Cold War spending on science and technology during the 1980s, driven, as the first boom had been, by national-security concerns. Yet with the end of the Cold War in the early 1990s, that second bubble burst, too. The second collapse sent droves of young physicists and mathematicians to Wall Street, where they helped to design complicated financial instruments like collateralized debt obligations: from one bursting bubble to another.[39]

Since that time, physicists have crafted new means to seek out and sustain the longshot efforts that might otherwise be lost or dissipated amid the discipline's boom-and-bust cycles. New centers like the Perimeter Institute near Toronto have sprung up, funded primarily by a local booster and wireless-technology billionaire. Established in 1999, the Perimeter has sponsored physicists who work on "foundational, non-directed research": topics like quantum gravity, the interpretation of quantum mechanics, and the latest puzzles in quantum information science. The Institute's founders sought to create a safe space in which young physicists could ask questions that might not fit easily within the discipline's prevailing fads or fashions. One can hear echoes of the earlier counterculture in the Institute's founding documents. Organizers sought "the flattest possible hierarchy" with a "youth-oriented focus." During the past decade, the Perimeter Institute has grown into an internationally recognized center, with an impressive Scientific Advisory

Committee and a full complement of local educational and outreach activities.[40]

Some physicists associated with the Perimeter Institute have gone further, launching a new initiative known as "FQXi," the "Foundational Questions Institute." Unlike the Perimeter, FQXi has no home in physical space; it is a web-based consortium of young and energetic physicists supported by a suite of private "angel" donors. Their goal is to "catalyze, support, and disseminate research on questions at the foundations of physics," particularly those that promise some "deep understanding of reality," but which are "unlikely to be supported by conventional funding sources." Since 2006 the group has disbursed some $5 million, about the cost of endowing one senior professorship at an elite university. The money has been parceled out in dozens of modest grants, most of them a few tens of thousands of dollars each.[41]

Many recipients of the FQXi funds are professional physicists with some university affiliation, but the group has branched out as well. Perhaps the best-known awardee is a young physicist, Garrett Lisi, who completed his physics training in the 1990s, just as the second Cold War bubble burst. Out of work, he wandered from ski resort to surfing village, backpacking from here to there and crashing on friends' couches, all while pursuing his vision of a new "theory of everything": some means of combining quantum theory with gravitation that might unify all the known forces of nature into a single überforce. Finding some way to treat gravity as a quantum phenomenon has long been a holy grail among theoretical physicists. David Finkelstein, former editor of the *International Journal of Theoretical Physics* and longtime participant in the Esalen workshops and the *est* foundation physics conferences, has dedicated most of his career to the effort. In recent years thousands of young physicists have joined the quest. Yet Lisi's approach seemed fresh; he pursued a complementary tack to the fashionable trends in string theory. Unlike the days of the Fundamental Fysiks Group, Lisi did not need his own Ira Einhorn to put his ideas into circulation: he was able to post his papers on the central web-based physics preprint server, arXiv.org. But they garnered little attention until journalists picked up the story of the unemployed surfer-dude with his just-might-be-right theory, supported by private donations funneled through the FQXi. Following a *New Yorker* profile of Lisi, the surfer-physicist became a sensation—indeed, a sensation with his own vaguely countercultural ideas about setting up scientist hostels or communes in which to incubate the big ideas of tomorrow.[42]

Few physicists are convinced that Lisi's model will be the last word on the topic. Like the interventions by Jack Sarfatti and Nick Herbert a generation ago, Lisi's efforts have stirred other leading experts to clarify the underlying physics and make progress toward the ultimate goal.[43] Yet whereas Sarfatti, Herbert, and the members of the Fundamental Fysiks Group had to create their perch from scratch, today a few more stable institutional bases exist to support young physicists like Lisi. Institutions like the Perimeter Institute, FQXi, and the web-based preprint server arXiv.org provide a safety net to catch those out-of-the-mainstream ideas that might otherwise have been lost to obscurity.

•

And so echoes of the Fundamental Fysiks Group continue to reverberate. We have overlooked contributions from collectives like the Fundamental Fysiks Group precisely because their efforts have been so smoothly reabsorbed within the mainstream, like so many once-radical innovations of the 1970s. Yoga, organic foods, networked personal computers, identity politics, even the U.S. Army's slogan of "Be all that you can be"—cribbed from the 1970s human potential movement—fail to raise an eyebrow anymore. These days the phrase "New Age" seems like more of a marketing ploy than an alternative worldview.[44]

Not every part of the Fundamental Fysiks Group's efforts has been absorbed into mainstream physics, of course. Turf battles continue to break out from time to time, reminders that the boundaries of physics can still be contested. As recently as spring 2010 a minor kerfuffle erupted when two young physicists in Britain rescinded invitations to former Fundamental Fysiks Group member Jack Sarfatti and Nobel laureate Brian Josephson. Sarfatti and Josephson had been invited to participate in a conference on interpretations of quantum theory. Perhaps under pressure from senior colleagues who wanted to distance the event from the taint of parapsychology, those invitations were revoked. The resulting hue and cry reached the pages of the *Times Higher Education* in Britain and countless blogs around the world.[45]

While few Nobelists besides Josephson pay much heed to out-of-body experiences or the quantum mechanics of parapsychology, however, the last decade has seen a sharp uptick of interest in the physics of consciousness, now seen as a legitimate borderland between theoretical physics and advanced neuroscience. A prominent physicist recently returned to some of the early work by Fundamental Fysiks Group member Henry Stapp and consciousness-as-hidden-variables theorist Evan Harris Walker—totally unaware of the context in which their work had been done—to bolster his own study of quantum processes in the brain.[46] Leading physicists and popular authors like Michio Kaku tackle the "physics of the impossible" these days, including excursions into telepathy, telekinesis, and teleportation, while keeping their heads held high and their books squarely on the *New York Times* best-seller list.[47] Most important, we may all benefit from the Fundamental Fysiks Group's legacy in the coming years. When our children send quantum-encrypted messages between their hyperfast quantum computers, they will be living in a world that a bunch of hippies helped to invent.

Acknowledgments

My memories of the 1970s are decidedly hazy—not because they were clouded by psychedelics, but because I was too young even to know what a "hippie" was at the time. My exploration of that era has been incomparably enriched by the generosity of many people who have shared their recollections and reminiscences with me. Several also provided correspondence, notes, and photographs from their personal collections. I am sincerely grateful to them: Alain Aspect, Larry Bartell, Charles H. Bennett, Fritjof Capra, A. Lawrence Chickering, John Clauser, Roger Cooke, Dennis Dieks, Freeman Dyson, Ira Einhorn, Werner Erhard, David Finkelstein, Robert W. Fuller, Uri Geller, GianCarlo Ghirardi, Jeffrey Goldstone, Alan Guth, David Harrison, Nick Herbert, David Hess, Roman Jackiw, Alwyn van der Merwe, Jeffrey Mishlove, Charles Petit, Norman Quebedeau, Elizabeth Rauscher, Jack Sarfatti, Saul-Paul Sirag, Henry Stapp, Barbara Stevenson, Bob Toben, George Weissmann, Stephen Wiesner, Fred Alan Wolf, H. Dieter Zeh, Anton Zeilinger, John Zipperer, and Wojciech Zurek.

I could not have completed this book without a team of tireless research assistants. My thanks to Sarah Rundall, Alma Steingart, Dan Volmar, Alex Wellerstein, Lambert Williams, and Benjamin Wilson. In addition to his dogged sleuthing through all manner of texts, Alex Wellerstein also created the terrific line drawings that appear throughout the book. Librarians and archivists at several institutions proved to be remarkably helpful, including the Albert Einstein Papers at Hebrew University in Jerusalem (especially Barbara Wolff); the Niels Bohr Library at the American Institute of Physics in College Park, Maryland; the American Philosophical Society in Philadelphia (especially Charles Greifenstein); the Institute Archives at the California Institute of Technology (especially Bonnie Ludt and Shelley Erwin); the California History Section of the California State Library in Sacramento; the Institute Archives at the Massachusetts Institute of Technology (especially Nora Murphy); and the University of California at Davis (especially Daniel Goldstein).

This project began while I enjoyed a year of sabbatical at the Susan and Donald Newhouse Center for the Humanities at Wellesley College. I am grateful to Tim Peltason, then director of the Newhouse Center, for creating such a warm and stimulating environment, and for his personal encouragement as I began marching down the road toward this book. My friends and colleagues in my home departments at the Massachusetts Institute of Technology, in the Program in Science, Technology, and Society and the Department of Physics, have continued to provide inspiration. My agent, Max Brockman, and my editor, Angela von der Lippe, each offered unstinting support and valuable insights. I am also indebted to Peter Skolnik for crash-course lessons on finer points of media and entertainment law.

Several friends, students, and colleagues offered comments on portions of the manuscript. My thanks to Ken Alder, Irinéa Batista, Joan Bromberg, Jimena Canales, K. C. Cole, Olival Freire, Michael Gordin, Stefan Helmreich, Caroline Jones, David S. Jones, Rebecca Lemov, Vincent Lépinay, Clapperton Mavhunga, Patrick McCray, Lisa Messeri, Latif Nasser, Chad Orzel, Mina Park, Heather Paxson, Sophia Roosth, Michael Rossi, Natasha Schüll, Sam Schweber, Hanna Rose Shell, David Singerman, Rebecca Slayton, Alma Steingart, Robin Wasserman, Lambert Williams, Rosalind Williams, Benjamin Wilson, and Nasser Zakariya. Though we might not agree on every point, I am deeply grateful for their generous input.

Writing this book has been one crazy ride. I have been lucky beyond measure to share the voyage with my beloved wife, Tracy Gleason. I began working on the book soon after our children, Ellery and Toby, were born. Now they are almost old enough to read it. My thanks to them for teaching me all about the unbreakable bonds of entanglement.

$\sqrt{}$

Notes

Abbreviations

AE. Albert Einstein Archives. Hebrew University of Jerusalem, Israel.

EAR. Elizabeth A. Rauscher papers. Apache Junction, Arizona, in Dr. Rauscher's possession.

GCG. GianCarlo Ghirardi papers. Trieste, Italy, in Professor Ghirardi's possession.

HM. Henry Margenau papers. Manuscripts and Archives, Yale University, New Haven, Connecticut.

HPS. Henry P. Stapp papers. Berkeley, California, in Dr. Stapp's possession.

JAW. John A. Wheeler papers. American Philosophical Society, Philadelphia, Pennsylvania.

JFC. John F. Clauser papers. Walnut Creek, California, in Dr. Clauser's possession.

NBL. Niels Bohr Library, American Institute of Physics, College Park, Maryland.

NH. Nick Herbert papers. Boulder Creek, California, in Dr. Herbert's possession.

RPF. Richard P. Feynman papers. Institute Archives, California Institute of Technology, Pasadena, California.

RWJ. Roman W. Jackiw papers. Cambridge, Massachusetts, in Professor Jackiw's possession.

SPS. Saul-Paul Sirag papers. Eugene, Oregon, in Mr. Sirag's possession.

TSK. Thomas S. Kuhn papers. Collection MC 240, Institute Archives and Special Collections, Massachusetts Institute of Technology, Cambridge, Massachusetts.

VFW. Victor F. Weisskopf papers. Collection MC 572, Institute Archives and Special Collections, Massachusetts Institute of Technology, Cambridge, Massachusetts.

WHZ. Wojciech H. Zurek papers. Los Alamos, New Mexico, in Dr. Zurek's possession.

Introduction

1 "World premiere: Bank transfer via quantum cryptography based on entangled photons," press release from the Institut für Experimentalphysik, University of Vienna, April 21, 2004; available at http://www.secoqc.net/html/press (accessed September 18, 2008). See also the research paper by physicist Anton Zeilinger's team, reporting on the Viennese bank transfer: Poppe et al. (2004).

2 "Geneva is counting on quantum cryptography as it counts its votes," press release from the Geneva State Chancellery, October 11, 2007; available at http://www.idquantique.com/news (accessed September 18, 2008).

3 On publishing trends in the field, see Bettencourt, Kaiser, Kaur, Castillo-Chavez, and Wojick (2008). Accessible introductions to the field include Monroe and Wineland (2008); and the special issues of *Scientific American* (January 5, 2005) and *Physics World* (March 2007). See also Lloyd (2006) and Orzel (2009).

4 Wayner (1999), Naik (2009), McKay (1999), and Hesseldahl (2006).

5 Central Intelligence Agency, unclassified "memorandum for the record," December 4, 1979, copy in *JAW*, Sarfatti folders; Wilson (1979). *Scientific American* also ran a feature article on the topic that same year: d'Espagnat (1979). As we will see in chapter 5, the *Scientific American* article's author crossed paths with the physicists I focus on here. On *Oui* magazine, see Anon. (1981).

6 See esp. Holton (1988), Feuer (1974), Beller (1999), and Gilder (2008).

7 Kevles (1995), Forman (1987), Schweber (1994), and Galison (1997).

8 Holton (1998), Pais (1991), Cassidy (1992), and Moore (1989).

9 Kaiser (2004, 2007a).

10 Kaiser (2002). See also Kevles (1995), chap. 25; Leslie (1993), chap. 9; and Moore (2008).

11 Gustaitus (1975). On changes at the magazine, see also Anon. (1975).

12 Sirag (1977a, b) and Sarfatti (1977a). On Leary's acceptance of Sirag's and Sarfatti's essays for *Spit in the Ocean*, see Sirag (2002), 111. On Leary's and Kesey's counterculture exploits, see Wolfe (1968), Lee and Shlain (1992), and Lattin (2010).

13 Gold (1993), 15–17, 38, 115.

14 Anon. (1974a), Woodward and Lubenow (1979), Garfinkel (1982), Anon. (1977a), Carroll (1981), and Roosevelt (1980). On the 1977 humanistic psychology conference, see also Jerry Diamond to Sarfatti, n.d. (ca. January 1977), copy in *JAW*, Sarfatti folders.

15 Heirich (1976), 697. Heirich referred to work by Jack Sarfatti, Fred Alan Wolf, Henry Stapp, and Evan Harris Walker, among others.

16 See esp. Pamplin and Collins (1975); Pinch (1979); Collins and Pinch (1979, 1982); and Collins (1992), chap. 5.

17 Popper (1963). See also Popper (1976), esp. chap. 8.

18 See esp. Collins (1992), Gieryn (1999), and Laudan (1983). See also Ross (1991), 23–30.

19 Reagan quoted in Braunstein and Doyle (2002), 6.

20 Rorabaugh (1989), chap. 4; Doyle (2001); Braunstein and Doyle (2002); Rossinow (2002); and Gosse (2005).

21 Lee and Shlain (1992), 119; and Nick Herbert, "Doctor Quantum drops acid," available at http://members.cruzio.com/~quanta/doctorquantum.html (accessed October 2, 2007). See also Novak (1997) and Wasserman (2000).

22 Rorabaugh (1989), 134, 135, 137; Lee and Shlain (1992), 154; and Braunstein (2002).

23 Novak (1997), 100.

24 Melton (1992), 20; and Kyle (1995), 157.

25 Lee and Shlain (1992), 148.

26 Lewis and Melton (1992); Kyle (1995), 13, 14, 65; and Schulman (2001), 96.

27 Roszak (1969), 48.

28 On similar intermingling of quintessential military-industrial sponsors with New Age and paranormal research, see also Burnett (2009, 2010), and Ronson (2004).

29 Cf. Peck (1985).

30 Nick Herbert as quoted in Physics/Consciousness Research Group, "A modest proposal to the Foundation for the Realization of Man," February 11, 1976, on p. 15; copy in *SPS*.

31 Cf. Forman (1971).

32 Lo, Popescu, and Spiller (1998), 15–17, 24, 77–88; Nielsen and Chuang (2000), 3, 24, 25, 528–531; and Jaeger (2007), 83, 147, 156, 157.

33 Cahill (1995).

34 Brisick (1995), Bernstein (1995), and Finn (1995).

Chapter 1: "Shut Up and Calculate"

1 Marin (1975), Heck and Thompson (1976), and Litwack (1976).

2 Erhard interview (2010).

3 Fred Alan Wolf, email to the author, November 12, 2007; Jack Sarfatti, email to the author, November 27, 2007 ("I think you're an asshole"); Wolf interview (2009); and Sarfatti interview (2009). Several sources documented the prevalence of the phrase "You're an asshole" in *est* trainings at the time: Heck and Thompson (1976), 20; Litwack (1976), 48, 50, 54; Fenwick (1976), 33, 34, 51, 96, 101, 165; and Hubner (1990a), 19.

4 Cf. Beller (1998).

5 The literature on the creation of quantum mechanics is vast. For an introduction, see esp. Jammer (1966); Darrigol (1992); Beller (1999); and Galison, Gordin, and Kaiser (2001).

6 Heisenberg (1971), 73–76.

7 Physicist N. David Mermin provides a fascinating and amusing genealogy of the phrase "shut up and calculate": Mermin (2004); see also Mermin (1989). I owe these references to Orzel (2009), 79, 80.

8 Bohr (1985), vol. 6; Bohr (1949), 215–20; and Heisenberg (1930). By no means did they always agree on what the double-slit experiment implied. See Beller (1999), chap. 11.

9 Moore (1989), 299.

10 Crease (2002), 19. For a classic presentation of the double-slit experiment, see Feynman with Leighton and Sands (1965), vol. 3, chap. 1.

11 Bohr (1949).

12 Albert Einstein to Erwin Schrödinger, June 17, 1935, as quoted in Fine (1986), 68 ("young whore"). On Einstein's use of the double-slit experiment for his criticisms of quantum theory, see Einstein to Schrödinger, April 26, 1926, in Przibram (1967), 28; and Jammer (1966), 360.

13 Wheaton (1983); Rodgers (2002), 15; and letters to the editor, *Physics World* 16 (May 2003): 20. For more recent experimental demonstrations, see Scully and Drühl (1981); Arndt et al. (1999); and Hillmer and Kwiat (2007).

14 Notes between Albert Einstein and Paul Ehrenfest, October 25, 1927, in *AE*, item 10-168. My translation; emphasis in original. See also Jammer (1966), 360. On Einstein's long-standing critique of quantum theory, see also Pais (1982), Fine (1986), Kaiser (1994), and Beller (1999).

15 Max Born introduced and refined his now-famous probability interpretation of the wavefunction in a series of short articles published in 1926. See, in particular, Jammer (1966), 283–290; and Beller (1990).

16 Albert Einstein to Erwin Schrödinger, May 31, 1928, in Przibram (1967), 31.

17 The same holds for the amount of energy exchanged in a given interaction and the time interval over which the interaction takes place. See Jammer (1966), chap. 7; Cassidy (1992), chap. 12; and Beller (1999), chaps. 4 and 5.

18 Heisenberg (1930), 76–79; see also Feynman with Leighton and Sands (1965), vol. 3, chap. 1, 6–9.

19 Einstein to Schrödinger, June 19, 1935 ("ridiculous little Talmudic philosopher," my translation of "talmudistische Philosoph"), in *AE*, item 22-047. On Bohr's complementarity, see esp. Holton (1988), 99–146; Folse (1985); Murdoch (1987); Kaiser (1992); and Beller (1999), chap. 6.

20 Albert Einstein to Max Born, undated, ca. January 1927, in Born (2005), 93; and Einstein (1934), 168, 169 ("transitory significance," "still believe").

21 Einstein to Born, December 4, 1926, in Born (2005), 88 ("Quantum mechanics is certainly

imposing"). See also Einstein to L. Cooper, October 31, 1949, in *AE*, item 411; Fine (1986); and Kaiser (1994).

22 Erwin Schrödinger (1935b), 812, as translated in Fine (1986), 65. On the correspondence between Einstein and Schrödinger that led to Schrödinger's article, see Fine (1986), chap. 5.

23 Wolfgang Pauli to Niels Bohr, February 11, 1924, in Pauli (1979), 143 ("very unphilosophical"). On Bohr's approach, see Kaiser (1992) and Faye (1991). On Heisenberg's philosophical pretensions, see also Carson (2010).

24 Bohr (1934, 1958); Born (1956), 107; and Pauli (1994). See also Beller (1998).

25 Albert Einstein to Erwin Schrödinger, June 17, 1935, as quoted in Fine (1986), 68 ("epistemology-soaked orgy"); Albert Einstein to Paul Bonofield, September 18, 1939, in *AE*, item 6-118-1 ("My own opinion"). See also Einstein to Maurice Solovine, April 10, 1938, in Einstein (1987), 85; Einstein (1949), esp. 671, 672.

26 Max Born to Albert Einstein, July 15, 1925, in Born (2005), 82; Erwin Schrödinger to Albert Einstein, May 30, 1928, in Przibram (1967), 30. See also Kragh (1999), 168–73.

27 Kuhn et al. (1967).

28 Haas (1928), chaps. 11 and 16; Heisenberg (1930), 65; Weyl (1931), 76; Born (1936), 82–85; and Sommerfeld (1930), 37, 257.

29 On American physicists' philosophical and pedagogical approaches to quantum mechanics in the 1930s, see Kaiser (2007a) and (forthcoming), chap. 4.

30 Einstein to Maurice Solovine, February 12, 1951, in Einstein (1987), 123.

31 Michael Cohen, entry of May 14, 1953, in Caltech "bone books," box 1, vol. 7 ("invested in analysis"), available in the Archives of the California Institute of Technology, Pasadena, California; Frederick Zachariasen, entry of May 27, 1953, ibid. ("usual spiel"); Romain (1960), 62 ("avoids philosophical discussion"); Falkoff (1952), 460, 461 ("philosophically tainted questions"); and Feshbach (1962), 514 ("musty atavistic to-do"). See also Kaiser (2007a).

32 Weiner (1969) and Rider (1984).

33 Edward Teller (with Robert F. Christy and Emil J. Konopinski), "Lecture notes on quantum mechanics," autumn 1945, on 79. A copy of these notes is available as part of "Notes on physics courses given at Los Alamos, 1943–1946," in *NBL*, call number AR31029. On effects of physicists' wartime projects, see esp. Forman (1987); Schweber (1994), chap. 3; and Galison (1997), chap. 4.

34 Smyth (1951); and Bureau of Labor Statistics report as quoted in Barton (1953), 6 ("If the research in physics").

35 Kaiser (2006a); cf. Forman (1987) and Leslie (1993).

36 Kaiser (2002).

37 Raymond T. Birge to E. W. Strong, August 30, 1950, in Raymond Thayer Birge correspondence and papers, call number 73/79c, Bancroft Library, University of California at Berkeley.

38 "Opinions of returning graduate students in physics," 86-pp report, 1948, call number UAV 691.448, in Harvard University Archives, Pusey Library, Cambridge, Massachusetts ("The classes are so large"). See also Kaiser (2004).

39 Gerjuoy (1956), 118 ("With these subjects"); and Uhlenbeck (1963), 886 ("easy to teach").

40 Eyvind Wichmann, "Comments on *Quantum Mechanics*, by L. I. Schiff (Second Edition)," n.d., ca. January 1965, in Leonard I. Schiff papers, call number SC220, Stanford University Archives, box 9, folder "Schiff: Quantum mechanics" ("The book kept me sufficiently busy").

41 Kaiser (2007a).

42 Hans Freistadt, who convened the group, thanked its members in Freistadt (1957), 65.

43 Schrecker (1986), 289, 290; and Wang (1999), chap. 7.
44 U.S. House, Committee on Un-American Activities (1953a), 190–212 and (1953b) 1795–99; Cattell (1960), s.v. "Darling, Byron T."; and Schrecker (1986), 207–9.
45 Freistadt (1953), 221, 229, 237 and (1957). Freistadt's review article received roughly one citation every other year over the next two decades, several of them in philosophy journals rather than physics ones: *Science Citation Index* (1961–). On political leanings of other group members, see Newman (2002); David Bohm to Melba Phillips, n.d., ca. June 1952, in David Bohm Archives, Birkbeck College, London, folder C46, and related correspondence in folders C3, C46, and C48; and John Stachel, email to the author, October 23, 2007.
46 Kevles (1995), chap. 25; Schweber (1988); Leslie (1993), chap. 9; and Moore (2008).
47 Kaiser (2002), 149–53.

Chapter 2: "Spooky Actions at a Distance"

1 Bernstein (1991a), 50, 51; Whittaker (2002), 14–17; and Gilder (2008), chaps. 27 and 28.
2 Bernstein (1991a), 53, 64, 65; Whittaker (2002), 17; cf. Born (1949). Von Neumann's proof appeared in von Neumann (1932), later translated as von Neumann (1955), on 305–24. On von Neumann's proof and the drawn-out debate it inspired, see Jammer (1966), 367–70; Pinch (1977); Bell (1982); and Jackiw and Shimony (2002), 84–87. Most of Bell's papers on the foundations of quantum mechanics were republished in Bell (2004b).
3 Scholars still debate the extent to which Bohm's political views shaped his own approach to physics, or the reception it received from others. See, e.g., Cross (1991); Olwell (1999); Mullet (1999, 2008); Kojevnikov (2002); and Freire (2005).
4 Bell (1982), 990 ("saw the impossible done"); and Bohm (1952a, b). See also Albert (1992), chap. 7; Cushing (1994); and Buchanan (2008). As is now well known, Bohm's 1952 papers bore strong similarity to a 1927 proposal by Louis de Broglie, which de Broglie had quickly abandoned in the face of criticism from Wolfgang Pauli. See, e.g., Jammer (1966), 291–93; and Cushing (1994), chaps. 7 and 8.
5 Bell (2002), 3–5; and Burke and Percival (1999), 9, 10. On the Bells' decision to leave Harwell, see Bernstein (1991a), 18, 20.
6 Bell (1966), 452 ("first ideas"); Bernstein (1991a), 67, 68; cf. Jauch and Piron (1963). On Bell's significant contributions to "mainstream" nuclear and particle theory, see Burke and Percival (1999), 4–9; and Jackiw and Shimony (2002), 100–112.
7 Einstein, Podolsky, and Rosen (1935).
8 Bohr (1935). On the conditions surrounding Bohr's response, see Rosenfeld (1967). The literature on the Einstein-Bohr debate, and on the EPR thought experiment in particular, is enormous. See esp. Fine (1986), esp. chap. 3; Kaiser (1994); and Beller (1999), chap. 7.
9 Einstein, Podolsky, and Rosen (1935), 779 ("since at the time"); Albert Einstein to Max Born, March 18, 1948 ("bristles"), reprinted in Born (2005), 162; Einstein to Born, March 3, 1947, in Born (2005), 154, 155 ("spooky actions at a distance"). See also Howard (1985).
10 Bohm (1951), 614–22.
11 On the discovery of quantum-mechanical spin and its significant conceptual break from ordinary angular momentum, see, e.g., Jammer (1966), 133–56; and Tomonaga (1997).
12 The device pictured in figure 2.1 is a simplified version of a Stern-Gerlach apparatus, first conceived by Otto Stern in 1921 and put to use by Stern with Walther Gerlach a few months later. See Friedrich and Herschbach (2003).
13 Bell (1964). Erwin Schrödinger introduced the term "entanglement" in Schrödinger (1935a), 555.

14 Several other physicists derived this particular expression for S, building on Bell's work. It is often referred to as the CHSH inequality based on the authors' initials: Clauser, Horne, Shimony, and Holt (1969).

15 Bell (1964), 199; and Bernstein (1991a), 84, 85. On the quantum-mechanical calculation, see, e.g., Sakurai (1985), 223–232. Several authors have since simplified Bell's original proof, and popular treatments abound. Among the best are those by N. David Mermin, several of which are republished in Mermin (1990b), esp. chaps. 10–12. Philosopher Tim Maudlin's treatment is also particularly clear: Maudlin (1994), chap. 1.

16 D'Espagnat (2003), 112–14 ("divisibility by thought").

17 Mermin (1981, 1985, 1990a, 1994); Bell (2004a); Jacobs and Wiseman (2005); and Kwiat and Hardy (2000).

18 Lloyd (2006), 120, 121 ("beer or whiskey?").

19 Nielsen and Chuang (2000), 117 ("iron"), emphasis in original. Citation data from *Science Citation Index* (1961–). On the rarity of accumulating so many citations, compare with data on top-cited publications within high-energy physics available at http://www.slac.stanford.edu/spires (accessed January 15, 2008) and the data in Redner (2005). For recent assessments of the significance of Bell's theorem, see Jackiw and Shimony (2002) and Bertlmann and Zeilinger (2002).

20 Bell (1966); Bernstein (1991a), 67, 68; and Jackiw and Shimony (2002), 87.

21 Anderson and Matthias (1964); Bernstein (1991a), 74, 75; Wick (1995), 89, 90. On physicists' postwar anxieties about information overload and its effects on their research journals, see Kaiser (forthcoming), chap. 3. On page fees at the *Physical Review*, see Scheiding (2009).

22 Other than Bell's own 1966 review article on hidden variables, completed before his 1964 paper but published after it, the first article to cite Bell's 1964 paper was Clark and Turner (1968), 447. Citation data from *Science Citation Index* (1961–).

23 Clauser interview with Joan Bromberg (2002), 34, 51, 52; and Selleri interview with Olival Freire (2003), 23; both transcripts available in *NBL*. One member of Selleri's group, the experimentalist Vittorio Rapisarda, died in a car accident while driving from Catania to Bari for one of the group's regular meetings: 48, 49. For an indication of the tight-knit community working on Bell's theorem at the time, see the acknowledgments in Vigier (1974); Bohm and Hiley (1976b); Garuccio and Selleri (1976); Lamehi-Rachti and Mittig (1976); Baracca, Cornia, Livi, and Ruffo (1978); and Selleri (1978).

24 Based on data in *Science Citation Index* (1961–).

Chapter 3: Entanglements

1 This chapter draws inspiration from historians' grappling with scientists' "self-fashioning" and personae. See esp. Biagioli (1993); Daston and Sibum (2003); and Shapin (2008).

2 Clauser interview with Bromberg (2002), 11, 12, 14, 15; and Clauser interview (2009). See also Wick (1995), 104, 105; and Clauser (2002), esp. 71, 77, 78.

3 J. S. Bell to John Clauser, March 5, 1969 ("shake the world"); see also Clauser to Bell, February 14, 1969, and Bohm to Clauser, February 25, 1969, all in *JFC*, folder "Random correspondence." David Wick notes that Clauser's letter was the first direct response Bell had received to his work: Wick (1995), 106n124.

4 Clauser (1969), 578.

5 Most of Shimony's papers on foundations of quantum theory are collected in Shimony (1993), vol. 2.

6 Shimony interview with Bromberg (2002), 39, 40 (receipt of Bell's preprint), 49 ("kooky

paper"), 51, 52 ("quantum archaeology"), and 53 ("whole thing on ice"). Shimony attributes the term "quantum archaeology" to his then–graduate student, Michael Horne (ibid., 51).

7 Shimony interview with Bromberg (2002), 54, 55, 71 ("civilized"); Clauser interview with Bromberg (2002), 34, 35; and Clauser (2002), 80, 81. See also Wick (1995), 103–13; and Aczel (2001), chap. 14.

8 Clauser interview with Bromberg (2002), 18, 35; and Clauser (2002), 62, 63, 70–72, 78. On the stigma at the time—especially its effects on graduate students and postdocs—see also Harvey (1980, 1981).

9 Clauser interview with Bromberg (2002), 35 ("some of which I picked up"); Clauser interview (2009); Shimony interview with Bromberg (2002), 73, 74; and Clauser, Horne, Shimony, and Holt (1969). The journal received the article submission on August 4, 1969.

10 On Townes, masers, lasers, and early skepticism about their compatibility with quantum mechanics, see Bromberg (1991), 17–19.

11 Clauser interview with Bromberg (2002), 12, 13, 69–71, 73, 74 ("Dumpster diving"); Freedman and Clauser (1972). On early experimental tests of Bell's theorem, see also Clauser and Shimony (1978); Freire (2006); and Gilder (2008), chaps. 29, 30.

12 Clauser interview with Bromberg (2002), 41, 42; Shimony interview with Bromberg (2002), 71, 74; Freedman and Clauser (1972); and *Science Citation Index* (1961–), s.v. "Bell, John S."

13 Shimony to Clauser, August 8, 1972, in *JFC*, reporting the attitude of the physics department chair at San Jose State College. See also Freire (2006), 604.

14 Rauscher interview (2008).

15 Rauscher interview (2008).

16 Rauscher interview (2008). See also Timothy Pfaff, "An interview with Elizabeth Rauscher," *California Monthly* (University of California Alumni Magazine), ca. 1979–80 (clipping in *EAR*). Rauscher's first article appeared as "Fundamentals of fusion" (1960). On cutbacks, see Anon. (1965), 16; and Clark (1966), 70.

17 Statistics on female physics degree recipients are calculated from data in Adkins (1975), 278–81.

18 Rauscher interview (2008). Rauscher likened her experiences during graduate school to those described in Keller (1977).

19 On the Livermore group, see E. A. Rauscher, flyer, "Ideals and purpose of the Tuesday night club" (1969), in *EAR*. On Rauscher's summer course, see *The Magnet* [Lawrence Berkeley Laboratory newsletter] 15 (December 1971): 5; "The philosophy of science," *The Magnet* 17 (June 1973): 1; "Philosophy of science course to start," *Beam Line* [Stanford Linear Accelerator Center newsletter] 3 (June 20, 1972): 2 ("rap sessions"); and R. B. Neal (acting director, SLAC), memo to "all hands," June 14, 1972, in *EAR*. On the late-1960s and early-1970s protests against physicists' facilities, see Kaiser (forthcoming), chap. 5. On more recent community-building efforts at Livermore, see Gusterson (1996).

20 Rauscher interview (2008). On Arthur Young and his institute, see also Mishlove (1975), 263–78.

21 Federici (1967), 5; and Anon. (1967a), back page.

22 See the photographs accompanying Pasolli (1966), 19; and Novick (1966), 17.

23 Saul-Paul Sirag, email to the author, July 18, 2010.

24 Saul-Paul Sirag, email to the author, December 11, 2007; Sirag (2002), esp. 97, 118. See also the list of seminars for the Institute for the Study of Consciousness, winter 1976, several of them led by Sirag; copy in *SPS*.

25 Nick Herbert, emails to the author, November 28, 2007, and December 1, 2007 ("no-

nonsense"); and Herbert, "Doctor Quantum drops acid," available at http://members
.cruzio.com/~quanta/doctorquantum.html (accessed October 2, 2007). On Schiff's textbook
and prevailing trends in teaching quantum mechanics at the time, see Kaiser (2007a) and
(forthcoming), chap. 4.

26 Nick Herbert, email to the author, July 15, 2009 ("looking like an insane hippy").

27 Nick Herbert, email to the author, July 13, 2010.

28 Heinz Pagels first told Herbert about Bell's theorem: Nick Herbert, email to the author,
December 1, 2007. Herbert's rederivation of Bell's theorem appears in Herbert (1975). The
paper was first submitted to the journal in June 1973. On the group's discussions with
Clauser, see also Rauscher interview (2008); Sirag (2002), 107–9; and Sirag (1977a), 13.

29 Stapp interview (2007). See also Miller (2009).

30 Stapp interview (1998); Henry Stapp, email to the author, October 1, 2007; and Stapp interview (2007) ("in a huff"). On Stapp's graduate work, see Stapp (1956) and Kaiser (2005), 109.
Stapp's 1968 preprint was later released as Stapp (1976). On his early lecturing about Bell's
theorem, see Kripal (2007), 310; and Stapp interview (2007). Many of Stapp's publications
on the topic are collected in Stapp (2004, 2007b).

31 George Weissmann, remarks transcribed from videotape of the Fundamental Fysiks Group
reunion, November 18, 2000, San Francisco. My thanks to George Weissmann for providing
videotapes of the reunion.

32 George Weissmann, email to the author, March 3, 2008; and Weissmann interview (2008).

33 Weissmann interview (2008).

34 Weissmann interview (2008); and Elizabeth Rauscher, remarks transcribed from videotape
of the Fundamental Fysiks Group reunion, November 18, 2000, San Francisco ("verboten,"
"very deep issues"). On Chew's pedagogical style and "secret seminars," see Kaiser (2005),
chap. 9.

35 Rauscher interview (2008); Weissmann interview (2008); and Capra (1988), 54–57.

36 Wolf interview (2009). On Project Orion, see also Dyson (2002).

37 Wolf interview (2009). Cf. Kaiser (2004). On Cold War "gadgeteering," see also Forman
(1987) and Mody (2008).

38 Wolf interview (2009).

39 Fred Alan Wolf, email to the author, September 24, 2007; Wolf interview (2007) ("Suddenly
all of politics," "apple was rotten"); and Gale Reference Team, "Biography: Wolf, Fred Alan
(1934–)," *Contemporary Authors Online*, available via http://gale.cengage.com (accesssed
September 25, 2007). See also Trombley (1967) and Anon. (1967b).

40 Wolf interview (2007) (strangers' office visits, "most physicists at that time"); and Wolf
(1991), 35 ("lost the magic").

41 Wolf interview (2007); and Wolf (1991), 39, 40.

42 Wolf interview (2009).

43 Jack Sarfatt to John A. Wheeler, March 31, 1971, in *JAW*, Sarfatti folders; University of
California online library catalog, s.v. "Sarfatt, Jack"; Sarfatti (2002a); and Jack Sarfatti,
email to the author, November 27, 2007. Sarfatti changed his name from "Sarfatt" in the
mid-1970s, restoring it to an ancestral spelling; see Sarfatt (1977b), 3. Cf. Sarfatt (1963,
1965, 1967a); Sarfatt and Stoneham (1967); Sarfatt (1967b, 1967c, 1967d, 1969); Cummings,
Herold, and Sarfatt (1970); Sarfatti (1972); and Sarfatt (1974a).

44 Wolf interview (2009) and Sarfatti interview (2009). My thanks to Wolf for sharing a copy
of the short film.

45 Jack Sarfatt to John A. Wheeler, May 26, 1973, in *JAW*, Sarfatti folders. See also Sarfatt (1974b).
46 Sarfatti (2002a), 45, 46 ("like Bob Hope"), 82 (meeting Salam). On Salam and his Centre, see de Greiff (2002).
47 Saul-Paul Sirag, email to the author, December 11, 2007; Sirag (2002), 96, 97, 109, 110; Sarfatti (2002a), 44n64, 48n81; Rauscher interview (2008) ("the idea"); George Weissmann, emails to the author, November 28, 2007, and March 3, 2008.
48 Elizabeth Rauscher, remarks transcribed from videotape of the Fundamental Fysiks Group reunion, November 18, 2000, San Francisco ("I had figured"). See also Elizabeth Rauscher, "List of lectures presented at the Fundamental 'Fysiks' Group, LBNL," in *EAR*. This list matches the one reproduced in Collins and Pinch (1982), 189n4.

Chapter 4: From Ψ to Psi

1 Anthropologist David Hess argues that parapsychology specialists should be distinguished from New Age enthusiasts: the former often worked in laboratories and tried to establish scientific protocols while the latter engaged with occult phenomena outside the trappings of academic research. While Hess's typology is helpful, individuals frequently moved seamlessly across those boundaries, including several members of the Fundamental Fysiks Group. Cf. Hess (1993).
2 Elizabeth Rauscher, "Possible topics of investigation," notes on the May 7, 1975, organizational meeting of the Fundamental Fysiks Group, in *EAR*. Rauscher was also fond of making maps like the one in fig. 4.1. See Rauscher (1979), 52. On the group's interest in quantum mechanics and psi, see also Collins and Pinch (1982), chap. 4.
3 See esp. Moore (1977); Oppenheim (1985); Owen (1990, 2004); Winter (1998); Gordin (2004), chap. 4; and Morrisson (2007).
4 Quoted in Moore (1989), 252; see also 169–77.
5 Jordan (1927b, 1955).
6 Wolfgang Pauli, as quoted in Miller (2009), 162; see also chaps. 8–10.
7 Pauli, "Science and Western thought" (1955), as translated and reprinted in Pauli (1994), 137–48, on 148. See also the other essays from the early 1950s reprinted in Pauli (1994).
8 Anon. (1955a, b) and Price (1955). Cf. Oppenheim (1985), chap. 4; and Mauskopf and McVaugh (1980), 298–306.
9 Members of the group routinely invoke Jung's notion of "synchronicity" when describing their experiences; e.g., Sarfatti (1997); Sirag (2002); and Weissmann interview (2008).
10 Sullivan (1973). At the time, *California Living* was the Sunday magazine associated with both the *San Francisco Chronicle* and the *San Francisco Examiner*.
11 Sarfatti (2002a), 47, 99. On the Stanford Research Institute, see Leslie (1993), 243–47.
12 Dewar (1977); Puthoff (1996), esp. 65; Schnabel (1997), 86, 87, 97, 100, 200; and Kripal (2007), 340–45. Cf. Pantell and Puthoff (1969). On controversies surrounding Scientology, see esp. Sappell and Welkos (1990) and Behar (1991).
13 Targ and Puthoff (1974) and (1977), chap. 7; Gilliam (1978a); and Schnabel (1997), chap. 9. For more skeptical accounts of these studies, see Randi (1975), esp. chaps. 3, 4, 11 and (1982), 131–45; Gardner (1976); and Marks and Kammann (1980), chaps. 6–10.
14 Petit (1973); Bess (1973); Sullivan (1973); Anon. (1973a); and Rensberger (1974).
15 Sarfatti (2002a), 47, 48. On Mitchell's moon-orbiting telepathy experiment, see the unsigned, redacted Central Intelligence Agency "Memorandum for the record," September 27, 1971, released in response to Freedom of Information Act request (CIA FOIA reference F-2009-

00079); Anon. (1971); Mitchell (1974); and Mitchell with Williams (1996). On the founding of Mitchell's institute, see also Anon. (1974a); Gardner (1976), 43; Schnabel (1997), 134; and Kripal (2007), 340, 341.

16 Gustaitus (1975) and Wilson (1979), 131. For a more recent account of Sarfatti's "Spectra" phone calls, see Sarfatti (2002a), 24–32. Cf. Puharich (1974).

17 Puharich (1979a), 10, 11 (on the BBC broadcast); and Hasted, Bohm, Bastin, and O'Regan (1975). A copy of Jack Sarfatti's original press release, entitled "Historic test of Uri Geller's psycho-kinetic powers by physicists at Birkbeck College, University of London," dated June 24, 1974, can be found in *JAW*, Sarfatti folders; it was circulated to two dozen recipients by Ira Einhorn in a package dated June 26, 1974. Most of the text was also published in Sarfatti (1974a).

18 Sarfatti (1974a), 46 ("My personal professional judgment"); Hasted et al. (1975), 471; Sarfatti (1975a), 355 ("ambiguity in the interpretation") and (1974b), 3 ("intrinsically nonlocal"). Sarfatti cited the February 1974 preprint of Bohm and Hiley (1975). Hasted and Bohm soon parted company on the matter of psi phenomena: Hasted delved deeper into experimental tests of phenomena like spoon bending, while Bohm wrote a few cautionary papers against over-extending his own ideas about hidden variables and nonlocality. See Hasted (1976, 1977, 1978/79, 1979); Randi (1982), 215–21; Bohm and Hiley (1976a); Sirag, Musès, and Bohm (1976), 29; and Peat (1997), 271, 272. British physicist John G. Taylor, who endorsed Geller's psychic powers on the BBC broadcast in November 1973, likewise had a change of heart. Compare Taylor (1975a, b) with (1980).

19 Wigner (1962). See also Jammer (1974), 498–500; and Stapp (2007a).

20 Wigner, "Remarks on the mind-body question," as reprinted in Wheeler and Zurek (1983), 173 ("Consciousness enters"), 176–78.

21 Seitz, Vogt, and Weinberg (1998), 367, 383, 384; Shimony interview with Bromberg (2002), 30–36; and Freire (2007).

22 The quotation is from I. I. Rabi. See Rigden (1987), 46.

23 Bernstein (1991b); and Wheeler with Ford (1998), chaps. 4–6.

24 Patton and Wheeler (1975), 560.

25 Patton and Wheeler (1975), 560–62. Wheeler also included his stick-figure "observer" and "participator" cartoon in at least four more talks given between August 1975 and March 1976, published as Wheeler (1977), 6; see ibid., 29, 30 for a list of venues at which he delivered this talk.

26 Wheeler (1978), 41. Based on a conference talk delivered in June 1977.

27 Wheeler introduced this quasar variant of the delayed-choice experiment in a talk at the American Philosophical Society in Philadelphia in June 1980; portions were reprinted in Wheeler (1983), quotation on 192 (emphasis in original).

28 Patton and Wheeler (1975), 564 ("gives the world the power"); and Wheeler (1983), 209 ("Acts of observer-participancy"). Wheeler's self-observing "U" universe cartoon reappeared in Wheeler (1980), 362 and (1983), 209.

29 On Sarfatti's and Wolf's repeated requests to spend their sabbaticals at Princeton, see Sarfatti to John A. Wheeler, February 28, 1973, April 29, 1973, and May 26, 1973; Fred Alan Wolf and Jack Sarfatti to Wheeler, May 7, 1973 ("We understand"), and May 25, 1973 (telegram); Wheeler to Sarfatti and Wolf, May 22, 1973, and Wheeler to Sarfatti, July 10, 1973 ("I hated so much to seem unwelcome"), all in *JAW*, Sarfatti folders. Wheeler first briefly introduced the idea of his "participatory universe" in Wheeler (1973), 244; and in Wheeler, Misner, and Thorne (1973), 1273. Sarfatti cited a preprint of Wheeler and Patton's 1974 Oxford talk in

   Sarfatti (1975b), 279, 334n9; and in a letter to the editor of *Physics Today*, dated April 17, 1974, copy in *JAW*, Sarfatti folders. On the paucity of citations to Wheeler's published conference talks, see *Science Citation Index* (1961–), s.v. "Wheeler, J. A." On Wheeler's interactions with experimentalists during the 1980s, see Bromberg (2008).

30 Sarfatti (1974b), 6, 7; Sarfatti, letter to the editor of *Physics Today*, April 13, 1974 ("Age of Aquarius" students), copy in *JAW*, Sarfatti folders.

31 Sarfatti (1974b), 6, 7. Sarfatti's thinking on the topic changed dramatically after he read Bohm and Hiley (1993). See, e.g., Levit and Sarfatti (1997); and Jack Sarfatti, email to the author, September 27, 2010.

32 O'Regan (1974), 9, 10. Cf. Sarfatti (2002a), 57.

33 See the letters collected under "Geller and magicians," *Science News* 106 (August 3, 1974): 78, 79. On earlier magicians' efforts at debunking spiritualist and other psychic claims, see Noakes (2008), 331; and Houdini (1924).

34 Sarfatti (1975a), 355 ("I do not think"). See also Rensberger (1975), 59; Gardner (1976), esp. 44, 46–49; and Randi (1975).

35 Naval Surface Weapons Center quoted in Straczynski (1977), 158 ("Geller has altered").

36 Straczynski (1977), 160 ("The enormous spate"); and Randi (1975), 121 ("digging his own grave"). See also, e.g., Weil (1974a, b).

37 Sarfatti (1975a), 355. Sarfatti elaborated upon his theoretical model in Sarfatti (1974b, 1975b, 1977b). On Sarfatti's more recent views, see Jack Sarfatti, email to the author, December 7, 2009. On Crookes's and Thomson's responses to allegations against individual mediums, see Oppenheim (1985), chaps. 4, 8.

38 Wolf interview (2007); Wolf (1991), 39, 40, 45–47; and Jack Sarfatti to John Wheeler, April 27, 1974, in *JAW*, Sarfatti folders.

39 Erhard interview (2010) and Wolf interview (2007). On the May Lectures, see also Capra (1988), 149, 150.

40 Wolf interview (2007); and Wolf (1991), 183, 184 ("I can't say"). See also Puharich, (1974, 1979a). Sarfatti credited many of the ideas in his *Psychoenergetic Systems* paper to joint effort with Wolf: Sarfatti (1974b), 5, 6, 8.

41 Nick Herbert, email to the author, December 1, 2007. On Walker's training, see his entry in the Dissertations and Theses database, available via http://proquest.umi.com.

42 Walker (1970, 1971, 1972, 1972/73, 1975, 1977, 1978/79). Walker later published an account of his work for nonspecialists: Walker (2000).

43 Nick Herbert, email to the author December 1, 2007 ("house organ"); Sirag, Musès, and Bohm (1976), 27–29 (quotations); and Sirag (1977b), 77, 78.

44 On Walker's presentation to the Fundamental Fysiks Group, see Elizabeth Rauscher, "The origins and operations of the era of the Fundamental Physics Group," prepared for the group's reunion in November 2000, in *EAR*. On the metaphase typewriter, see Sirag and Babbs (1977), 53 ("as a joke"); and Nick Herbert, "Metaphase typewriter," available at http://www2.cruzio.com/~quanta/meta.html (accessed May 29, 2008) ("spirits of the dead"). Jeffrey Mishlove, who attended one of Herbert's demonstrations of the metaphase typewriter, confirmed that many details of Sirag's fictionalized account were accurate: Mishlove interview (2008).

45 Sirag and Babbs (1977), 53 ("realm of mind"); Nick Herbert, email to the author, May 29, 2008.

46 Nick Herbert, "Application for radioactive material license," n.d. (ca. October 1973), in *NH*.

47 Nick Herbert, "Metaphase typewriter: An exercise in free assembly," unpublished manu-

script (January 1974), on 7, 8; copy in *NH*. See also Herbert (1976) and (1993), 199–208. On the thallium bans, see Anon. (1973b), 26.

48 Herbert, "Metaphase typewriter: An exercise in free assembly," copy in *NH*.

49 Sirag and Babbs (1977), 53 ("rather suspect communication").

50 Herbert, "Metaphase typewriter: An exercise in free assembly," copy in *NH*; and (1993), 199–208.

51 Sirag and Babbs (1977), 57 ("anininfinitime"); Sirag (2002), 111 ("byjung").

52 Weissmann interview (2008); and George Weissmann, email to the author, March 3, 2008. Cf. Teilhard de Chardin (1959), published posthumously.

53 LaMothe (1972), quotations on xi. See also Kress (1999). Kress is the CIA agent who arranged Puthoff and Targ's first CIA grants. See also Puthoff (1996); Targ (1996); and Schnabel (1997), 94. On earlier fears over the missile and manpower gaps, see Kaiser (2006a).

54 Kress (1999), 71; Schnabel (1997), 97; and Kripal (2007), 340, 341.

55 Puthoff and Targ (1976), 335. Puthoff expanded upon the protocol in a February 1978 letter to scientists who sought to replicate the remote-viewing findings at the U.S. Army Aberdeen Proving Ground in Maryland. Puthoff's letter was reprinted as "Appendix B: SRI Protocol" in U.S. Army Material Systems Analysis Activity (1979), 57–62.

56 Puthoff and Targ (1976), 344; see also Targ and Puthoff (1974), 606.

57 Targ and Puthoff (1974) and Puthoff and Targ (1976).

58 Gilliam (1978a), 4 ("If a man"); see also Petit (1973), 1 ("mystic powers," "supernatural phenomena").

59 Puthoff and Targ published several descriptions of their remote-viewing work in the open literature, in addition to filing dozens of classified reports. See esp. Targ and Puthoff (1974, 1975, 1977); Puthoff and Targ (1976, 1979); Puthoff (1996); and Targ (1996). On "ESPionage," see Kripal (2007), 340–45; and Kress (1999). Several detailed, if credulous, books on the remote-viewing projects have been published since the first wave of declassification in the mid-1990s. See esp. Schnabel (1997) and Smith (2004). See also Ronson (2004).

60 Rauscher interview (2008).

61 Rauscher interview (2008); Rauscher correspondence with John Wheeler, in *JAW*, series I, box "Princeton University, Mathematical - RA," folder "Rauscher, Elizabeth A"; and series II, box "Princeton University, Dept of Physics 3 - Reg," folder "Rauscher, Elizabeth A."

62 Rauscher's early articles on relativity included Rauscher (1972, 1973). These grew out of her long unpublished technical report: Rauscher (1971).

63 Rauscher applied her ideas about multidimensional geometry to psi phenomena such as remote-viewing in Rauscher (1979; 1983a, b); and Ramon and Rauscher (1980). Although the motivations are obviously rather different, Rauscher's early multidimensional models bear some resemblance to recent work at the forefront of physics and cosmology on universes with large extra dimensions. See esp. Arkani-Hamed, Dimopoulos, and Dvali (1998); Randall and Sundrum (1999); and Guth and Kaiser (2005).

64 Rauscher interview (2008). Puthoff and Targ explicitly cited Bell's theorem and quantum nonlocality as a likely explanation in Puthoff and Targ (1976), 349 and (1979), 42—although in each case they cited the wrong paper by Bell (his 1966 review article on hidden variables, rather than his 1964 paper on nonlocality), thus skewing citation searches based on Bell's 1964 article. Had they cited the correct Bell paper, the proportion of articles on Bell's theorem contributed by participants in the Fundamental Fysiks Group would have been even higher.

65 Rauscher interview (2008). Rauscher is described as the house theorist for Puthoff and

Targ in Dewar (1977), 12. See also the discussion following Targ and Puthoff's presentation in Oteri (1975), 175–79. A follow-up report, prepared by other members of the Fundamental Fysiks Group, continued in Rauscher's vein. See Hubbard and May (1982).

66 Rauscher, "Is ESP a science? Open meeting Wed.," *The Magnet* (August 28, 1975), page not numbered ("any subject"), clipping in *EAR*; and Rauscher interview (2008). On particle physicists' "golden events," see esp. Pickering (1984) and Galison (1987, 1997).

67 Rauscher, Weissmann, Sarfatti, and Sirag (1976); and Rauscher interview (2008).

Chapter 5: New Patrons, New Forums

1 David Marks and Richard Kammann, two psychologists in New Zealand, emerged as the most persistent critics of the Puthoff-Targ remote-viewing work, and carried on an extended debate with the SRI researchers. See Marks and Kammann (1978) and (1980), chaps. 2 and 3; Tart, Puthoff, and Targ (1980); Marks (1981, 1986); Puthoff and Targ (1981); and Marks and Scott (1986). See also Hyman (1977, 1995); and Alcock (1981, 1988). My thanks to Elizabeth Scalia for providing a copy of the 1995 American Institutes for Research report. All of these interactions mirrored earlier run-ins between experimental psychologists and parapsychology researchers in the 1930s, which had likewise turned on abstruse statistical arguments: Mauskopf and McVaugh (1980), chap. 9.

2 Romm (1977), 15 (comparison to Lysenko); and Randi (1982), 131–60.

3 Portions of Wheeler's AAAS talk, along with his petition to the AAAS leadership, are reproduced in Gardner and Wheeler (1979). See also Anon. (1979), C2; and Randi (1982), 228–32. On "moonshine," see also Jack Sarfatti, "The quantum mechanics of superluminal communication," unpublished typescript received in John Wheeler's office July 2, 1979, on p. 6, in *JAW*, Sarfatti folders; Sarfatti, "Research bulletin #5," October 6, 1979, in *JAW*, Sarfatti folders; and Jack Sarfatti to John Wheeler, January 20, 1982, in *JAW*, Sarfatti folders.

4 On CSICOP, see Pinch and Collins (1984), 539 ("scientific-vigilante"); Hess (1993), 11–13; Rensberger (1976), 19; Dewar (1977), 11; and Gilliam (1978b), 4. On ASTOP, see materials in *JAW*, series III, box "D - Extrasensory Perception 1," folder 1, and box "Extransensory Perception 2 - Hi," folders 2–5.

5 On funding for remote viewing from CIA and Pentagon sources, see Kress (1999), 71, 79–81; and Schnabel (1997), 97, 142, 205, 206, 220, 221, 320, 321, 380. On the Aberdeen tests, see U.S. Army Material Systems Analysis Activity (1979), quotations on 48, 49. See also Ronson (2004).

6 Geiger (1986), chaps. 4 and 5; Owens (1990); Kohler (1991); Lowen (1997), chap. 1; and Lécuyer (2010).

7 Young (1972), 45 ("science can best serve"). See also Young (1979) and Mishlove (1975), 263–278.

8 Nick Herbert, email to the author, June 2, 2008 ("wonderful intellectual salon"); Saul-Paul Sirag, emails to the author, June 5 and 6, 2008. See also the list of seminar series for the winter 1976 session at the Institute for the Study of Consciousness, copy in *SPS*; and Saul-Paul Sirag, "Consciousness Theory Group," unpublished report dated March 1977, in *SPS*.

9 Herbert, email to the author, June 2, 2008 ("psychic healing," "weird vibrations," "We would take any drug"); Sirag, emails to the author, June 5 and 6, 2008. See also Beverly Rubik and Nick Herbert, "A meeting of minds in Big Sur," n.d. (ca. February 1987), in *NH*. Young presented his own theories in Young (1976a and b).

10 Fred Alan Wolf, email to the author, November 12, 2007; and Jack Sarfatti, email to the author, November 27, 2007.

11 Marin (1975); Litwack (1976); Pressman (1993), 2, 50, 89, 90, 93, 212; and Fuller interview (2007). Sympathetic biographies of Erhard include Bartley (1978) and Self (1992).

12 Erhard interview (2010).

13 Rupert (1992), 129 ("excessively confrontational"); and Glass, Kirsch, and Parris (1977), 245 ("employ a confrontational, authoritarian model"). See also Kirsch and Glass (1977).

14 Anon. (1977b), 95.

15 Critical coverage included Marin (1975); Fenwick (1976); Heck and Thompson (1976); Kornbluth (1976); Litwak (1976); and Woodward (1976). More positive accounts from the time included Bry (1976) and Bartley (1978).

16 Erhard interview (2010).

17 Pressman (1993), 89, 90 ("San Francisco's most dazzling salon," "With a formal education"); and Fuller interview (2007).

18 Fred Alan Wolf, email to the author, September 24, 2007; Wolf interview (2007); Jack Sarfatti, email to the author, November 27, 2007; Saul-Paul Sirag, email to the author, December 11, 2007; and "Articles of incorporation of Physics/Consciousness Research Group," corporation number C0729058 (January 7, 1975), available from California Secretary of State, Business Programs Division; quotation from p. 1. Sarfatti noted that he filed the legal documents for incorporation himself in Physics/Consciousness Research Group, "A modest proposal to the Foundation for the Realization of Man," unpublished grant proposal, February 11, 1976, on 4 (copy in *SPS*).

19 Barbara Stevenson (former chief of staff, Office of Werner Erhard), email to the author, July 15, 2010; Erhard interview (2010); and Physics/Consciousness Research Group, "A modest proposal to the Foundation for the Realization of Man," February 11, 1976, on 4 (copy in *SPS*).

20 Sirag (1977b), 79, 80 ("communicate the excitement"); and Sarfatti (1974b).

21 Jack Sarfatti, "Excerpts from 'The Time Traveller's Handbook,'" dated May 31, 1976, on 3 ("Be here–Now!"), 5 (Beatles), 9 ("The only thing"), 10 ("Scientific speculation"), copy in *JAW*, Sarfatti folders. Cf. Dass (1971). On Alpert's and Leary's exploits at Harvard and beyond, see Wasserman (2000) and Lattin (2010).

22 Robert Anton Wilson, "Scientists confess strange encounters," undated clipping, ca. 1976, in *SPS*. Sirag suggests that the article might have originally appeared in a local underground paper such as the *Berkeley Barb*: Sirag, email to the author, July 29, 2010. Wilson expanded on his interactions with Sarfatti, Sirag, and the rest in Wilson (1977, 1979).

23 "Physics/Consciousness Program De Anza - Foothill College, Spring Quarter, 1976"; Jack Sarfatti, "Distribution list—Prepublication copies," May 22, 1975, in *JAW*, Sarfatti folders. See also Saunders and Norton (1976), 22. On the central place devoted to Puthoff and Targ's remote viewing in Sarfatti's presentations on quantum theory, see the lecture outline for the "Philosophy of quantum mechanics" course in the De Anza - Foothill College circular; and Jack Sarfatti, "Einstein: On quanta and human awareness," lecture notes for another PCRG seminar, n.d., ca. 1976, in *JAW*, Sarfatti folders.

24 "Physics/Consciousness Research Group" flyer (n.d., ca. 1976), ("the metaphors arising"), in *JAW*, Sarfatti folders.

25 Sarfatti (1977b), 75 ("warm personal relationship").

26 Physics/Consciousness Research Group, "A modest proposal to the Foundation for the Realization of Man," February 11, 1976, on 4 ("personal account of Werner Erhard"), copy in *SPS*; Robert W. Fuller to Saul-Paul Sirag, April 20, 1976, in *SPS*. See also Sarfatti (2002a), 44n64; Sarfatti, email to the author, November 27, 2007; and Sirag, email to the author, July 13, 2010.

27 Fowler (1989), B8 (on Vietnam-era intelligence work); and Sirag (2002), 120.

28 Delores M. Nelson (Information and Privacy Coordinator, CIA), letter to the author, November 4, 2008, in response to FOIA request F-2009-00110; David M. Hardy (Records Management Division, FBI), letter to the author, January 28, 2009, in response to FOIPA No. 1121034-000; Pamela N. Phillips (Acting Initial Denial Authority, National Security Agency), letter to the author, October 28, 2008, in response to FOIA case 57343; and Alesia Y. Williams (Chief, Freedom of Information Act Staff, Defense Intelligence Agency), letter to the author, November 30, 2009, in response to FOIA case number 0019-2009.

29 Fowler (1989), B8, mentions Koopman's work on the military training films. See also file 46-11705, FBI Los Angeles field office, dated November 8, 1976, concerning the allegations of "fraud against the government," released in response to FOIPA No. 1121034-000.

30 Fowler (1989), B8; and Sirag (2002), 120.

31 Fowler (1989), B8; Sirag (2002), 120, 121; and Sirag, email to the author, July 13, 2010. Koopman was listed as a participant in the Berkeley PCRG "Tao-of-Physics Discussion Group," October 27, 1976, in *SPS*. See also Leary with Wilson and Koopman (1977).

32 Jack Sarfatti, "Physics Consciousness Research Group: Financial statements for the ten month period ended 1/31/77," in *JAW*, Sarfatti folders.

33 Kripal (2007), 12 (on 1962 catalog cover); and Goleman, (1985), C6 ("valhala"). See also Anderson (1983).

34 Kripal (2007), chap. 2.

35 Erhard interview (2010). On Erhard's and Murphy's relationship, see also Anderson (1983), 272–77.

36 Sarfatti, email to the author, November 27, 2007; Herbert, email to the author, November 28, 2007; Sirag, email to the author, December 11, 2007; and Michael Murphy and Jack Sarfatti, "Physics and consciousness," 1976 Esalen catalog, quotations on 3, copy in *NH*.

37 Jack Sarfatti to Richard Feynman, September 5, 1975, in *RPF* 29:1. On Feynman's visits to Esalen, see Feynman to Esalen Institute, March 18, 1974, in *RPF* 24:16; and Feynman with Leighton (1985), 301–8. On John Lilly's work and connection to Esalen, see Burnett (2010).

38 Feynman and Hibbs (1965), 23 ("in the nature"). On Feynman's college philosophy course, see Feynman with Leighton (1985), 45–52.

39 Richard Feynman to Jack Sarfatti, September 22, 1975, in *RPF* 29:1.

40 David Finkelstein interview (2007); David Finkelstein to Richard Feynman, June 9, 1975, in *RPF* 25:5. Finkelstein demonstrated that a black hole's event horizon acts as a one-way membrane: matter can fall in, but cannot pass back out. See Finkelstein (1958).

41 Finkelstein interview (2007). See also Finkelstein (1969; 1972a, b; 1974); and Finkelstein, Frye, and Susskind (1974).

42 Karl H. Pribram to Werner Erhard, February 4, 1976, copy in *SPS*.

43 Nick Herbert, invitation to the February 1988 Esalen workshop (on audiovisual aids), and Beverly Rubik and Nick Herbert, "A meeting of minds in Big Sur," report on the February 1987 Esalen workshop (crystals; "break the old conference mold"), both in *NH*. On LSD trips at Esalen, see Sirag (2002), 110, 111.

44 On reserving baths and other quoted regulations, see Nancy Kaye Lunney (Programs office, Esalen Institute) to Richard Feynman, n.d., ca. January 1983, in *RPF* 24:16.

45 Nick Herbert, report on "Bell's theorem and the nature of reality," January 1980, in *NH*. On the Esalen workshops, see also Nick Herbert, email to the author, November 28, 2007; and Saul-Paul Sirag, email to the author, December 11, 2007.

46 Herbert, "Bell's theorem and the nature of reality," January 1980, in *NH*.

47 Kripal (2007), 302, 309, 310; Capra (1988), 117–19. Nick Herbert filed detailed (and quite amusing) reports after each year's Esalen workshop (copies in *NH*); my thanks to him for sharing these reports with me.

48 Richard Feynman to Faustin Bray, February 25, 1983 ("primitive drums playing"), in *RPF* 24:16; on Feynman's adventures at Esalen, see Feynman with Leighton (1985), 301–8.

49 Sirag (1977b), 81 (on fund-raising for Clauser's experiments); Ed Fry to Olival Freire, August 5, 2005, in Freire's possession. On Fry's experiments, see Freire (2006), 593, 594; and Fry and Walther (2002). My thanks to Olival Freire for sharing his correspondence with me.

50 Herbert, email to the author, November 28, 2007; Nick Herbert to John S. Bell, March 29, 1982 ("merry"), in *NH*; and Clauser (2002), 73. I saw Clauser's plaque while interviewing him at his home: Clauser interview (2009).

51 H. Dieter Zeh, email to the author, May 10, 2008; Nick Herbert, annual reports on Esalen workshops for 1980, 1981, and 1983, in *NH*. On the slow recognition of Zeh's work on decoherence, see also Zeh interview with Fábio Freitas (2008) and Zeh (2006).

52 D'Espagnat (1971b). D'Espagnat organized workshops on Bell's theorem for the Italian summer schools at Varenna (1970) and Erice (1976); see d'Espagnat (1971a); and Freire (2004), 1755, 1756 and (2006), 592, 603.

53 Bernard d'Espagnat to John A. Wheeler, February 16, 1982, in *JAW*, series II, folder "D'Espagnat, Bernard." See also Nick Herbert to John Bell, March 29, 1982, in *NH*.

54 Nick Herbert to John S. Bell, March 29, 1982, in *NH*; and Nick Herbert, emails to the author, November 28, 2007, and March 7, 2008.

55 Stapp (1980), 794; Clauser interview with Bromberg (2002), 38, 39 ("bunch of nuts," "open discussion forum"), 84, 85 ("wealthy Los Angelinos"). Materials pertaining to the twenty-fifth reunion, held in November 2000, can be found in *EAR*; photographs from the event are available at http://quantumtantra.com/reunion.html (accessed March 7, 2008).

Chapter 6: Spreading (and Selling the Word)

1 Goudsmit (1973), 357; John Clauser emails to the author, July 8, 2009; and Clauser (2002), 72.

2 Stapp interview (2007); and Freire (2006), 602, 603. My thanks to Olival Freire for sharing a copy of the tables of contents from the difficult-to-find *Epistemological Letters*.

3 Biographical information on Einhorn is drawn primarily from McCormick (1970); Walter (1979); and Levy (1988).

4 Kuhn (1962). On the book's impact, see Garfield (1987); and Owen Gingerich, email to the author, November 6, 2009 (on the book's dominance across Harvard's curriculum). See also Ira Einhorn to Thomas Kuhn, July 11, 1966, in *TSK* 11:10. An editor at the University of Chicago Press reports that the second (1970) and third (1996) editions of Kuhn's book together have sold more than half a million copies to date; sales figures for the original (1962) edition, also published by Chicago, are more difficult to reconstruct: Karen Darling, email to the author, November 2, 2009.

5 Kuhn (1962), chaps. 6–8. On X-rays and nuclear fission as examples of anomalies, see 57–61.

6 Ira Einhorn to Thomas S. Kuhn, January 16, 1964, in *TSK* 11:10.

7 Kuhn to Einhorn, January 24, 1964, in *TSK* 11:10.

8 Masterman (1970); and Kuhn (1970), 174–210.

9 See the correspondence between Einhorn and Kuhn from 1964 through 1966 in *TSK* 11:10.

10 Einhorn to Kuhn, September 27, [1964], and Kuhn to Einhorn, October 26, 1964, in *TSK* 11:10.

11 Einhorn to Kuhn, November 1, 1964, in *TSK* 11:10.

12 Kuhn to Einhorn, November 11, 1965, and Einhorn to Kuhn, November 18, 1965, in *TSK* 11:10.

13 Einhorn to Kuhn, November 18, 1965 ("sterile") and July 11, 1966 (from California), in *TSK* 11:10; Thompson (1971), 26–39; and Levy (1988), 128.

14 MacNamara (1967), 147.

15 MacNamara (1967), 147–49.

16 Einhorn (1970/71), 57 ("Today's mysticism"); Einhorn to Kuhn, November 27, [1966], in *TSK* 11:10; and Levy (1988), 76, 77 ("Intro to Hippiedom").

17 Einhorn to Kuhn, November 27, [1966], in *TSK* 11:10.

18 Einhorn (1966); see also Einhorn (1970/71).

19 McCormick (1970), 58; and Levy (1988), 81. On the San Francisco Be-In, see Rorabaugh (1989), 141; and Lee and Shlain (1992), 159–64.

20 McCormick (1970), 58.

21 Einhorn to Kuhn, November 27, [1966], in *TSK* 11:10; and Levy (1988), 93, 94. On Alpert's and Leary's psychedelic exploits at Harvard, see Lee and Shlain (1992), 84–89; Wasserman (2000); and Lattin (2010). On Hoffman, Rubin, and the "Yippies," see Lee and Shlain (1992), chap. 8; and Gitlin (1997), chap. 9.

22 Kuhn to Einhorn, December 19, 1966; see also Einhorn to Kuhn, February 26, 1968, in *TSK* 11:10; and Freeman Dyson, email to the author, April 3, 2008.

23 McCormick (1970); and Walter (1979), 148. On controversy over who deserved credit for organizing the first Philadelphia Earth Day events, see Levy (1988), 114–20; and the open letter from the "Earth Week Committee of Philadelphia," dated November 27, 1998, and published as a letter to the editor in the *Philadelphia Inquirer*, available at http://www.amgot .org/ einhorn/eday.htm (accessed February 5, 2009).

24 On Puharich and his alleged military and CIA ties, see Lee and Shlain (1992), 158; and Levy (1988), 128–30. Puharich received at least a dozen patents in the United States for medical inventions (most of them concerning hearing-aid improvements) between 1957 and 1983; he received duplicate patents for the same inventions throughout Europe. See http://www .google.com/patents and http://ep.espacenet.com (both accessed January 26, 2010). My thanks to Alex Wellerstein for his help with the patents search.

25 Levy (1988), 128–30.

26 Levy (1988), 131, 132; see also Wilson (1981).

27 The Bell executive was Moses Hallett; see Walter (1979), 145.

28 Moses Hallett, quoted in Walter (1979), 146.

29 Walter (1979), 145, 146; Levy (1988), 121–25. Cf. Turner (2006), chap. 3.

30 Walter (1979), 145, 146; and Levy (1988), 132.

31 Moses Hallett quoted in Walter (1979), 145.

32 On the phrase "internet before the internet," see Lopez (1997) and Baker (1999).

33 Einhorn (1977), 74; and Walter (1979), 146.

34 Walter (1979), 146–48; Levy (1988), 156–61.

35 Henderson (1979), 89; and Moses Hallett quoted in Walter (1979), 146.

36 Freeman Dyson, email to the author, April 3, 2008.

37 The Diebold Corporate Issues Program, "The emergence of personal communication networks among people sharing the new values and their possible use in sensitizing operating management," n.d. (1978), copy available in the John Diebold papers, acc. #8234-84-01-30, box 90, American Heritage Center, University of Wyoming, Laramie. My thanks to Alex Wellerstein and Sarah Rundall for locating and retrieving this report.

38 Smith (1976), 103, 104. See also Rauscher interview (2008); and Nick Herbert, email to the author, February 26, 2008.

39 Jack Sarfatti and Fred Alan Wolf, "A Dirac equation description of a quantized Kerr space-time," unpublished preprint (1973), copy in *JAW*, Sarfatti folders. See also Bearden (1988).

40 Rosenberg (1978); Anon. (1978a); and Levy (1988), 315–18.

41 Drucker (1978).

42 Levy (1988), 189, 190, 196, 253–55.

43 Ira Einhorn, letter to the author, June 3, 2008; Anon. (1987a); Anon., "William Grant White-head," obituary posted on a website maintained by the Princeton University class of 1965, available at http://www.pu65.com/memorials/whitehead.html (accessed January 26, 2010); and Levy (1988), 130, 131.

44 Puharich (1973), with introduction by Einhorn on vii–xii, and (1974), 8.

45 Toben with Sarfatti and Wolf (1975); Fred Alan Wolf, email to the author, April 10, 2008; Sarfatti interview (2009); Wolf interview (2009); Toben interview (2010); and Sarfatti (2002b), 28 ("altered state of consciousness"). Editions appeared in German (Synthesis-Verlag, 1980 and 1984; Fischer-Taschenbuch-Verlag, 1990); and in Japanese (Seidosha, 1985).

46 LeShan (1974); and Toben interview (2010). On the Esalen Institute Publishing Program, see Anderson (1983), 191, 192.

47 Sarfatti (2002a), 44n64, 48n81; Levy (1988), 196; and Capra (1988), 44–46.

48 Stapp interview (2007).

49 Ira Einhorn, open letter dated September 1, 2002, entitled "A snapshot of my 70's," reprinted in Sarfatti (2002b), 160–64, on 161; and Walter (1979), 146. Cf. Young (1976b), vii, and (1976a).

50 Leonard (1978), ix, x; and Levy (1988), 159. On Leonard's connections to Esalen and humanistic psychology, see also Kripal (2007), chap. 9.

51 Mishlove (1975) and Mishlove interview (2008). On Geller's performance in Zellerbach Auditorium, see also Puharich (1974), 231–34.

52 See the list of "members and associates of PCRG," in Physics/Consciousness Research Group, "A modest proposal to the Foundation for the Realization of Man," February 11, 1976, on 5, in *SPS*.

53 Zukav (1979), 7, 8 ("godfathers"), 31 ("I had spoken often").

54 Zukav (1979), 31–34.

55 Zukav (1979), 52–54.

56 Zukav (1979), 56 ("Physics has become a branch"), 136 ("The exact sciences," "If the new physics"), 327 ("Do not be surprised"). On Zukav's reduction of physics to psychology, and his rampant use of anthropomorphic metaphors, see also Leane (2007), 95–105.

57 Zukav (1979), 176 (on Ram Dass), 309–17 (on connections to parapsychology). On the use of Ram Dass's *Be Here Now*, see also Jack Sarfatti, "Excerpts from 'The Time Traveller's Handbook,'" May 31, 1976, on 3, in *JAW*, Sarfatti folders.

58 Physics/Consciousness Research Group, "A modest proposal to the Foundation for the Realization of Man," February 11, 1976, on 32 ("This is the kind of experience"), 35 (photo and caption of Sarfatti and Huang); copy in *SPS*.

59 Jack Sarfatti, "Book proposal, *Faster than Light: Breaking the Space Time Barrier*," n.d. (ca. March 1979), copy in *JAW*, Sarfatti folders.

60 March (1979), 55; and Sokolov (1979), BR5. Not all reviews were so negative. The first review in the *New York Times*, for example, lauded the book, singling out the final chapter (centered around Sarfatti's ideas about Bell's theorem and consciousness) as the book's main "payoff": Lehmann-Haupt (1979), C23.

61 Zukav (2001), ix, 4, 267n, 328n. The 2001 HarperCollins edition is a reprint of the amended 1979 edition.

62 Jack Sarfatti, "Excerpts from forthcoming book, *Faster-than-Light?*" n.d. (ca. late June 1979), on 7; "Research bulletin #5," October 6, 1979, on 2; "For the public record," November 3, 1980, on 2; and "Higher Intelligence Agency," October 28, 1981, on 3; all in *JAW*, Sarfatti folders. See also Sarfatti (2002a), 131, 132.

63 Anon. (1980), 14; and Appelbaum (1983), 39, 40. Cf. Hofstadter (1979).

64 Appelbaum (1983), 39, 40.

65 Leane (2007), chap. 1; and Lewenstein (1992).

66 Levy (1988), 196, 206. The conference proceedings appeared as Puharich (1979b).

67 Walter (1979), 147; Levy (1988), chaps. 9, 10.

68 Walter (1979), 144, 149; Baker (1999); and Levy (1988), 21–25.

69 Rauscher interview (2008); Walter (1979), 271; and Levy (1988), 25. The heiress was Barbara Bronfman, who had married into the Seagram's liquor family fortune.

70 Walter (1979), 147, 149; and Levy (1988), 251.

71 Jack Sarfatti, "Quantum Communications Network, distribution list for May 16, 1980 disclosure of EPR COMM device," May 22, 1980, copy in *JAW*, Sarfatti folders.

72 Jack Sarfatti, "Jubilee for Zarathustra," n.d. (summer 1979), copy in *JAW*, Sarfatti folders; and Levy (1988), 254 ("He certainly didn't *act*").

73 Levy (1988), 255, 256, 265.

74 Matza (1997), E1.

75 Slobodzian and Sataline (1997), A8; and Baker (1999).

76 Associated Press (1997), B1; Molotsky (1998), A22; and Loyd (1999), B2.

77 "Post Conviction Relief Act," 42 Pa. Cons. Stat. §9543(c), available at http://www.lexisnexis .com (accessed February 8, 2010); Moran (1988), B1; Molotsky (1998), A22; and Soteropoulos (2002a), B2.

78 Baker (1999); see also Rubin (1999), A1. Jack Sarfatti reproduced some email messages from Einhorn (including his username) in Sarfatti (2002b), 164.

79 Whitney (1999), A14; and Anon. (2001), A17.

80 "Ira Einhorn's 1999 statement on his innocence," reproduced in Sarfatti (2002b), 165–67; and Einhorn, "An American travesty: My story," 19pp typescript mailed to the author on December 3, 2009. Cf. Ditzen (1999), A8; and Slobodzian (2003), B3.

81 Loyd (1999a), A1; Baker (1999).

82 My research assistant, Alex Wellerstein, and I tried for more than two years to locate a copy of the 1979 arraignment-hearing transcript. After contacting the city and county clerks, the Philadelphia district attorney's office, and all of Einhorn's previous attorneys, I wrote to Einhorn to ask if he knew where copies might be. (His prison writings reveal close familiarity with other transcripts from his various trials.) Einhorn replied that he receives similar queries every few months: Einhorn, letter to the author, December 3, 2009. Steven Levy quoted extensively from the 1979 hearing transcripts in Levy (1988), 21–25.

83 Lin (2002a), A10.

84 Levy (1988), chap. 10; and Einhorn, letters to the author, June 3, 2008, and December 3, 2009. See also Baker (1999); and Soteropoulos (2002b), B1.

85 Sarfatti (2002b), 156, 157, 164n, 172; and Rauscher interview (2008). See also Lin (2002b), A10.

86 Judge William Mazzola, quoted in Soteropoulos (2002c), A1; and Einhorn, "An American travesty," mailed to the author on December 3, 2009.

Chapter 7: Zen and the Art of Textbook Publishing

1 Elizabeth Rauscher, "List of lectures presented at the Fundamental 'Fysiks' Group, LBNL," in *EAR*. Capra spoke on May 16 and 21, 1975; the group's first (organizational) meeting was held on May 7, 1975. On the 1976 discussion series, see the flyer "Tao-of-Physics Discussion Group," ca. October 27, 1976, in *SPS*. On Sarfatti's and Capra's meetings in Europe, see Sarfatti (2002a), 44n64, 48n81; on Capra's Esalen workshops, see Capra (1988), 117–20.

2 Capra (1988), 22–25. The Santa Cruz physicist who invited Capra was Michael Nauenberg; see Nauenberg interview with Jarrell (1994), 37.

3 Capra (1988), 23 ("schizophrenic life"), 27 (on Alan Watts). On Watts's connections with Esalen, see Kripal (2007), 59, 73, 76, 99, 121–25.

4 Capra (1988), 34. Capra opened *The Tao of Physics* by recounting his "Dance of Shiva" experience on the beach: Capra (1975), 11.

5 Capra (1988), 34.

6 Fritjof Capra to Victor F. Weisskopf, November 12, 1972, in *VFW*, box NC1, folder 26.

7 Capra to Weisskopf, November 12, 1972, in *VFW*. On Weisskopf's career, see Kaiser (2007c); and Weisskopf (1991). The oft-stolen textbook was Blatt and Weisskopf (1952).

8 Capra to Weisskopf, January 11, 1973 (quotations); Capra to Weisskopf, March 23, 1973, both in *VFW*, box NC1, folder 26.

9 Weisskopf to Capra, April 19, 1973, in *VFW*, box NC1, folder 26.

10 Capra (1988), 44, 45 ("rather hard-headed"), 53, 54. Capra's early essays included Capra (1972, 1974). On Chew's bootstrap program, see Kaiser (2005), chaps. 8 and 9.

11 Capra (1988), 46; and Appelbaum (1983), 39, 40.

12 Capra to Weisskopf, May 7, 1976; and Weisskopf to Capra, June 21, 1976 (quotations), both in *VFW*, box NC1, folder 26.

13 On sales, see Capra to Weisskopf, July 8, 1976, in *VFW*, box NC1, folder 26; and Appelbaum (1983), 39, 40. On subsequent editions and translations, see the full list at http://www.fritjofcapra.net (accessed June 12, 2008).

14 Several years later, two comparative-religion scholars scoffed that Capra's book "seemed to misinterpret Asian religions and cultures on almost every page": Diem and Lewis (1992), 49.

15 De Witt (1977), C1 ("Tall and slim"); and Capra (1975), quotations on 307.

16 Capra (1975), 19 (sutras, Vedas), 24 ("Our tendency to divide"), 25 ("The further we penetrate").

17 Capra (1975), 57.

18 Capra (1975), 141 ("The idea of 'participation'"). Although Capra emphasized quantum entanglement and interconnectedness throughout chapters 4 and 10, he did not mention Bell's theorem by name in the first edition of his book. Many of the physicists' articles from which he quoted in his first edition, however, were devoted to Bell's theorem and the nature of quantum nonlocality, especially Stapp (1971), quoted by Capra on 132, 136, and 139; and Bohm and Hiley (1975), quoted by Capra on 138. Capra likewise drew upon Wheeler (1973), quoted by Capra on 141. These were precisely the same sources cited by Sarfatti (some in preprint form) in Sarfatti (1974b), written while Sarfatti and Capra were both in London during the fall of 1973. Capra added a lengthy discussion of Bell's theorem in the afterword to the book's second edition: Capra (1984), 299–303.

19 Capra (1975), 160 ("this notion," coat of arms); see also 114, 115, and chaps. 11–13.

20 Capra (1975), 34 (quotations); see also 305, 306.

21 Miles (1982), N8 ("sold amazingly well"); Dull (1978), 387 ("pleasing way"); Jonathan Westphal, in Clarke, Parker-Rhodes, and Westphal (1978), 294 ("Capra is clearly in earnest"); and

Shimony (1981), 436 ("quotes charming *haikus*"). For other reviews, see Kauffman (1977); White (1979); Clarke, Parker-Rhodes, and Westphal (1978); Restivo (1978, 1982); and Clifton and Regehr (1990).

22 Dull (1978), 387, 389; and Chrisopher Clarke, in Clarke, Parker-Rhodes, and Westphal (1978), 289–91.

23 Restivo (1978), 151–55; and Scerri (1989), 688.

24 On inaccuracies of lumping all Eastern traditions into a single worldview, see Esbenshade (1982), 225; and Scerri (1989), 688. On diversity of interpretations among physicists, see Restivo (1978), 156–58; Shimony (1981), 436 ("dissolves the precise meaning"); and Clifton and Regehr (1990), 82, 85–88 ("muddling," 88), 101n17 ("schizophrenic"). On Capra's notion of parallel "empiricism," see Westphal, in Clarke, Parker-Rhodes, and Westphal (1978), 296; and Clifton and Regehr (1990), 77.

25 On charges of circular reasoning, see Esbenshade (1982), 226. On oscillations between complementary and confirmatory arguments, see Restivo (1978), 165–67; and Clifton and Regehr (1990), 77, 78. On accounting for similarities, see Westphal, in Clarke, Parker-Rhodes, and Westphal (1978), 298; and Parker-Rhodes, ibid., 291, 292 ("basic tendencies of the human mind"). On possible "contamination" of terms between the two traditions, see Restivo (1978), 153.

26 Asimov (1978), 19; and Bernstein (1978/79), 6–9.

27 Capra (1975), 25; and Mansfield (1976), 56.

28 Fritjof Capra to Victor Weisskopf, July 8, 1976, in *VFW*, box NC1, folder 26; Harrison (1979b), 779 ("This leads naturally"); and Scerri (1989), 688 ("Anyone involved"). Jack Sarfatti likewise adopted Capra's book as a textbook for one of his popular seminars on science and religion, run by the Physics/Consciousness Research Group: Sarfatti, "Physics/Consciousness Program De Anza-Foothill College, Spring Quarter 1976," on 4, 5, in *JAW*, Sarfatti folders.

29 Harrison (1979b), 782 (on "Son of Zen of Physics"); David Harrison, emails to the author, July 3, 2007, January 5, 2008, and April 17, 2008. Harrison included the quoted homework assignment with the January 5, 2008, email message. LeShan's "interesting test" originally appeared in LeShan (1969) and reappeared as a chapter in LeShan (1974). On Harrison's other interests at the time, see Harrison (1978, 1979a); and Harrison and Prentice (1980).

30 Harrison (1982b), 873, 874 (on physics majors); Harrison (1982a), 811 (on standard curricula) and 815 (acknowledgment to Herbert); Nick Herbert, email to the author, April 16, 2008 ("spirited" correspondence); and Harrison, email to the author, April 17, 2008. The first quantum mechanics textbook to include any material on Bell's theorem was Sakurai (1985), 223–32; see Ballentine (1987), 787.

31 Clifton and Regehr (1990), 73, 74.

32 Pedagogical critiques include Esbenshade (1982); and Scerri (1989); cf. Harrison (1982b), 873 ("most of these students"); and Harrison, email to the author, July 3, 2007 ("bums in the seats").

33 Several reviewers highlighted this "ideological" use of Capra's book: physicists could use it as a hedge against antiscientific sentiments of the day. See Kauffman (1977), 461; Restivo (1982), 39, 43, 45, 47, 53; and Scerri (1989), 688.

Chapter 8: Fringe?!

1 John Bell to Robert A. McConnell, May 20, 1986, reprinted in McConnell (1987), 51, 52. Jeremy Bernstein quoted from the same letter in his profile of Bell: Bernstein (1991a), 79, 80.

2   Margenau (1936); Lindsay and Margenau (1936); Margenau (1950, 1953, 1961, 1978).

3   Margenau (1966), 214, 215. See also Margenau (1956, 1957); and Margenau's correspondence with Gardner Murphy, president of the American Society for Psychical Research, ca. 1965–1967, in *HM*, series I, folder 1:9.

4   Henry Margenau and Lawrence LeShan to the editor of *Science*, April 24, 1978, in *HM*, series I, folder 1:6.

5   Henry Margenau and Lawrence LeShan, untitled letter to the editor (submission), ca. April 1978, in *HM*, series I, folder 1:6; emphasis in original.

6   Henry Margenau and Lawrence LeShan to the editor of *Science*, December 15, 1978, in *HM*, series I, folder 1:6.

7   Henry Margenau to Philip Abelson, January 22, 1979, in *HM*, series I, folder 1:6. See also Christine Karlik to Henry Margenau, February 12, 1979, in the same folder. Margenau and LeShan expanded upon their brief, unsuccessful submission to *Science* in their book: LeShan and Margenau (1982).

8   Ramon and Rauscher (1983); Rauscher (1983c); and Wigner (1983), 1479.

9   Hall, Kim, McElroy, and Shimony (1977), reprinted in Shimony (1993), 2: 323–31. Cf. Clauser and Shimony (1978).

10 Feinberg (1967). See also Anon. (1992), D26.

11 Feinberg (1969b); see also Feinberg (1969a, 1971). Feinberg later joined the advisory board of the Foresight Institute, founded in 1986 by nanotechnology enthusiast Eric Drexler. See the interview with Feinberg in the institute's newsletter, *Foresight Update 9*, originally published June 30, 1990, available at http://www.foresight.org/ Updates/Update09 (accessed December 21, 2009). On Einhorn's interactions with futurist Alvin Toffler and the Congressional Clearinghouse for the Future during the 1970s, see Levy (1988), 9, 159, 313.

12 Puharich (1979a), 10.

13 Feinberg (1975), 54–73. Rauscher thanked Feinberg for helpful discussions in several of her parapsychology papers: Rauscher (1979), 79, and (1983b), 115; and Ramon and Rauscher (1980), 669. Evan Harris Walker also thanked Feinberg for "constructive and valuable comments": Mattuck and Walker (1979), 140.

14 Rauscher delivered a seminar for Bohm's group at Birkbeck College in London on November 23, 1977; see flyer announcing the talk in *EAR*. Bohm met with the Fundamental Fysiks Group in Berkeley a few months later, in April 1978. See Elizabeth Rauscher to John A. Wheeler, n.d. [received in Wheeler's office on March 29, 1979], including enclosure of typed notes entitled "David Bohm Lectures," April 9 and 10, 1978, in *JAW*, series I, folder "Rauscher, Elizabeth"; and Elizabeth Rauscher, "The origins and operations of the era of the Fundamental Fysiks Group from 1975 to 1978," unpublished memo ca. November 2000, in *EAR*. See also Bohm (1981); and Peat (1997), 195–200, 225–31.

15 Costa de Beauregard (1967); cf. Bergmann (1969), 85–87. See also, e.g., Costa de Beauregard (1977). On Costa de Beauregard's 1975 Berkeley visit, see Sirag (2002), 109, 110; see also Rauscher, "The origins and operations of the era of the Fundamental Fysiks Group from 1975 to 1978," unpublished memo ca. November 2000, in *EAR*.

16 Compare Costa de Beauregard (1976) with (1978). See also Costa de Beauregard (1975, 1979).

17 Collery (1981). *Tonus*, the magazine in which the interview originally appeared, was a weekly newspaper aimed at medical professionals: Solange Collery, email to the author, December 23, 2009.

18 Mattuck (1967); see also *Science Citation Index* (1961–), s.v. "Mattuck, Richard D." On the importance of Mattuck's textbook, see Kaiser (2005), 243, 244, 268–70.

19 Mattuck (1977, 1978/79) and Mattuck and Walker (1979). On Mattuck's later work, see *Science Citation Index* (1961–), s.v. "Mattuck, Richard D."

20 Josephson (1992).

21 Wallace (1976), 4. See also Brian Josephson, "Foreword," in Puharich (1979b), 4, 5.

22 Josephson, Mattuck, Walker, and Costa de Beauregard (1980), 48–51, in response to Gardner and Wheeler (1979): 39, 40.

23 Durrani (2000), 5; and Cartlidge (2002), 10, 11. See also Professor Josephson's website at http://www.tcm.phy.cam.ac.uk/~bdj10 (accessed April 28, 2008).

24 Daniel Greenberger to John Clauser, February 6, 1985, in *JFC*, folder "Random correspondence." See also the related correspondence in the same folder. The conference proceedings, entitled *New Ideas and Techniques in Quantum Measurement Theory*, were published as a special issue of *Annals of the New York Academy of Sciences* 480 (December 1986): 1–632. See Greenberger (1986), xiii, xiv.

25 Participants who had won or would go on to win the Nobel Prize included Willis Lamb, Anthony Leggett, and Roy Glauber. See also Stapp (1986); Herbert (1986); and correspondence between Daniel Greenberger and John Clauser, ca. February–November 1985, in *JFC*, folder "Random correspondence."

26 Sarfatti quoted in Browne (1986), C3. See also Sarfatti interview (2009); and Wolf interview (2009).

27 Nick Herbert, email to the author, March 11, 2009.

28 See, e.g., Bernstein (1991a), 6, 7; Greene (2004), 113–23; Clegg (2006), chap. 3; and Davies and Gribben (2007), 224, 225. Although Aspect's experiment was certainly a crucial milestone, too much can be made of its role in jump-starting interest in foundations of quantum mechanics, a point to which I return in the coda. See also Freire (2004).

29 John F. Clauser to Christian Imbert, November 16, 1973, in *JFC*, folder "Random correspondence."

30 Clauser interview with Joan Bromberg (2002), 33, 34; and Clauser interview (2009).

31 D'Espagnat interview with Freire (2001), 11, 12. Recall from chapter 5 that d'Espagnat attended one of the Esalen workshops, holding forth on Bell's theorem and nonlocality in Esalen's famous hot tubs.

32 Aspect interview (2009); see also Aczel (2001), 177–80; and Freire (2006), 606–8. Before leaving for Cameroon, Aspect completed his "doctorat de troisième cycle" in 1971, a degree that lay somewhere between a master's degree and a PhD from a U.S. university. The textbook that inspired Aspect was Cohen-Tannoudji, Diu, and Laloë (1973).

33 John F. Clauser to John S. Bell, February 14, 1969, in *JFC*, folder "Random correspondence." Cf. Freedman and Clauser (1972).

34 Aspect interview (2009); Aspect (1976), 1949; and Freire (2006), 607.

35 Aspect (2002), 119, and Aspect interview (2009).

36 Aspect interview (2009); see also Freire (2006), 608.

37 Aspect (1976); and Aspect, Dalibard, and Roger (1982). See also Leggett (2009).

38 Aspect (1976, 1981). See also Jack Sarfatti to Nick Herbert, October 28, 1981, and Jack Sarfatti to John Wheeler, January 20, 1982, in *JAW*, Sarfatti folders.

39 Rauscher, "The origins and operations of the era of the Fundamental Fysiks Group from 1975 to 1978," unpublished memo ca. November 2000, in *EAR*; Mrs. Evan G. "Bootsie" Galbraith

to Jack Sarfatti, May 7, 1982, reprinted in Sarfatti (2002a), 152, 153; and Chickering interview (2009).

40 D'Espagnat (1979), 178, 179; Jack Sarfatti, "Memorandum for the record," April 7, 1982, in *JAW*, Sarfatti folders; and Saul-Paul Sirag, email to the author, April 7, 2009.

41 Alain Aspect to John Clauser, June 29, 1981, in *JFC*, folder "Random correspondence"; and John Clauser, email to the author, December 24, 2009. See also Alain Aspect to John Clauser, May 7, [1981?], in *JFC*, folder "Random correspondence"; and Clauser interview with Bromberg (2002), 33; Clauser interview (2009). The paper that Clauser edited was published as Aspect, Grangier, and Roger (1981). This was the first of Aspect's papers to report data collected with his new apparatus; he had not yet gotten the time-varying switches to work.

42 Aspect, Dalibard, and Roger (1982).

43 Aspect interview (2009); Clauser interview (2009); Jack Sarfatti to John Wheeler, n.d. (ca. October 15, 1984), in *JAW*, Sarfatti folders; and Sirag (2002), 109.

44 Fuller interview (2007). See also Fuller (1961); Fuller and Wheeler (1962); and Byron and Fuller (1969).

45 Fuller interview (2007). See also Fuller and Wallace (1975).

46 Fuller interview (2007). On the Solvay conferences, see, e.g., Mehra (1975).

47 Erhard interview (2010) ("actually make some difference"); and Fuller interview (2007). See also Sidney Coleman to Roman Jackiw, July 26, 1976, in *RWJ*.

48 Jackiw interview (2007).

49 Jackiw interview (2007). On Coleman, see also Marquard (2008); and Glashow (2008), 69. Coleman served on my physics dissertation committee at Harvard in the mid-1990s.

50 Sidney Coleman, form-letter invitation dated July 26, 1976; copies in *RWJ*, and in *RPF*, folder 25:1.

51 Original invitees included Curtis Callan, Geoffrey Chew, Roger Dashen, Ludvig Fadeev, Richard Feynman, David Finkelstein, Alfred Goldhaber, Jeffrey Goldstone, David Gross, T. D. Lee, Stanley Mandelstam, Yoichiro Nambu, Alexander Polyakov, Claudio Rebbi, Leonard Susskind, Gerard 't Hooft, and Ken Wilson. (See Coleman form-letter invitation, July 26, 1976.) Of these, Feynman, Gross, Lee, Nambu, 't Hooft, and Wilson have received the Nobel Prize. Almost all of those invited agreed to participate; they were joined by Steven Weinberg, John Wheeler, and Edward Witten. See Sidney Coleman and Robert Fuller to Richard Feynman, December 30, 1976, in *RPF*, folder 25:1.

52 Jackiw interview (2007). See also correspondence in *RWJ*, and in *RPF*, folder 25:1.

53 Howard Sherman to Roman Jackiw, December 23, 1976, in *RWJ*; see also Sherman to Richard Feynman, December 30, 1976, in *RPF*, folder 25:1.

54 Erhard interview (2010).

55 Sidney Coleman to Roman Jackiw, December 17, 1976 ("conscious policy"), in *RJW*; and Erhard interview (2010) ("I couldn't follow"). See also Jackiw interview (2007) and Fuller interview (2007).

56 Handwritten postscript on Robert Fuller to Richard Feynman, April 22, 1977, in *RPF*, folder 25:1; see also Erhard interview (2010).

57 Richard Feynman to Howard Sherman, February 4, 1977, in *RPF*, folder 25:1.

58 Sidney Coleman to Robert Fuller, March 1, 1977, in *RPF*, folder 25:1.

59 Roy Ascott [Acting President and Dean of the College, San Francisco Art Institute] to Jack Sarfatti, February 17, 1977, in *JAW*, Sarfatti folders.

60 Jack Sarfatti to David Finkelstein, February 15, 1977, in *JAW*, Sarfatti folders.

61 Jack Sarfatti, "Plato's anticipation of quantum logic," 16pp typed manuscript, April 8, 1977,

copy in *JAW*, Sarfatti folders. Wheeler wrote across his copy that he had responded to Sarfatti on February 28, 1977, and suggested Lowe (1927).

62 Victor Weisskopf to Jack Sarfatti, March 7, 1977, in *JAW*, Sarfatti folders. See also Weisskopf to Sarfatti, February 11, 1977, also in *JAW*, Sarfatti folders.

63 See, e.g., Heck and Thompson (1976); Kornbluth (1976); Litwack (1976); Woodward (1976); Fenwick (1976); and Glass, Kirsch, and Parris (1977).

64 Victor Weisskopf to Jack Sarfatti, March 15, 1977, in *JAW*, Sarfatti folders. Roman Jackiw does not recall Weisskopf ever raising the issue with him: Jackiw, email to the author, December 1, 2009.

65 Weiskopf to Sarfatti, March 7, 1977.

66 Martin Gardner to Jack Sarfatti, July 1, 1977, in *JAW*, Sarfatti folders. See also Gardner to Sarfatti, June 23, 1977, in *JAW*, Sarfatti folders.

67 Jack Sarfatti to Robert R. Curtis, Jr. [*est* house counsel], July 11, 1977, and circulated widely in photocopy form; copy in *JAW*, Sarfatti folders.

68 Sidney Coleman to Robert Fuller, March 1, 1977, in *RPF* folder 25:1; Roman Jackiw to Robert Fuller, May 5, 1977, in *RWJ*.

69 Jack Sarfatti, handwritten message dated July 11, 1977, written on top of Weisskopf to Sarfatti, March 7, 1977, and distributed widely in photocopy form; copy in *JAW*, Sarfatti folders.

70 Robert R. Curtis, Jr. [*est* house counsel] to Jack Sarfatti, July 8, 1977, in *JAW*, Sarfatti folders.

71 Sarfatti to Curtis, July 11, 1977; and Jack Sarfatti, typed memo "To: All physicists invited to *est* 'Quantum Gravity' conference and to the press," July 29, 1977, copy in *JAW*, Sarfatti folders.

72 Jackiw interview (2007) and Fuller interview (2007).

73 Norman Quebedeau, email to the author, December 2, 2009 ("well-heeled"). See also http:// www.imdb.com/name/nm2611476 (accessed December 11, 2009). Sarfatti sent a copy of Quebedeau's cartoon of "Dr. Jack Sarfatti and the Temple of *est*" to John Wheeler, prob- ably enclosed with Sarfatti's letter of July 11, 1977 (written on top of an earlier letter from Weisskopf to Sarfatti); copy in *JAW*, Sarfatti folders.

74 Kim Burrafato, email to the author, December 1, 2009 ("shouting with almost mad glee"); Michael Sarfatti (Jack's brother), email to the author, December 1, 2009; and Jack Sarfatti, email to the author, December 1, 2009.

75 The recording is available at http://www.qedcorp.com/book/psi/hitweapon.htm (accessed November 30, 2009). On Stephen Hill's *Hearts of Space* radio show, see http://www.hos.com (accessed December 11, 2009).

76 Fuller interview (2007); Anon. (1978b), 22; Gordon (1978), 43, 44.

77 Fuller interview (2007); and Gordon (1978), 42. See also Pressman (1993), 155–67. Cf. the Hunger Project website: http://www.thp.org (accessed June 30, 2010).

78 Roman Jackiw to Robert Fuller, May 5, 1977, in *RWJ*.

79 Susskind (2008).

80 McCoy, Perk, and Wu (1981), 760.

81 Among them was MIT physicist Jeffrey Goldstone, who argued strenuously in private with Coleman against getting involved with Erhard: Jeffrey Goldstone, personal communication, December 4, 2009.

82 Sidney Coleman to Laurel Sheaf, March 8, 1979, with cc to Steven Adler, Alan Luther, Roman Jackiw, and Werner Erhard; copy in *RWJ*.

83 Petit (1981), 32. A brief description of the article appeared on the newspaper's front page that day. My thanks to Charles Petit and Daniel Goldstein for supplying copies of the article.

84 Sidney Coleman, "Press reports of Franklin House conferences," memorandum dated February 3, 1981, in *RWJ*.

85 Carroll (1981), 23. Charles Petit, author of the January 1981 article on the "secretive" *est*-sponsored physics conference, confirmed that he was in regular contact with Jack Sarfatti at the time: Charles Petit, email to the author, December 1, 2009.

86 Sidney Coleman to Steven Adler, Werner Erhard, Roman Jackiw, et al., February 3, 1981, in *RWJ*. See also Roman Jackiw to Werner Erhard, February 6, 1981; and Doug Bell ["Office of Werner Erhard"] to Roman Jackiw, February 16, 1981, in *RWJ*.

87 Sidney Coleman to Steven Adler, Werner Erhard, Roman Jackiw, et al., March 23, 1982, in *RWJ*.

88 Michael Turner to Alan Guth, August 3, 1983, in *RWJ*. Guth was the organizer for the 1984 meeting. Guth was also the main supervisor of my physics dissertation, and coauthor on more recent work.

89 Michael Turner to So-Young Pi, August 22, 1983, in *RWJ*. Turner included copies of his letter to Guth of August 3, 1983, when writing to the other invited participants.

90 Roman Jackiw to Werner Erhard, August 16, 1983, in *RWJ*. On Turner's position at Fermilab, see Overbye (1991), 206–11.

91 Sidney Coleman to Steven Adler, Werner Erhard, Roman Jackiw et al., November 6, 1985, in *RWJ*. Ed Witten was listed as a confirmed participant in advance of the first meeting that Coleman and Jackiw organized: see Sidney Coleman and Robert Fuller to Richard Feynman, December 30, 1976, in *RPF*, folder 25:1.

92 Alan Guth, personal communication, October 3, 2007; Jackiw recalled similar behavior: Jackiw interview (2007).

93 Fenwick (1976), 43 ("zombielike faces") and 75 ("catatonic stares").

94 Sidney Coleman memo to Steven Adler, Werner Erhard, Roman Jackiw et al., February 3, 1981, in *RWJ*.

95 Greenberger (1986), xiii.

96 Roman Jackiw to Werner Erhard, April 5, 1985, in *RWJ*. See also Erhard's memo about his impending divorce, March 13, 1983; Erhard to Jackiw, n.d., ca. March 1983; and the other personal correspondence in *RWJ*. See also the correspondence between Feynman and Erhard in *RPF*, folder 25:1; Jackiw interview (2007); and Fuller interview (2007).

97 Anderson (1991), B10.

98 Zeman (1991), 8 ("embattled *est* guru," "his whereabouts"); Walker (1991), 31 ("pulled up stakes"); Macintyre (1992); cf. Pressman (1993), 263–75.

99 See, e.g., Lattin (1990), A4; Hubner (1990a, b); Richard Rapaport, "Respect," *San Francisco Focus*, ca. 1990, clipping in *RWJ*; and Gelman, Abramson, and Leonard (1991), 72. See also Pressman (1993), 142, 143, 253–58; cf. Self (1992) and Snider (2003).

100 Gonneke Spits to Roman Jackiw and So-Young Pi, January 24, 1991, in *RWJ*.

101 Roman Jackiw to Werner Erhard, May 29, 1991, in *RWJ*; correspondence between Jayne Sillari and Jackiw, spring 1991, in *RJW*; and Jackiw interview (2007).

Chapter 9: From FLASH to Quantum Encryption

1 Herbert (1975), 316.

2 Griffiths (2005), 427.

3 For a concise and accessible description of the various paradoxes, see Herbert (1988), quotation on 4 ("change yesterday today," attributed to Nic Harvard).

4 Nick Herbert, postcard to John Clauser, February 6, 1975 ("intrinsically almost obscenely

non-local"); and Clauser to Herbert, February 11, 1975, both in *JFC*, folder "Random correspondence."

5 Elizabeth Rauscher, "List of lectures presented at the Fundamental 'Fysiks' Group, LBNL," in *EAR*. Entries for September 12, 1975 (Sarfatti) and October 10, 1975 (Herbert).

6 Stapp (1977a), 202. As noted on p. 191, the article was based on lectures originally delivered in Trieste, Italy, in December 1975, though Stapp's analysis clearly benefited from the Fundamental Fysiks Group as well: he thanked John Clauser, George Weissmann, and other members of the group for "discussions that contributed significantly to the development of this paper" (p. 204).

7 Jack Sarfatti, "Disclosure document describing a 'faster-than-light quantum communication system," May 8, 1978, copy in *JAW*, Sarfatti folders.

8 Ibid.

9 Ibid., quotations on 2. See also Sarfatti's press release, "On fractals and the possibility of faster-than-light quantum communication," February 2, 1978, copy in *JAW*, Sarfatti folders.

10 For example, Eberhard spoke about "quantum physics, external reality, and tests of quantum theory" in February and March 1976: Elizabeth Rauscher, "List of lectures presented at the Fundamental 'Fysiks' Group, LBNL," in *EAR*.

11 Nick Herbert, email to the author, May 3, 2009.

12 Eberhard (1978), 410, 411.

13 Eberhard (1978), esp. 403, 404, 408, and 416.

14 Stated more formally, Eberhard's proof indicated that the superluminal connections inherent in individual events would always be masked or hidden from view when considering averages over many such events. So much for Sarfatti's signaling device. His scheme depended on detecting interference patterns at B. But any interference pattern, whether sharp or washed out, would be composed of many individual photons striking the screen, one at a time. Interference patterns, in other words, are a kind of statistical average.

15 *Science Citation Index* (1961–). Sometimes scientists repeated Eberhard's argument without citing his paper, perhaps because they had come up with the gist of the argument on their own, e.g. Bartell (1980b), 1358, 1359; and Aspect (1981), 78, 79.

16 Pagels (1982), chaps. 12 and 13, quotation on 174. See also Herbert (1988), 159, 161. On Pagels and Herbert's relationship, see Pagels (1982), vii; and Nick Herbert, emails to the author, November 28, 2007; December 1, 2007; and February 26, 2008.

17 Eberhard (1978), 410.

18 Jack Sarfatti, "Response to Stapp, 6/26/79," handwritten notes, June 29, 1979; copy in *JAW*, Sarfatti folders.

19 Jack Sarfatti, "Research bulletin #5: Seeds of superluminal quantum physics," unpublished, October 6, 1979; copy in *JAW*, Sarfatti folders.

20 Lloyd G. Carter, raw feed of untitled UPI article, dated November 28, 1979; copy in *JAW*, Sarfatti folders. The article ran in several newspapers across the country, including as Carter (1979), 9, and (1980), 44. Nowadays, Sarfatti acknowledges that he probably should have conceded earlier to Stapp on this matter: Sarfatti interview (2009).

21 Jack Sarfatti, "Corporate vision, prepared by i^2 Associates," January 1979, on 1–3; copy in *JAW*, Sarfatti folders.

22 Sarfatti, "Corporate vision," copy in *JAW*, Sarfatti folders.

23 Sarfatti, "Corporate vision," copy in *JAW*, Sarfatti folders.

24 Jack Sarfatti, "The quantum mechanics of superluminal communication," draft from late June 1979, on 7 ("subsist at the poverty level"); copy in *JAW*, Sarfatti folders.

25 Central Intelligence Agency, unclassified "memorandum for the record," December 4, 1979, on 3–5; copy in *JAW*, Sarfatti folders. Sarfatti underlined the word "subliminal" and added a handwritten note on the copies of the memo that he circulated: "Note amusing synchronicity—misspelling of 'superluminal' to 'subliminal.'"

26 In general other states of polarization are possible as well. In fact, any linear combination of the linear and circular polarization states is possible, leading to the most general state of elliptical polarization.

27 Buchwald (1989), chap. 2. On early attributions of polarization states to individual photons, see Beck (1927); Jordan (1927a); and Ruark and Urey (1927).

28 Nick Herbert, email to the author, July 14, 2009.

29 Beth (1936), 115–17, 125. Reprinted in American Association of Physics Teachers (1963), 27–37.

30 Beth (1936), 115–17.

31 Nick Herbert, email to the author, July 14, 2009.

32 Copies of Nick Herbert's QUICK preprint are no longer extant. This description comes from two sets of handwritten notes by Jack Sarfatti from June 1979: Sarfatti, "Analysis of Nick Herbert's design for faster-than-light communicator," dated June 26, 1979; and Sarfatti, "Response to Stapp, 6/26/79," dated June 29, 1979, both in *JAW*, Sarfatti folders. Herbert's QUICK preprint was also analyzed in detail in Ghirardi and Weber (1979), 599–603; and Ghirardi, Rimini, and Weber (1980), 293–98.

33 Sarfatti, "Analysis of Nick Herbert's design," and Sarfatti, "Response to Stapp," both in *JAW*, Sarfatti folders.

34 GianCarlo Ghirardi, letter to the author, January 22, 2009.

35 GianCarlo Ghirardi, email to the author, January 23, 2009.

36 Selleri interview with Freire (2003), 6, 41.

37 Ghirardi cited two unpublished papers by Selleri, in addition to Herbert's QUICK preprint: Selleri, "Einstein locality and the quantum-mechanical long-distance effects," 1979 preprint based on his presentation at the Udine meeting; and a preprint of Cufaro-Petroni, Garruccio, Selleri, and Vigier (1980), 111–14.

38 Ghirardi and Weber (1979), 602, 603. Ghirardi's and Weber's argument relied on the so-called Wigner-Araki-Yanase theorem, a variant or extension of Heisenberg's uncertainty principle. The theorem specifies limits to how precisely a given quantity can be measured if a distinct, incompatible quantity (such as total angular momentum) is conserved. See Wigner (1952); Araki and Yanase (1960); and Yanase (1961).

39 Shimony (1984), 225–30. See also GianCarlo Ghirardi, letter to the author, January 22, 2009.

40 Ghirardi, Rimini, and Weber (1980), 298.

41 Herbert, "FLASH—A superluminal communicator based upon a new kind of quantum measurement," preprint (January 1981); copy in *JFC*, folder "OLT correspondence." "OLT" refers to "objective local theories," Clauser's term for the type of theory that Bell's theorem seemed to rule out. See Clauser and Horne (1974).

42 Siegman (1971), 59–64, 183.

43 Herbert, FLASH preprint (January 1981). The same phrase appears in the published version: Herbert (1982), 1177.

44 Actually, the results at station 2 would show some statistical scatter—perhaps twenty-three photons in state *R* and twenty-seven in state *L* during one round, rather than always twenty-five in each.

45 Herbert and Karush (1978).

46 Herbert, "FLASH—A superluminal communicator based upon a new kind of quantum

measurement," preprint (January 1981); copy in *JFC*, folder "OLT correspondence." See also Nick Herbert, email to the author, February 26, 2008.

47 Herbert, "Esalen sessions on the problem of reality (Feb 2–6, 1981)," on 7; copy in *NH*.

48 Nick Herbert to Philippe Eberhard, March 15, 1981, in *JFC*, folder "OLT correspondence." Eberhard's original letter is no longer extant. From Herbert's reply, it appears that Eberhard had asked about unitarity, that is, whether the probabilities for various outcomes in the FLASH device would add up to one.

49 E.g., Sarfatti, unpublished "Jubilee for Zarathustra: A quantum epic that describes itself," n.d., ca. June 1979; Sarfatti, "Research bulletin #5: Seeds of superluminal quantum physics," October 6, 1979; and Jack Sarfatti to Nick Herbert, October 28, 1981; copies in *JAW*, Sarfatti folders.

50 Alwyn van der Merwe (editor, *Foundations of Physics*) to GianCarlo Ghirardi, March 17, 1981, in *GCG*.

51 The reviewer was Asher Peres, a French-born physicist who had emigrated to Israel as a young boy and who later studied with Einstein's collaborator, Nathan Rosen. See Peres (2003), 458. Peres composed a moving, brief autobiography on the occasion of his seventieth birthday, which was published posthumously: Peres (2006). See also Anon. (2006).

52 GianCarlo Ghirardi, letter to the author, January 22, 2009 ("I must confess"); and GianCarlo Ghirardi to Alwyn van der Merwe, April 22, 1981, in *GCG*.

53 GianCarlo Ghirardi, "Referee's report on the paper 'FLASH—A superluminal communicator based upon a new kind of quantum measurement' by Nick Herbert," April 22, 1981, in *GCG*.

54 Ibid. Symbolically, Herbert had assumed that the laser gain tube would create a quantum state with n copies of the original polarization, or $|R\rangle \rightarrow |nR\rangle = |RRRRRRR\ldots\rangle$. The initial state, however, can be rewritten as $|R\rangle = 2^{-1/2}\,[|H\rangle + i\,|V\rangle]$, where i is the imaginary number (square root of -1). Any laser tube that functioned as Herbert's scheme required would also need to transform $|H\rangle \rightarrow |nH\rangle$ and $|V\rangle \rightarrow |nV\rangle$. So the action of the tube would actually be $|R\rangle \rightarrow 2^{-1/2}\,[|nH\rangle + i\,|nV\rangle]$, a superposition composed of all H or all V, rather than a state of half H and half V. Herbert seems to have tacitly assumed the transformation would be $|R\rangle \rightarrow \{2^{-1/2}\,[|H\rangle + i\,|V\rangle]\}^n$, in which the nonlinearity is made most manifest.

55 Compare the published version of Herbert (1982), 1175, 1176, with the original preprint from January 1981 (copy in *JFC*, folder "OLT correspondence"), on 6, 7.

56 Nick Herbert to John Bell, March 29, 1982, in *NH*; copy also in *JFC*, folder "OLT correspondence."

57 Tullio Weber to Alwyn van der Merwe, March 8, 1982 (including referee report as enclosure), in *GCG*.

58 Stapp interview (2007); and Nick Herbert to John Bell, March 29, 1982, in *NH*.

59 John A. Wheeler to Thomas A. Griffy, April 2, 1980, in *JAW*, series II, folder "Zurek, Wojciech."

60 See, e.g., Wojciech H. Zurek to Fritz Rohrlich, November 11, 1982: "I credit John Wheeler for my interest in the fundamental issues raised by quantum theory of measurements"; copy in *JAW*, series II, folder "Zurek, Wojciech."

61 Wojciech Zurek to John Wheeler, August 10, 1979, in *JAW*, series II, folder "Zurek, Wojciech."

62 Wojciech Zurek, email to the author, September 16, 2007.

63 Wojciech Zurek, email to the author, August 4, 2007. See also correspondence in *JAW*, series II, folder "Zurek, Wojciech," regarding Zurek's and Wootters's participation in teaching a follow-up graduate seminar. Wheeler taught a version of his seminar while visiting at Columbia University in the early 1980s; see the description in Bernstein (1991b), 93–95.

64 Wootters and Zurek (1979); see also Wojciech Zurek email to the author, August 4, 2007.

65 John A. Wheeler to C. Kennel, February 13, 1985, in *JAW*, series II, folder "Zurek, Wojciech." See also Wheeler and Zurek (1983).

66 Wheeler's archives contain thousands of pages from Sarfatti, spanning the period 1971 through 1988; Herbert and Einhorn also mailed dozens of items to Wheeler during the 1970s and early 1980s. See *JAW*, Sarfatti, Herbert, and Einhorn folders.

67 Wojciech Zurek, email to the author, March 16, 2009.

68 Larry Bartell, email to the author, April 14, 2009 ("anything that is interesting"); see also Bartell, emails to the author, April 16 and 17, 2009; and Bartell (1980a). On Bartell's impressive and lengthy career, see Kuczkowski (1999).

69 See, e.g., Larry Bartell to Jack Sarfatti, February 19, 1980; and Larry Bartell to John A. Wheeler, May 22, 1980, both in *JAW*, Sarfatti folders. See also Bartell, emails to the author, April 14, 16, and 17, 2009; Nick Herbert, email to the author, April 14, 2009; and Jack Sarfatti, email to the author, July 22, 2009.

70 Bartell (1980b), esp. 1358, 1359.

71 Bartell, email to the author, April 14, 2009; see also Nick Herbert, "Fourth Esalen seminar on the nature of reality: Brief impressions," February 7, 1983, in *NH*.

72 Bartell (1980a), as reprinted in Wheeler and Zurek (1983), 455, 456.

73 Group member John Clauser, for example, received Nick Herbert's preprints (copies of which remain in his files) and talked about each iteration with Herbert at the annual Esalen workshops. At the same time, Clauser was in frequent contact with Marlan Scully, an expert in quantum optics then at the University of New Mexico in Albuquerque. Scully, in turn, began to cross paths with Zurek and Wootters at conferences during 1981 and 1982. See *JFC*, folders "OLT correspondence" and "Random correspondence"; Nick Herbert's annual reports on the Esalen workshops, 1980–1985, in *NH*; Marlan O. Scully to John Clauser, September 28, 1981, in *JFC*, folder "Random correspondence"; and the proceedings of the 1981 NATO summer school, published as Meystre and Scully (1983). On the March 1982 San Antonio meeting at which Zurek, Wootters, and Scully also met, see Zurek, unpublished research notebook, entry for March 17, 1982, in *WHZ*. On Scully's work during this period, see also Bromberg (2006).

74 Wojciech Zurek, email to the author, March 9, 2008; see also Zurek (1983).

75 Zurek research notebook, entries for March 17 and 29, 1982, in *WHZ*.

76 See Georges Lochak and Simon Diner (Perugia conference organizers) to Wojciech Zurek, October 22, 1981; and Zurek to Lochak and Diner, November 10, 1981, in *JAW*, series II, folder "Zurek, Wojciech."

77 Nick Herbert, email to the author, February 25, 2009.

78 Selleri (1984), 101–28, esp. 119; and Gozzini (1984), esp. 129, 137.

79 Scarani, Iblisdir, and Gisin (2005), 1228n5 (attributed to Gisin).

80 In addition to the published papers by Selleri and Gozzini in the conference proceedings, see also Tarozzi (1984); and Srinivas (1984), esp. 375nn18–20.

81 Wojciech Zurek research notebook, entry for April 27, 1982, in *WHZ*.

82 Wojciech Zurek, email to the author, March 9, 2008.

83 Zurek, email to the author, March 9, 2008. See also Misner, Thorne, and Zurek (2009), 46.

84 Wootters and Zurek (1982), 802, 803. For accessible introductions to the no-cloning theorem, see Buzek and Hillery (2001) and Wootters and Zurek (2009).

85 Dennis Dieks, email to the author, April 8, 2008; and Roger Cooke, email to the author, April 9, 2008.

86 Dieks, email to the author, April 8, 2008, and Dieks (1982), 271.

87 Dieks, "Communication by EPR devices" (1982), 272n2.

88 Dieks, email to the author, April 8, 2008. Most physics journals at the time (as now) used a "single-blind" reviewing system: referees were told the names of authors who had written the submissions under consideration, but identities of referees were not revealed to the authors.

89 Ghirardi, letter to the author, January 22, 2009.

90 GianCarlo Ghirardi and Tullio Weber to Alwyn van der Merwe, April 22, 1983, in *GCG*.

91 Alwyn van der Merwe to GianCarlo Ghirardi and Tullio Weber, June 24, 1983, in *GCG*.

92 Wojciech Zurek, email to the author, September 16, 2007.

93 Van der Merwe to Ghirardi and Weber, June 24, 1983, in *GCG*. See also Alwyn van der Merwe, unpublished memo, March 7, 2002, in *GCG*; and van der Merwe, email to the author, February 26, 2009.

94 Van der Merwe to Ghirardi and Weber, June 24, 1983, in *GCG*.

95 Ghirardi, letter to the author, January 22, 2009; and Ghirardi, email to the author, January 23, 2009.

96 Rudolf Peierls to Adriano Gozzini, June 26, 1983, in *GCG*; see also Adriano Gozzini to GianCarlo Ghirardi, June 28, 1983, in *GCG*.

97 Ghirardi and Weber (1983).

98 For an accessible introduction, see Singh (1999). In fact, RSA encryption had been invented in secret—appropriately enough—a few years earlier by Clifford Cocks, a mathematician working for the British Government Communications Headquarters, a continuation of the wartime organization that had tackled the Enigma code. Cocks's work remained classified until 1997. See Singh (1999), 279–92. See also MacKenzie (2001), chap. 5.

99 See, e.g., Brassard (1988), chap. 4. In one standard scenario, users A and B share, in public, two large positive integers, n and g (with $g < n$). User A then selects a secret number, x, and calculates $X = g^x \bmod n$, where "mod" means keeping the remainder after X has been divided by n. (For example, $3^4 \bmod 5 = 1 \cdot 3^4 = 81$, and 81 can be divided evenly by 5 sixteen times, leaving a remainder of 1.) Meanwhile, user B selects her own secret number, y, and calculates $Y = g^y \bmod n$. Next users A and B exchange X and Y in public, while keeping x and y to themselves. User A can then calculate $Y^x \bmod n = g^{xy} \bmod n$. Likewise, B can calculate $X^y \bmod n = g^{xy} \bmod n$. By sharing only some of their large numbers, they have arrived at identical values with which to construct a secret encryption key.

100 Singh (1999), chap. 6. See also Brassard (1988).

101 MIT News Office (1994).

102 Stephen Wiesner, emails to the author, March 10 and 11, 2009.

103 Charles Bennett, email to the author, February 25, 2009.

104 Wiesner, email to the author, March 11, 2009; and Clauser interview (2009).

105 Wiesner, email to the author, March 10, 2009. On the Columbia protests, see Avorn et al. (1969) and Matusow (1984), 331–35.

106 Wiesner (1983), 85. On Wiesner's "quantum money" scheme, see also Singh (1999), 334–37.

107 Stephen Wiesner, email to the author, March 10, 2009.

108 Nearly fifteen years later, Wiesner's paper was published in the newsletter of the "Special Interest Group on Algorithms and Computation Theory" (SIGACT) of the Association for Computing Machinery (ACM): Wiesner (1983).

109 Wiesner, email to the author, March 10, 2009; and Clauser interview (2009).

110 Nick Herbert, email to the author, March 11, 2009.

111 Charles Bennett, email to the author, February 25, 2009. See also Singh (1999), 338, 339.

112 Bennett, Brassard, Breidbart, and Wiesner (1983).

113 See John A. Wheeler to Mirdza Berzins Anderson, April 10, 1985, in *JAW*, series II, folder "Bennett, Charles H."

114 Bennett, email to the author, February 25, 2009.

115 Bennett and Brassard (1984).

116 Bennett and Brassard (1984), 175.

117 Ekert (1991). In Ekert's protocol, experimenters A and B announce the detector settings they had used for each run after all of their measurements are complete. For all those cases in which their detector settings happened not to match, they check to make sure that Bell's inequality was violated—that is, they redo John Clauser's old Berkeley experiment, to confirm that the quantum states had been unperturbed between source and detectors. With that security check complete, the experimenters can then focus attention, as in the original BB84 protocol, on those cases for which the detector settings did happen to match.

118 Hiskett et al. (2006); see also Nikbin (2006).

119 Schmitt-Manderbach et al. (2007).

120 On DARPA funding, see Oullette (2004/2005), 22–25. On Los Alamos's group, see http://www.lanl.gov/science/centers/quantum/index.shtml; on the program at NIST, see http://qubit.nist.gov; and on comparable worldwide collaborations, see http://www.quantiki.org (all accessed July 29, 2009).

121 Private firms include idQuantique in Geneva (http://idquantique.com), MagiQ Technologies in New York (http://magiqtech.com), Quintessence Labs in Australia (http://www.quintessencelabs.com), and SmartQuantum in France (http://www.smartquantum.com). Major corporations with active research groups in quantum encryption include IBM (http://www.almaden.ibm.com/st), Hewlett-Packard (http://www.hpl.hp.com/research/qip), Toshiba (http://www.toshiba-europe.com/research), and Mitsubishi and NEC (http://global.mitsubishielectric.com/news/news_releases/2006/mel0655.pdf); all accessed July 29, 2009.

122 Stix (2005), Nikbin (2006), Biever (2004), Kanellos (2004), Hesseldahl (2006), Pease (2008), and Naik (2009).

123 Jack Sarfatti, "The future machine," unpublished memo, February 7, 1983, on 10; copy in *JAW*, Sarfatti folders.

124 Van der Merwe to Ghirardi and Weber, June 24, 1983, in *GCG*.

125 Milonni and Hardies (1982), 321, 322, esp. n4 (Peter Milonni was at the Perugia conference); Mandel (1983), 188; and Glauber (1986), esp. 338, 339, 362–366. For a useful overview of these calculations, see also Chyba and Abraham (1985). Mandel presented similar work in Friberg and Mandel, "Photon statistics of the linear amplifier," paper presented at the Fifth Rochester Conference on Coherence and Quantum Optics, June 13–15, 1983; bound abstracts from the conference available in *JFC*. John Clauser was the chair of the session in which Mandel spoke, and recalls that Mandel told him there that he had undertaken the research because of Herbert's FLASH paper: Clauser interview (2009).

126 Herbert, "Fourth Esalen seminar on the nature of reality: Brief impressions," February 1, 1983, on 3; and "Esalen reality seminar 1985," n.d., on 1; both in *NH*. See also Herbert (1986), 578 and (1988), 176, 177.

127 Charles Bennett, email to the author, February 25, 2009.

128 See esp. Kuhn (1959), Ord-Hume (1977), Gardner (1991), Schaffer (1995), and Smith (1998).

129 Glauber (1986), 336. See also Herbert (1988), 177, 178.

130 Popescu and Rohrlich (1998), 45; Jozsa (1998), 64. See also Spiller (1998), 5; Nielsen and Chuang (2000), 56, 57; and Jaeger (2007), 153–57.

131 Based on data in *Science Citation Index* (1961–).

132 De Angelis, Nagali, Sciarrino, and De Martini (2007). In the preprint version (arXiv: 0705.1898v2 [quant-ph]), they thank both Zurek and Ghirardi; available at http://arxiv.org/abs/0705.1898 (accessed July 31, 2009). See also GianCarlo Ghirardi, letter to the author, January 22, 2009. The Italian team conducted the test in order to contribute to "a deeper, more complete understanding of the quantum cloning process and of the distribution of quantum information among the generated [imperfect] clones" (De Angelis et al. [2007], 193601-1).

133 Barbara Honegger, "Summary: Big Sur realism conference, Esalen 1985," in *NH*.

Chapter 10: The Roads from Berkeley

Epigraph. My thanks to George Weissmann for providing videotapes of the Fundamental Fysiks Group reunion, November 18, 2000, San Francisco, from which these remarks were transcribed.

1 Rauscher interview (2008); Weissmann interview (2009); Elizabeth Rauscher, remarks transcribed from videotape of the Fundamental Fysiks Group reunion, November 18, 2000, San Francisco.

2 Jack Sarfatti to Saul-Paul Sirag, January 30, 1980, in *JAW*, Sarfatti folders.

3 Nick Herbert, email to the author, June 2, 2008. On the Consciousness Theory Group, see also Saul-Paul Sirag, email to the author, March 3, 2010.

4 Jack Sarfatti, "Psi wars!" personal ad, *The Daily Californian* (May 18, 1982): 12 ("duel of wits"). See also Sarfatti, "For the record," memorandum, November 3, 1980, in *JAW*, Sarfatti folders; and "Higher Intelligence Agency, Physical Science Institute," memorandum June 28, 1982, in *JAW*, Sarfatti folders.

5 Carroll (1981), 23.

6 Jack Sarfatti to John Wheeler, November 21, 1980, in *JAW*, Sarfatti folders.

7 Deers (1986). See also Chickering interview (2009).

8 A. Lawrence Chickering to Richard D. DeLauer, March 12, 1982, in *JAW*, Sarfatti folders.

9 Chickering to DeLauer, March 12, 1982.

10 Jack Sarfatti to Reagan Administration, "Request for funds," November 9, 1981, in *JAW*, Sarfatti folders.

11 Chickering to DeLauer, March 12, 1982. Such requests are relatively common in the world of Pentagon discretionary spending, which rarely (if ever) needs to pass through peer review. One sociologist has likened this funding approach to Pascal's wager, named for the seventeenth-century philosopher Blaise Pascal. Pascal reasoned that it is better to believe in God than not to, even if the likelihood of God's existence seems slim, because the cost of faith is small and the price of being wrong much higher. See Collins (2004), 338–44. Cf. Weinberger (2006).

12 Richard D. DeLauer to A. Lawrence Chickering, March 29, 1982, in *JAW*, Sarfatti folders; Domenic A. Maio (colonel, U.S. Air Force, and military assistant to Defense Undersecretary for Research and Engineering) to A. Lawrence Chickering, July 12, 1984, reprinted in Sarfatti (2002a), 151; and Jack Sarfatti to John Wheeler, February 8, 1982, describing the reaction of Assistant Secretary of Defense Frank Carlucci, in *JAW*, Sarfatti folders. On the JASON group, see Finkbeiner (2006).

13 Chickering interview (2009); and Mrs. Evan G. ("Bootsie") Galbraith to Jack Sarfatti, May 7, 1982, reprinted in Sarfatti (2002a), 152, 153.

14 Jack Sarfatti to Alwyn van der Merwe, November 9, 1981, in *JAW*, Sarfatti folders; emphasis in original.

15 Jack Sarfatti, "Summary of the book: Faster-than-Light," July 1, 1982, in *JAW*, Sarfatti folders. See also, e.g., Jack Sarfatti, "For the public record," November 3, 1980; Sarfatti to John Wheeler, February 25, 1982; and Sarfatti, "The Future Machine," February 7, 1983, all in *JAW*, Sarfatti folders; and Sarfatti (2002a).

16 Sarfatti (1986). See also Sarfatti (1987), 118–20. Sarfatti's idea was to use quantum entanglement to create coherent beams of W bosons, the carriers of the weak nuclear force. The W bosons, in turn, could stimulate neutrons inside the enemy warheads to undergo radioactive decay into protons, changing the nuclear composition of the weapons. Ordinarily W bosons decay after traveling only fractions of a millimeter. Sarfatti thought he could cook up a stable beam by exploiting entanglement.

17 The architect Lee Porter Butler, chair of Ekose'a, Inc., gave Sarfatti a modest grant in 1980, which quickly ran out; see Jack Sarfatti, "Basic research from the Galois Institute of Mathematical Physics," May 16, 1980, in *JAW*, Sarfatti folders. In the 1990s, computer scientist and entrepreneur Joe Firmage established the International Space Sciences Organization (ISSO) in San Francisco, which supported Sarfatti's work for several years: Sarfatti interview (2009). See also Butler (1980) and Sarfatti (2002b), 121.

18 Sarfatti interview (2009). For more on Sarfatti's work on "metric engineering"—ways one might control the metric of space-time for purposes of interstellar travel—see his recent books: Sarfatti (2002a, b).

19 Jack Sarfatti, email to the author, July 16, 2010.

20 Saul-Paul Sirag, email to the author, March 3, 2010. See also Nick Herbert's annual reports on the Esalen workshops, ca. 1979–1988, in *NH*.

21 Sirag, email to the author, March 3, 2010. See also Kripal (2007), 302.

22 Sirag, email to the author, March 3, 2010. See also Sirag (1977c, 1979, 1983, 1982, 1989). Sirag expanded upon his ideas in a long appendix to the second edition of Jeffrey Mishlove's book *The Roots of Consciousness*, which replaced Jack Sarfatti's appendix in the first edition: Sirag (1993). Sirag also gave updates on his research each year at the Esalen workshops, as described in Nick Herbert's annual reports in *NH*.

23 Sirag, email to the author, March 3, 2010; and Sirag (2002).

24 Johnson (1988), 27. Capra maintains a list of editions of his various books on his website: http://www.fritjofcapra.net/publishers.html (accessed March 11, 2010).

25 Fritjof Capra, email to the author, March 18, 2010; and Capra (1982); on other editions, see http://www.fritjofcapra.net/ publishers.html (accessed March 11, 2010).

26 Johnson (1988); and Romero (2006), C16. See also Kyle (1995), 88, 94, 95.

27 On *Mindwalk*, see Canby (1992), C22.

28 Capra with Spretnak and Lutz (1984); Capra (1988, 1996, 2002, 2007); and Capra with Steindl-Rast and Matus (1991). On the Center for Ecoliteracy, see http://www.ecoliteracy.org (accessed March 11, 2010). See also Anon. (2002/2003); Pisani (2007); and Fritjof Capra's remarks videotaped at the Fundamental Fysiks Group reunion, November 18, 2000, San Francisco.

29 Wolf interview (2009).

30 Wolf interview (2009); Cassutt (1980), 50–52. My thanks to John Zipperman for providing a copy of the *Future Life* article. See also Huerta (1979), I10 (on Wolf's and Leary's performances). For more on Leary's lectures that summer, see Sullivan (1979), F1.

31 Wolf interview (2009) and Wolf (1981).

32 My thanks to Dr. Wolf for providing a copy of his appearance on the *Seattle Today* show from spring 1981, originally broadcast on KING-TV, an NBC affiliate. See also Cassutt (1980), 52 ("youniverse").

33 Wolfe (1976a), reprinted in Wolfe (1976b), 126–67.

34 Wolf interview (2009); Fred Alan Wolf, email to the author, April 10, 2008; and McDowell (1982a), C18 and (1982b), C9.

35 Wolf interview (2009); Toben with Wolf (1982); Wolf, (1984, 1985, 1986, 1988).

36 Wolf (1991, 1994, 1996, 2001, 2004, 2005).

37 Overbye (2006), F3; Fred Alan Wolf, email to the author, March 13, 2010. The cartoons can be found by searching for "Dr. Quantum" at http://www.youtube.com (accessed March 13, 2010).

38 Nick Herbert, emails to the author, November 28, 2007, May 4, 2009, July 15, 2009, and July 16, 2009; and Herbert (1985). The literary agent with whom Herbert began to work was John Brockman, who also served as Fritjof Capra's agent beginning with Capra's *Turning Point* (1982); Brockman also began representing Fred Alan Wolf in the late 1980s. They had all crossed paths at the Esalen workshops in the late 1970s, where Brockman would sometimes go to scout out promising new topics or talent. (Nick Herbert, email to the author, April 14, 2009; Fritjof Capra, email to the author, March 18, 2010; and Wolf interview [2009].) I worked with Brockman's agency while preparing this book, though not with John Brockman himself.

39 Herbert (1985), 17, 18, 21, 24, 52, 162, 170, 174, 181, 182, 219–26, 239–41, 257.

40 Lehmann-Haupt (1985), C13.

41 Stein (1987), 478, 479. The first textbook to treat Bell's theorem appeared in the same year as Herbert's *Quantum Reality*: Sakurai (1985), on 223, 232, a book aimed at advanced graduate students. See also Ballentine (1987), 787. During the mid-1990s, Herbert's *Quantum Reality* was assigned in the course Physics 121 at Harvard University ("History and philosophy of physics"), taught by Professor Peter Galison, which was aimed at physics majors; I served as a teaching assistant.

42 Nick Herbert, email to the author, July 15, 2009. See also Herbert (1987, 1990).

43 Herbert (1988, 1989).

44 Nick Herbert, email to the author, July 16, 2009; and Herbert (1993).

45 Nick Herbert, email to the author, December 8, 2007; and Herbert, "Entanglement tele-graphed communication avoiding light-speed limitation by Hong Ou Mandel effect," unpub-lished preprint arXiv:0712.2530 (December 2007), available from http://www.arXiv.org. In the second version of his paper (uploaded on December 19, 2007, four days after the first version), Herbert added an appendix to describe the critiques by physicist Lev Vaidman—an expert in foundations of quantum mechanics and quantum information science—who had emailed Herbert to describe faults in Herbert's scheme.

46 Nick Herbert, emails to the author, April 16, 2008, and May 4, 2009. Herbert's blog is avail-able at http://quantumtantra.blogspot.com. See also Herbert (2000). My thanks to Herbert for sharing a copy of his book.

47 Rauscher interview (2008).

48 Rauscher interview (2008). On John F. Kennedy University, see http://www.jfku.edu (accessed March 12, 2010).

49 Rauscher interview (2008). A list of Rauscher's grants and contracts can be found in *EAR*. See also Fülling, Bruch, Rauscher, Neill, Träbert, Heckmann, and McGuire (1992).

50 Rauscher interview (2008). See also Elizabeth A. Rauscher and William L. Van Bise, "Non-superconducting apparatus for detecting magnetic and electromagnetic fields," United

States Patent no. 4,724,390 (issued February 9, 1988); Elizabeth A. Rauscher and William L. Van Bise, "External magnetic field impulse pacemaker non-invasive method and apparatus for modulating brain through an external magnetic field to pace the heart and reduce pain," United States Patent no. 4,723,536 (issued February 9, 1988); and Elizabeth A. Rauscher and William L. Van Bise, "Non-invasive method and apparatus for modulating brain signals through an external magnetic or electric field to reduce pain," United States Patent 4,889,526 (issued December 26, 1989), all available at http://www.google.com/patents (accessed March 12, 2010).

51 See, e.g., Rauscher and Targ (2001) and Rauscher (2005, 2006).

52 Institute of HeartMath, "The Global Coherence Project," press release (2008), available at http://www.heartmath.org (accessed March 15, 2010).

53 Weissmann interview (2009). See also the corporate filings for Padma Marketing Corporation (C1104386), dated November 6, 1981, and February 19, 1982, available from the California Secretary of State, Business Programs Division.

54 Weissmann interview (2009); Anon. (1987b), 38; Anon. (1991), 43, 44; and Luis Treacy, "Food, drugs, or frauds?" originally published by Consumers Union and available at http://www .healthguidance.org/entry/9895/1/Food-Drugs-or-Frauds.html (accessed January 8, 2010).

55 Young (1988), 6, 7; and Anon. (1991), 43, 44 ("defendants were aware").

56 Weissmann interview (2009); corporate filings for George Weissmann, Incorporated (updated name for "Padma Marketing Corporation," still under C1104386), dated January 3, 1989, and June 15, 1990, available from California Secretary of State, Business Programs Division. See also the advertisements for "Adaptrin" in *Yoga Journal* (September/October 1988): 87; and in Walker (1989), 243. On recent research, see Jenny et al. (2005), and references therein.

57 Compton (1998), 5; Anon. (1999); and Anon. (2000). The people's choice awards were sponsored by the National Nutritional Foods Association; Veat products won "best in show" as well as awards in individual categories both years.

58 Weissmann interview (2009).

59 Many of Stapp's essays and articles were republished in Stapp (2004, 2007). See also, e.g., Schwartz, Stapp, and Beauregard (2005).

60 Stapp interview (2007); see also Stapp (1995), 78, 79. The parapsychologist who approached Stapp was Helmut Schmidt, who has been active in the field since the late 1960s. Many members of the Fundamental Fysiks Group have interacted with Schmidt over the years: Sarfatti (2002a), 96. In fact, the psychokinesis experiments that George Weissmann tried to replicate during his postdoctoral fellowship in Zurich, ca. 1979/1980, had originally been conducted by Schmidt: Weissmann interview (2009).

61 Stapp interview (2007).

62 Stapp interview (2007). Stapp elaborated on his method in an unpublished draft of his paper, "Theoretical model of a purported empirical violation of the predictions of quantum theory," preprint dated September 17, 1993, in *HPS*.

63 Stapp interview (2007); and Stapp (1995), 79.

64 Stapp interview (2007). The original version of Stapp's paper is no longer extant, though later versions can be found in *HPS*. Stapp built upon an idea by Steven Weinberg to adopt a nonlinear generalization of the Schrödinger equation: Weinberg (1989).

65 Anonymous referee report in response to the September 1993 revision of Stapp's paper, in *HPS*. See also Stapp interview (2007).

66 Stapp interview (2007); and Stapp (1994). As per the journal editor's instructions, Stapp cited Schmidt's account of the experiment: Schmidt (1993).

67 Benjamin Bederson to Henry P. Stapp, December 22, 1994, in *HPS*.

68 Dowling (1995), 78; Stapp (1995), 78, 79; and Berezin, Malin, and Dowling (1996), 15, 81.

69 Stapp interview (2007); see also Schmidt and Stapp (1993).

70 Edwin T. Jaynes to Peter L. Scott (Chair, Board of Studies in Physics, University of California at Santa Cruz), January 31, 1973, in *JFC*; see also Abner Shimony to John Clauser, May 19, 1972, and August 8, 1972, in *JFC*. Also quoted in Freire (2006), 604. See also Clauser interview with Bromberg (2002), Shimony interview with Bromberg (2002), and Clauser interview (2009).

71 Clauser interview (2009).

72 Clauser interview with Bromberg (2002); and Clauser interview (2009).

73 Clauser interview (2009). See also John F. Clauser, "Rotation, acceleration, and gravity sensors using quantum-mechanical matter-wave interferometry with neutral atoms and molecules," United States Patent 4,874,942 (issued October 17, 1989), available at http://www.google.com/patents (accessed March 18, 2010).

74 Clauser interview (2009). See also John F. Clauser, "Ultrahigh resolution interferometric x-ray imaging," United States Patent 5,812,629 (issued September 22, 1998), available at http://www.google.com/patents (accessed March 18, 2010).

75 Anon. (2010).

76 Rauscher interview (2008); Weissmann interview (2009); and Weissmann's remarks at the Fundamental Fysiks Group reunion, November 18, 2000, San Francisco.

77 Elizabeth Rauscher and George Weissmann, form letter invitation dated October 16, 2000, in *EAR*.

78 Rauscher and Weissman, invitation dated October 16, 2000, in *EAR*.

79 Videotape of the Fundamental Fysiks Group reunion, November 18, 2000, San Francisco.

80 Elizabeth Rauscher, remarks transcribed from videotape of the Fundamental Fysiks Group reunion, November 18, 2000, San Francisco.

Coda: Ideas and Institutions in the Quantum Revival

1 Fölsing (1997), 98–100; and Galison (2003), 235–40, 326, 327.

2 The physics department chair at Berkeley saved thick files of correspondence from "cranks" during the 1950s, a practice that members of the department continue today. See the folders labeled "Scientific insanity," folders 6:41–42, in Department of Physics, University of California, Berkeley, *Records, ca. 1920–1962*, call number CU-68, Bancroft Library. On more recent collections, see Kahn (2002).

3 Lavigne (1994) and Scott (1997).

4 Laudan (1983), 111–27; Collins (1992); and Gieryn (1999).

5 Roszak (1969). On the etymology of "counterculture," see Braunstein and Doyle (2002), 6, 7. On the 1967 Be-In and Pentagon protest, see Lee and Shlain (1992), 160–65, 203, 204.

6 Roszak (1969), 7, 8 ("scientific world-view of the Western tradition"), 50 ("subversion"), 215 ("plague"). On McNamara, cf. Engelhardt (2007), part III; and Amadae (2003), chap. 1.

7 Kyle (1995), chaps. 6, 7; Carroll (1990), 236–38; and Schulman (2001), 96–101.

8 Gleick (1987), 243–72. On Ralph Abraham's presentation at the 1976 Esalen workshop, see Physics/Consciousness Research Group, "A modest proposal to the Foundation for the Realization of Man," February 11, 1976, on pp. 24, 25, in *SPS*; and Sirag (2002), 110, 111. On

the Santa Cruz group, see also Bass (1985); Williams and Thomas (2009); and Williams (forthcoming), chap. 2.

9 Moy (2004), Markoff (2005), and Turner (2006). Cf. Schumacher (1973).

10 Wisnioski (2003), Kirk (2007), and Vettel (2006). See also Ross (1991), 17.

11 See, e.g., Moore (2008).

12 Clauser and Shimony (1978); cf. *Science Citation Index* (1961–), s.v. "Clauser."

13 Shimony (1984), 225–30.

14 Gilder (2008).

15 Gell-Mann (1994); Gottfried and Yan (2003); Thacker, Leff, and Jackson (2002); Carr and McKagan (2009); and Baily and Finkelstein (2010). My thanks to Charles Baily for bringing Carr and McKagan's work to my attention.

16 See, e.g., Jaeger (2010), 94; and Bokulich and Jaeger (2010). On Nobelist Gerard 't Hooft's editorship of *Foundations of Physics*, see http://www.springer.com/physics/journal/10701 (accessed May 27, 2010).

17 Lloyd (1997), Shor (2002), and Devetak (2005).

18 In addition to the books cited in chapter 8, footnote 28, see Carr and McKagan (2009); and Wojciech Zurek, email to the author, August 1, 2007. The best historical work on the new experiments includes Freire (2006); Bromberg (2006, 2008); and Gilder (2008).

19 Galison (1995). See also Woit (2006) and Smolin (2006).

20 Olival Freire makes a similar point: Freire (2004). I have in mind similar debates within the history of technology over "technological determinism." See esp. Bijker, Hughes, and Pinch (1987); and Smith and Marx (1994).

21 Ed Fry to Olival Freire, August 5, 2005, in Freire's possession; see also Fry and Walther (2002).

22 Aspect interview (2009).

23 Anton Zeilinger, email to the author, June 2, 2009.

24 Annual citations to John Bell's article on Bell's theorem remained flat between 1982 and 1990, averaging forty-one citations per year. Only during the mid-1990s did the number of citations per year begin to rise exponentially. From 1982 through 1990, well over half (56 percent) of the physics articles on Bell's theorem were contributed by authors who had already published on the topic ("old authors"), rather than by new authors to the field. Moreover, the vast majority of old authors (69 percent) had published on the topic prior to publication of Alain Aspect's articles on his new experiments. Based on data in *Science Citation Index* (1961–). That pattern is starkly at odds with the usual pattern for growing fields, in which the number of new authors per year grows exponentially: Bettencourt, Kaiser, Kaur, Castillo-Chavez, and Wojick (2008). The citation pattern for Bell's theorem, on the other hand, matches the same pattern of citations to the 1935 Einstein-Podolsky-Rosen article in the *Physical Review*, which likewise remained flat and small until a burst and exponential rise in the mid-1990s: Redner (2005).

25 See, e.g., Nauenberg (2007); and Orzel (2009), chap. 10.

26 Jack Sarfatti, "Research Bulletin #5," October 6, 1979 ("'scientific laborers'"); Jack Sarfatti to Robert J. Curtis, Jr. [*est* house counsel], July 11, 1977 ("sold out"), circulated widely in photocopy form; both in *JAW*, Sarfatti folders.

27 Sirag (2002), 112, 113.

28 Cf. Kevles (1995), chaps. 24, 25; Leslie (1993), chap. 9; and Moore (2008), chap. 6.

29 Physics/Consciousness Research Group, "A modest proposal to the Foundation for the Real-

ization of Man," February 11, 1976, appendix III, on 21 ("Mythos," "Logos," "Werner makes a distinction," "confuse the symbol"), copy in *SPS*. Several historians have likewise noted the unusually pragmatic attitude of most U.S. physicists during the 1950s and 1960s. See esp. Cini (1980); Forman (1987); Schweber (1989, 1994); Pickering (1989); Leslie (1993); Galison (1997); and Kaiser (2005).

30 Physics Survey Committee (1972/1973), vol. 2, 1149 ("Since the pursuit," "quality of a quest"), 1150 ("humanistic subject").

31 Physics Survey Committee (1972/1973), vol. 2, 1211 ("discriminated," "thoughtful theorist"), 1232 ("self-renewal").

32 Kaiser (2007a), 32, 33 and (forthcoming), chap. 4. See also French and Taylor (1978), 100 ("speculative question").

33 W. E. Meyerhof, memorandum to graduate students in Stanford's physics department, September 29, 1972, in Felix Bloch papers, call number SC303, Stanford University Archives, folder 12:10. On the Harvard seminar, see Freire (2006), 596.

34 Harrison (1982a); on Capra's book in physics classrooms, see Clifton and Regehr (1990), 73, 74.

35 Morrison (1990), 86, 87, 619; and Miller (2008), 454.

36 I am still grateful to my long-suffering lab partner, Michael Gordin, for his patience and good humor during that fateful course in the fall of 1997.

37 Harvard University, Physics 191/247, *Advanced Laboratory Manual* (Cambridge: Harvard University, 1997), privately printed for classroom use, quotations on B-3-5.

38 The emergence of particle cosmology—at the time a speculative field mixing ideas about the smallest and largest features of the universe—likewise grew from physicists' reappraisal after the Cold War bubble had burst. See Kaiser (2006b, 2007b).

39 Bernstein (2008); and Overbye (2009), D1.

40 See http://www.perimeterinstitute.ca/en/About/History/History (accessed March 29, 2010). See also Seife (2003).

41 See http://www.fqxi.org/about (accessed March 29, 2010); and Max Tegmark, personal communication, March 11, 2010.

42 Merali (2007), 8–10; Wallace-Wells (2008); and A. Garrett Lisi, personal communication, September 10, 2009. See also Lisi (2007).

43 See, e.g., Distler and Garibaldi (2009).

44 See esp. Kripal (2007), chap. 19; Ronson (2004), 26; Ross (1991), 21; and Gosse (2005), chap. 13.

45 Reisz (2010); Don Troop, "A subatomic explosion," *Tweed* (blog affiliated with the *Chronicle of Higher Education*), posted April 29, 2010, available at http://chronicle.com/blogAuthor/ Tweed (accessed May 5, 2010); Daniel Cressey, "No place for the paranormal at physics conference," *The Great Beyond* (blog affiliated with *Nature* magazine), posted April 29, 2010, available at http://blogs.nature.com (accessed May 27, 2010); Chad Orzel, "Conference organizers should not live in caves," *Uncertain Principles* blog, posted April 29, 2010, available at http://scienceblogs.com/principles (accessed May 27, 2010); and Peter Woit, "Bohmian spat," *Not Even Wrong* blog, posted April 29, 2010, available at http://www.math .columbia.edu/~woit/wordpress (accessed May 27, 2010).

46 Tegmark (2000); and Tegmark, personal communication, March 11, 2010.

47 Kaku (2008); Anon. (2008), 26; and Dixler (2009), 24.

Interviews

Unless otherwise noted, all interviews were conducted by the author.

Aspect, Alain. April 22, 2009. Cambridge, Massachusetts.
Chickering, A. Lawrence. May 4, 2009, via telephone.
Clauser, John. May 20–23, 2002. Interview by Joan Lisa Bromberg; transcript in *NBL*.
———. May 29, 2009. Walnut Creek, California.
d'Espagnat, Bernard. October 26, 2001. Interview by Olival Freire; transcript in *NBL*.
Erhard, Werner. July 26, 2010, via telephone.
Finkelstein, David. October 1, 2007, via telephone.
Fuller, Robert W. November 8, 2007, via telephone.
Jackiw, Roman. October 3, 2007. Cambridge, Massachusetts.
Mishlove, Jeffrey. April 24, 2008, via telephone.
Nauenberg, Michael. July 12, 1994. Interview by Randall Jarrell; transcript available at http://physics.ucsc.edu/~michael/oral2.pdf.
Rauscher, Elizabeth. January 4, 2008, via telephone.
Sarfatti, Jack. May 22, 2009. San Francisco, California.
Selleri, Franco. June 24, 2003. Interview by Olival Freire; transcript in *NBL*.
Shimony, Abner. September 9, 2002. Interview by Joan Lisa Bromberg; transcript in *NBL*.
Stapp, Henry P. August 21, 1998. Berkeley, California.
———. October 4, 2007, via telephone.
Toben, Bob. January 29, 2010, via telephone.
Weissmann, George. March 13, 2008. Cambridge, Massachusetts.
———. August 27, 2009. Natick, Massachusetts.
Wolf, Fred Alan. September 26, 2007, via telephone.
———. May 22, 2009. San Francisco, California.
Zeh, H. Dieter. July 25, 2008. Interview by Fábio Freitas; transcript in *NBL*.

Bibliography

Aczel, Amir D. 2001. *Entanglement: The Greatest Mystery in Physics*. New York: Four Walls Eight Windows.

Adkins, Douglas. 1975. *The Great American Degree Machine: An Economic Analysis of the Human Resource Output of Higher Education*. Berkeley: Carnegie Commission on Higher Education.

Albert, David Z. 1992. *Quantum Mechanics and Experience*. Cambridge: Harvard University Press.

Alcock, James E. 1981. *Parapsychology: Science or Magic? A Psychological Perspective*. New York: Pergamon.

———. 1988. "A comprehensive review of major empirical studies in parapsychology involving random event generators or remote viewing." In National Research Council, *Enhancing Human Performance: Issues, Theories, and Techniques*, chapter 10. Washington, DC: National Academy Press.

Amadae, Sonja M. 2003. *Rationalizing Capitalist Democracy: The Cold War Origins of Rational Choice Liberalism*. Chicago: University of Chicago Press.

American Association of Physics Teachers. 1963. *Quantum and Statistical Aspects of Light: Selected Reprints*. New York: American Institute of Physics.

Anderson, P. W., and B. T. Matthias. 1964. "Editorial foreword." *Physics* 1 (June): i.

Anderson, Susan Heller. 1991. "Chronicle: Seeking to maximize his potential, Erhard sells *est* empire." *New York Times*, February 13: D10.

Anderson, Walter T. 1983. *The Upstart Spring: Esalen and the American Awakening*. Reading, MA: Addison-Wesley.

Anon. 1955a. "World-wide research moves ahead." *Newsletter of the Parapsychology Foundation* 2 (September/October): 1, 2.

———. 1955b. "The 'Science' magazine controversy." *Newsletter of the Parapsychology Foundation* 2 (September/October): 3–9.

———. 1965. "1,000 atomic workers face loss of jobs." *Los Angeles Times*, January 7: 16.

———. 1967a. "Hippie haven." *New York Daily News*, June 2: back page.

———. 1967b. "Budget: New Reagan total breaks record." *Los Angeles Times*, April 2: K5.

———. 1971. "Astronaut tells of E.S.P. tests." *New York Times*, June 22: 22.

———. 1973a. "The magician and the think tank." *Time* (March 12): 110.

———. 1973b. "Poison banned 7 years ago still being sold, E.P.A. says." *New York Times*, March 26: 26.

———. 1974a. "Boom times on the psychic frontier." *Time* (March 4): 56 (cover story).

———. 1974b. "Geller and magicians." *Science News* 106 (August 3): 78, 79.

———. 1975. "Citizen Coppola." *Time* (June 30): 49.

———. 1977a. "Humanistic psychology conference set." *Valley News* [Van Nuys, California], May 8: 19.

———. 1977b. "*est*-erical behavior?" *Newsweek* (May 9): 95.

———. 1977c. "Margolis is acquitted on 5 of 6 counts remaining in federal tax-fraud case." *New York Times*, October 1: 40.

———. 1978a. "Henderson discusses changes in socio-economic structure." *Harvard Crimson* 168, no. 67 (December 6): 1.

———. 1978b. "Administration plans to set up commission on world hunger." *New York Times*, February 4: 22.

———. 1979. "Association urged to bar paranormal investigators." *New York Times*, January 9: C2.

———. 1980. "Book award winners listed for six categories." *New York Times*, May 3: 14.

———. 1981. "*Playboy* plans sale of *Oui* magazine." *New York Times*, May 19: D4.

———. 1987a. "William Grant Whitehead, 44, former top editor at Dutton." *New York Times*, October 10: 40.

———. 1987b. "Food additives." *FDA Consumer* 21 (November): 38.

———. 1991. "Injunction actions." *FDA Consumer* 25 (March): 43, 44.

———. 1992. "Gerald Feinberg, 58, physicist; Taught at Columbia University." *New York Times*, April 23: D26.

———. 1999. "People's choice award winners announced at NNFA's Marketplace '99." *Business Wire*, July 12. Available at http://www.lexisnexis.com.

———. 2000. "People's choice award winners announced at NNFA's Marketplace 2000." *PR Newswire*, August 7. Available at http://www.lexisnexis.com.

———. 2001. "France agrees to extradition of culprit in killing in U.S." *New York Times*, July 13: A17.

———. 2002/3. "Connecting with Fritjof Capra." *Ecotecture: The Journal of Ecological Design*. Available at http://www.ecotecture.com.

———. 2006. "Asher Peres: List of publications." *Foundations of Physics* 36: 157–73.

———. 2008. "Best sellers." *New York Times Book Review*, April 13: 26.

———. 2010. "Entanglement pioneers bag Wolf Prize." *Physics World* (February 4). Available at http://physicsworld.com/cws/article/news/41633 (accessed February 4, 2010).

Appelbaum, Judith. 1983. "Paperback talk: A science with mass appeal." *New York Times*, March 20: 39–40.

Araki, Huzihiro, and Mutsuo Yanase. 1960. "Measurement of quantum mechanical operators." *Physical Review* 120: 622–26.

Arkani-Hamed, Nima, Savas Dimopoulos, and G. R. Dvali. 1998. "The hierarchy problem and new dimensions at a millimeter." *Physics Letters B* 429: 263–72.

Arndt, Markus et al. 1999. "Wave-particle duality of C_{60} molecules." *Nature* 401 (October 14): 680–82.

Asimov, Isaac. 1978. "Scientists and sages." *New York Times*, July 27: 19.

Aspect, Alain. 1976. "Proposed experiment to test the nonseparability of quantum mechanics." *Physical Review D* 14: 1944–51.

———. 1981. "Expériences basées sur les inégalités de Bell." *Journal de Physique* 42, colloque C2 (March): 63–80.

———. 2002. "Bell's theorem: The naive view of an experimentalist." In *Quantum [Un]speakables: From Bell to Quantum Information*, ed. R. A. Bertlmann and A. Zeilinger, 119–54. New York: Springer.

Aspect, Alain, Jean Dalibard, and Gérard Roger. 1982. "Experimental test of Bell's inequalities using time-varying analyzers." *Physical Review Letters* 49: 1804–7.

Aspect, Alain, Philippe Grangier, and Gérard Roger. 1981. "Experimental tests of realistic local theories via Bell's theorem." *Physical Review Letters* 47: 460–63.

Associated Press. 1997. "Letter calls for return of Einhorn." *Philadelphia Inquirer*, December 22: B1.

Avorn, Jerry et al. 1969. *Up Against the Ivy Wall: A History of the Columbia Crisis*. New York: Atheneum.

Baily, Charles, and Noah D. Finkelstein. 2010. "Teaching and understanding of quantum interpretations in modern physics courses." *Physical Review Special Topics: Physics Education Research* 6 (January): 010101.

Baker, Russ. 1999. "A touch of eden." *Esquire* 132 (December): 100–107.

Ballentine, L. E. 1987. "Resource Letter IQM-2: Foundations of quantum mechanics since the Bell inequalities." *American Journal of Physics* 55: 785–92.

Baracca, A., A. Cornia, R. Livi, and S. Ruffo. 1978. "Quantum mechanics, 1st-kind states, and local hidden variables: Three experimentally distinguishable situations." *Nuovo Cimento B* 43: 65–72.

Bartell, L. S. 1980a. "Complementarity in the double-slit experiment: On simple realizable systems for observing intermediate particle-wave behavior." *Physical Review D* 21: 1698, 1699.

———. 1980b. "Local realism and the Einstein-Podolsky-Rosen paradox: On concrete new tests." *Physical Review D* 22: 1352–60.

Bartley, William W., III. 1978. *Werner Erhard: The Transformation of a Man, the Founding of est*. New York: C. N. Potter.

Barton, Henry A. 1953. "AIP 1952 annual report." *Physics Today* 6 (May): 4–9.

Bass, Thomas A. 1985. *The Eudaemonic Pie*. Boston: Houghton-Mifflin.

Bearden, Tom. 1988 [1978]. *The Excalibur Briefing: Explaining Paranormal Phenomena*. San Francisco: Strawberry Hill Press.

Beck, Guido. 1927. "Über einige Folgerungen aus dem Satz von der Analogie zwischen Lichtquant und Elektron." *Zeitschrift für Physik* 43: 658–74.

Beers, David. 1986. "Buttoned-down bohemians: Welcome to San Francisco's New Age Right." *Image* [Sunday magazine of the *San Francisco Examiner*], August 3: 14–18, 36.

Behar, Richard. 1991. "The thriving cult of greed and power." *Time* 137 (May 6): 50 (cover story).

Bell, John S. 1964. "On the Einstein Podolsky Rosen paradox." *Physics* 1: 195–200.

———. 1966. "On the problem of hidden variables in quantum mechanics." *Reviews of Modern Physics* 38: 447–52.

———. 1982. "On the impossible pilot wave." *Foundations of Physics* 12: 989–99.

———. 2004a [1987]. "Bertlmann's socks and the nature of reality." In *Speakable and Unspeakable in Quantum Mechanics*, 2nd ed., 139–158. New York: Cambridge University Press.

———. 2004b [1987]. *Speakable and Unspeakable in Quantum Mechanics*, 2nd ed. New York: Cambridge University Press.

Bell, Mary. 2002. "Some reminiscences." In *Quantum [Un]speakables: From Bell to Quantum Information*, ed. R. A. Bertlmann and A. Zeilinger, 3–5. New York: Springer.

Beller, Mara. 1990. "Born's probabilistic interpretation: A case study of 'concepts in flux.'" *Studies in History and Philosophy of Science* 21: 563–88.

———. 1998. "The Sokal hoax: At whom are we laughing?" *Physics Today* 51 (September): 29–34.

———. 1999. *Quantum Dialogue: The Making of a Revolution*. Chicago: University of Chicago Press.

Bennett, Charles H., and Gilles Brassard. 1984. "Quantum cryptography: Public key distribution and coin tossing." In *Proceedings of IEEE International Conference on Computers, Systems, and Signal Processing, Bangalore, India*, 175–79. New York: IEEE.

Bennett, Charles H., Gilles Brassard, Seth Breidbart, and Stephen Wiesner. 1983. "Quantum cryptography, or unforgeable subway tokens." In *Advances in Cryptology: Proceedings of Crypto '82*, ed. David Chaum, Ronald Rivest, and Alan Sherman, 267–75. New York: Plenum.

Berezin, Alexander A., Shimon Malin, and Jonathan P. Dowling. 1996. "More spirited debate on physics, parapsychology, and paradigms." *Physics Today* 49 (April): 15, 81.

Bergmann, Peter G. 1969. "Foundations of quantum theory." *Physics Today* 22 (May): 85–87.

Bernstein, Jeremy. 1978/79. "A cosmic flow." *American Scholar* 48: 6–9.

———. 1991a. "John Stewart Bell: Quantum engineer." In *Quantum Profiles*, 3–91. Princeton: Princeton University Press.

———. 1991b. "John Wheeler: Retarded learner." In *Quantum Profiles*, 93–142. Princeton: Princeton University Press.

———. 2008. *Physicists on Wall Street and Other Essays on Science and Society*. New York: Springer.

Bernstein, Richard. 1995. "Who saved civilization? The Irish, that's who!" *New York Times*, April 5: C23.

Bertlmann, R. A., and A. Zeilinger, eds. 2002. *Quantum [Un]speakables: From Bell to Quantum Information*. New York: Springer.

Bess, Donovan. 1973. "Demonstrations by Israeli psychic: His 'mind-force' bends metal." *San Francisco Chronicle*, June 2: 4.

Beth, Richard. 1936. "Mechanical detection and measurement of the angular momentum of light." *Physical Review* 50: 115–25.

Bettencourt, Luís, David Kaiser, Jasleen Kaur, Carlos Castillo-Chavez, and David Wojick. 2008. "Population modeling of the emergence and development of scientific fields." *Scientometrics* 75: 495–518.

Biagioli, Mario. 1993. *Galileo, Courtier: The Practice of Science in the Culture of Absolutism*. Chicago: University of Chicago Press.

Biever, Celeste. 2004. "First quantum cryptography network unveiled." *New Scientist* (June 4). Online edition available at http://www.newscientist.com/article/dn5076-first-quantum-cryptography-network-unveiled.html (accessed July 12, 2010).

Bijker, Wiebe, Thomas Hughes, and Trevor Pinch, eds. 1987. *The Social Construction of Technological Systems: New Directions in the Sociology and History of Technology*. Cambridge: MIT Press.

Blatt, J. M., and V. F. Weisskopf. 1952. *Theoretical Nuclear Physics*. New York: John Wiley.

Bohm, David. 1951. *Quantum Mechanics*. Englewood Cliffs, NJ: Prentice-Hall.

———. 1952a. "A suggested interpretation of the quantum theory in terms of 'hidden' variables, Part I." *Physical Review* 85: 166–79.

———. 1952b. "A suggested interpretation of the quantum theory in terms of 'hidden' variables, Part II." *Physical Review* 85: 180–93.

———. 1981. *Wholeness and the Implicate Order*. Boston: Routledge.

Bohm, David, and Basil Hiley. 1975. "On the intuitive understanding of nonlocality as implied by quantum theory." *Foundations of Physics* 5: 93–109.

———. 1976a. "Some remarks on Sarfatti's proposed connection between quantum phenomena and the volitional activity of the observer-participator." *Psychoenergetic Systems* 1: 173–79.

———. 1976b. "Nonlocality and polarization correlations of annihilation quanta." *Nuovo Cimento B* 35: 137–44.

———. 1993. *The Undivided Universe: An Ontological Interpretation of Quantum Theory*. New York: Routledge.

Bohr, Niels. 1934. *Atomic Theory and the Description of Nature*. Cambridge: Cambridge University Press.

———. 1935. "Can quantum-mechanical description of physical reality be considered complete?" *Physical Review* 48: 696–702.

———. 1949. "Discussion with Einstein on epistemological problems in atomic physics." In *Albert Einstein: Philosopher-Scientist*, ed. P. A. Schilpp, 201–41. Evanston, IL: Library of Living Philosophers.

———. 1958. *Atomic Physics and Human Knowledge*. New York: Wiley.

———. 1985. *Niels Bohr Collected Works*, vol. 6, *Foundations of Quantum Physics I (1926–1932)*, ed. Erik Rüdinger and Jørgen Kalckar. New York: North-Holland.

Bokulich, Alissa, and Gregg Jaeger, eds. 2010. *Philosophy of Quantum Information and Entanglement*. New York: Cambridge University Press.

Born, Max. 1936. *Atomic Physics*. Trans. John Dougall. New York: G. E. Stechert.

———. 1949. *Natural Philosophy of Cause and Chance*. Oxford: Clarendon.

———. 1956. *Physics in My Generation: A Selection of Papers*. New York: Pergamon.

———. 2005 [1971]. *The Born-Einstein Letters, 1916–1955: Friendship, Politics, and Physics in Uncertain Times*, rev. ed. New York: Macmillan.

Brassard, Gilles. 1988. *Modern Cryptology: A Tutorial*. New York: Springer.

Braunstein, Peter. 2002. "Forever young: Insurgent youth and the sixties culture of rejuvenation." In *Imagine Nation: The American Counterculture in the 1960s and '70s*, ed. Peter Braunstein and Michael William Doyle, 243–73. New York: Routledge.

Braunstein, Peter, and Michael William Doyle. 2002. "Introduction: Historicizing the American counterculture of the 1960s and '70s." In *Imagine Nation: The American Counterculture of the 1960s and '70s*, ed. Peter Braunstein and Michael William Doyle, 5–14. New York: Routledge.

Brisick, William C. 1995. "A tribute to the Irish monks whose literacy blesses us today." *Los Angeles Daily News*, March 26: L20.

Bromberg, Joan Lisa. 1991. *The Laser in America, 1950–1970*. Cambridge: MIT Press.

———. 2006. "Device physics vis-à-vis fundamental physics in Cold War America: The case of quantum optics." *Isis* 97: 237–59.

———. 2008. "New instruments and the meaning of quantum mechanics." *Historical Studies in the Natural Sciences* 38: 325–52.

Browne, Malcolm W. 1986. "Quantum theory: Disturbing questions remain unresolved." *New York Times*, February 11: C3.

Bry, Adelaide. 1976. *Est: 60 Hours That Transform Your Life*. New York: Harper and Row.

Buchanan, Mark. 2008. "Quantum determinism: Is there such a thing as pure chance?" *New Scientist* (March 22): 28–31.

Buchwald, Jed Z. 1989. *The Rise of the Wave Theory of Light: Optical Theory and Experiment in the Early Nineteenth Century*. Chicago: University of Chicago Press.

Burke, Philip G., and Ian C. Percival. 1999. "John Stewart Bell, 28 July 1928–1 October 1990." *Biographical Memoirs of Fellows of the Royal Society* 45: 3–17.

Burnett, D. Graham. 2009. "Games of chance." *Cabinet* 34 (Summer): 59–65.

———. 2010. "A mind in the water." *Orion* 29 (May/June): 38–51.

Butler, Lee Porter. 1980. *Ekose'a Homes: Natural Energy Saving Design*. San Francisco: Ekose'a.

Buzek, Vladimir, and Mark Hillery. 2001. "Quantum cloning." *Physics World* 14 (November): 25–29.

Byron, Frederick W., Jr. and Robert W. Fuller. 1969. *Mathematics of Classical and Quantum Physics*. Reading, MA: Addison-Wesley.

Cahill, Thomas. 1995. *How the Irish Saved Civilization: The Untold Story of Ireland's Heroic Role from the Fall of Rome to the Rise of Medieval Europe*. New York: Anchor.

Canby, Vincent. 1992. "Engaging in conversation on the Normandy coast." *New York Times*, April 8: C22.

Capra, Fritjof. 1972. "The dance of Shiva: The Hindu view of matter in the light of modern physics." *Main Currents in Modern Thought* 29: 15–20.

———. 1974. "Bootstrap and Buddhism." *American Journal of Physics* 42: 15–19.

———. 1975. *The Tao of Physics: An Exploration of the Parallels Between Modern Physics and Eastern Mysticism*. Boulder, CO: Shambhala.

———. 1982. *The Turning Point: Science, Society, and the Rising Culture*. New York: Simon and Schuster.

———. 1984 [1975]. *The Tao of Physics*, 2nd ed. New York: Bantam.

———. 1988. *Uncommon Wisdom: Conversations with Remarkable People*. New York: Simon and Schuster.

———. 1996. *The Web of Life: A New Synthesis of Mind and Matter*. New York: Doubleday.

———. 2002. *The Hidden Connections: Integrating the Biological, Cognitive, and Social Dimensions of Life into a Science of Sustainability*. New York: Doubleday.

———. 2007. *The Science of Leonardo: Inside the Mind of the Great Genius of the Renaissance*. New York: Doubleday.

Capra, Fritjof, with David Steindl-Rast and Thomas Matus. 1991. *Belonging to the Universe: Explorations on the Frontiers of Science and Spirituality*. San Francisco: Harper.

Capra, Fritjof, with Charlene Spretnak and Rüdiger Lutz. 1984. *Green Politics*. New York: Dutton.

Carr, L. D., and S. B. McKagan. 2009. "Graduate quantum mechanics reform." *American Journal of Physics* 77 (April): 308–19.

Carroll, Jerry. 1981. "Another eccentric genius in North Beach?" *San Francisco Chronicle*, May 11: 23.

Carroll, Peter N. 1990 [1982]. *It Seemed Like Nothing Happened: America in the 1970s*, 2nd ed. New Brunswick, NJ: Rutgers University Press.

Carson, Cathryn. 2010. *Heisenberg in the Atomic Age: Science and the Public Sphere*. New York: Cambridge University Press.

Carter, Lloyd G. 1979. "Contested possibilities of things faster than light," *Ukiah Daily Journal*, December 12: 9.

———. 1980. "He sheds light on ESP." *Daily Herald* [Chicago], January 20: 44.

Cartlidge, Edwin. 2002. "Pioneer of the paranormal." *Physics World* (May): 10, 11.

Cassidy, David C. 1992. *Uncertainty: The Life and Science of Werner Heisenberg*. San Francisco: W. H. Freeman.

Cassutt, Michael. 1980. "Captain Quantum: Physics superhero." *Future Life* 21 (September): 50–52.

Cattell, Jacques, ed. 1960. *American Men of Science*, 10th ed. Tempe, AZ: Jacques Cattell Press.

Chyba, T. H., and N. B. Abraham. 1985. "Amplification of a polarized single-photon state and determination of resulting polarization correlations." *Journal of the Optical Society of America B* 2: 377–82.

Cini, Marcello. 1980. "The history and ideology of dispersion relations: The pattern of internal and external factors in a paradigm shift." *Fundamenta Scientiae* 1: 157–72.

Clark, Evert. 1966. "War costs cut spending on science and technology." *New York Times*, January 17: 70.

Clark, P. M., and J. E. Turner. 1968. "Experimental tests of quantum mechanics." *Physics Letters A* 26: 447.

Clarke, Christopher, Frederick Parker-Rhodes, and Jonathan Westphal. 1978. "Review discussion: *The Tao of Physics* by F. Capra." *Theoria to Theory* 11: 287–300.

Clauser, John F. 1969. "Proposed experiment to test local hidden-variable theories." *Bulletin of the American Physical Society* 14: 578.

———. 2002. "Early history of Bell's theorem." In *Quantum [Un]speakables: From Bell to Quantum Information*, ed. R. A. Bertlmann and A. Zeilinger, 61–98. New York: Springer.

Clauser, John F., and Michael Horne. 1974. "Experimental consequences of objective local theories." *Physical Review D* 10: 526–35.

Clauser, John F., Michael A. Horne, Abner Shimony, and Richard A. Holt. 1969. "Proposed experiment to test local hidden-variables theories." *Physical Review Letters* 23: 880–84.

Clauser, John F., and Abner Shimony. 1978. "Bell's theorem: Experimental tests and implications." *Reports on the Progress of Physics* 41: 1881–927.

Clegg, Brian. 2006. *The God Effect: Quantum Entanglement, Science's Strangest Phenomenon*. New York: Macmillan.

Clifton, Robert K., and Marilyn G. Regehr. 1990. "Toward a sound perspective on modern physics: Capra's popularization of mysticism and theological approaches reexamined." *Zygon* 25: 73–104.

Cohen-Tannoudji, Claude, Bernard Diu, and Franck Laloë. 1973. *Mécanique quantique*. Paris: Hermann.

Collery, Solange. 1981. "D'Einstein à la télépathie: Interview d'Olivier Costa de Beauregard." *Tonus* 611 (November 2). Available at http://auriol.free.fr/parapsychologie/Costa-de-Beauregard.htm (accessed December 19, 2009).

Collins, Harry. 1992 [1985]. *Changing Order: Replication and Induction in Scientific Practice*, 2nd ed. Chicago: University of Chicago Press.

———. 2004. *Gravity's Shadow: The Search for Gravitational Waves*. Chicago: University of Chicago Press.

Collins, Harry, and Trevor Pinch. 1979. "The construction of the paranormal: Nothing unscientific is happening." In *On the Margins of Science: The Social Construction of Rejected Knowledge*, ed. Roy Wallis, 237–270. Keele: University of Keele.

———. 1982. *Frames of Meaning: The Social Construction of Extraordinary Science*. Boston: Routledge.

Compton, Laura. 1998. "Something to chew on: Veat." *San Francisco Chronicle*, October 30: 5.

Costa de Beauregard, Olivier. 1967. *Précis de Mécanique Quantique Relativiste*. Paris: Dunod.

———. 1975. "Quantum paradoxes and Aristotle's twofold information concept." In *Quantum Physics and Parapsychology*, ed. Laura Oteri, 91–102. New York: Parapsychology Foundation.

———. 1976. "Time symmetry and interpretation of quantum mechanics." *Foundations of Physics* 6: 539–59.

———. 1977. "Two lectures on the direction of time." *Synthese* 35: 129–54.

———. 1978. "S-matrix, Feynman zigzag, and Einstein correlation." *Physics Letters A* 67: 171–74.

———. 1979. "The expanding paradigm of the Einstein theory." In *The Iceland Papers: Select*

Papers on Experimental and Theoretical Research on the Physics of Consciousness, ed. Andrija Puharich, 161–91. Amherst, WI: Essentia Research Associates.

Crease, Robert P. 2002. "The most beautiful experiment." *Physics World* 15 (September): 19.

Cross, Andrew. 1991. "The crisis in physics: Dialectical materialism and quantum theory." *Social Studies of Science* 21: 735–59.

Cufaro-Petroni, Nicola, Augusto Garruccio, Franco Selleri, and Jean-Pierre Vigier. 1980. "Sur la contradiction entre la théorie quantique classique (idéalisée) de la measure et la conservation du carré du moment angulaire total dans le paradoxe d'Einstein, Podolski et Rosen." *Comptes Rendus* 290 B: 111–14.

Cummings, F. W., J. S. Herold, and J. Sarfatt. 1970. "Beyond the Hartree-liquid model of superfluid bosons." *Physica* 50: 15–26.

Cushing, James T. 1994. *Quantum Mechanics: Historical Contingency and the Copenhagen Hegemony*. Chicago: University of Chicago Press.

d'Espagnat, Bernard, ed. 1971a. *Foundations of Quantum Mechanics*. New York: Academic Press.

d'Espagnat, Bernard. 1971b. *Conceptual Foundations of Quantum Mechanics*. Menlo Park, CA: W. A. Benjamin.

———. 1979. "The quantum theory and reality." *Scientific American* 241 (November): 158–181.

———. 2003. *Veiled Reality: An Analysis of Present-Day Quantum Mechanical Concepts*. Boulder, CO: Westview Press.

Darrigol, Olivier. 1992. *From c-Numbers to q-Numbers: The Classical Analogy in the History of Quantum Theory*. Berkeley: University of California Press.

Dass, Baba Ram [Richard Alpert]. 1971. *Be Here Now, Remember*. New York: Crown.

Daston, Lorraine, and Otto Sibum, eds. 2003. *Scientific Personae and Their Histories*. Special issue of *Science in Context* 16 (June): 1–269.

Davies, Paul, and John Gribben. 2007. *The Matter Myth: Dramatic Discoveries That Challenge Our Understanding of Physical Reality*. New York: Simon and Schuster.

De Angelis, Tiziano, Eleonora Nagali, Fabio Sciarrino, and Francesco De Martini. 2007. "Experimental test of the no-signaling theorem." *Physical Review Letters* 99: 193601.

de Greiff, Alexis. 2002. "The tale of two peripheries: The creation of the International Centre for Theoretical Physics in Trieste." *Historical Studies in the Physical Sciences* 33: 33–60.

Devetek, Igor. 2005. "The private classical capacity and quantum capacity of a quantum channel." *IEEE Transactions on Information Theory* 51: 44–55.

Dewar, Elaine. 1977. "In search of the mind's eye: In the weird world of ESP, seeing is not believing." *Winnipeg Free Press Weekend Magazine*, July 30: 8–12.

de Witt, Karen. 1977. "Quantum theory goes East: Western physics meets yin and yang." *Washington Post*, July 9, C1.

Dieks, D. 1982. "Communication by EPR devices." *Physics Letters A* 92: 271, 272.

Diem, Andrea Grace, and James R. Lewis. 1992. "Imagining India: The influence of Hinduism on the New Age movement." In *Perspectives on the New Age*, ed. James R. Lewis and J. Gordon Melton, 48–58. Albany: State University of New York Press.

Distler, Jacques, and Skip Garibaldi. 2009. "There is no 'theory of everything' inside E8." *Communications in Mathematical Physics* (in press). Preprint arXiv:0905.2658 [math.RT], available at http://arXiv.org.

Ditzen, Stuart. 1999. "Prosecutor's zeal brought scrutiny: Barbara Christie almost always won her murder cases." *Philadelphia Inquirer*, November 13: A8.

Dixler, Elsa. 2009. "Paperback row." *New York Times Book Review*, April 26: 24.

Dowling, Jonathan P. 1995. "Parapsychological Review A?" (letter to the editor). *Physics Today* 48 (July): 78.

Doyle, Michael William. 2001. "Debating the counterculture: Ecstasy and anxiety over the hip alternative." In *The Columbia Guide to America in the 1960s*, ed. David Farber and Beth Bailey, 143–56. New York: Columbia University Press.

Drucker, Linda. 1978. "Institute Fellow Einhorn: Yippie turned teacher." *Harvard Crimson* 168, no. 44 (November 4): 1.

Dull, A. 1978. "The Tao of Physics" (book review). *Philosophy East and West* 28: 387–90.

Durrani, Matin. 2000. "Physicists probe the paranormal." *Physics World* (May): 5.

Dyson, George. 2002. *Project Orion: The True Story of the Atomic Spaceship*. New York: Henry Holt.

Eberhard, Philippe H. 1978. "Bell's theorem and the different concepts of locality." *Il Nuovo Cimento* B 46: 392–419.

Einhorn, Ira. 1966. "Philadelphia Letter." *Innerspace*, no. 3 (December): no pagination.

———. 1970/71. "The sociology of the now." *Psychedelic Review* 11 (Winter): 48–58.

———. 1977. "A disturbing communique." *The CoEvolution Quarterly* 16 (December 21): 74–76.

Einstein, Albert. 1934. "On the method of theoretical physics." *Philosophy of Science* 1 (April): 163–69.

———. 1949. "Reply to criticisms." In *Albert Einstein: Philosopher-Scientist*, ed. Paul A. Schilpp, 665–88. Evanston, IL: Library of Living Philosophers.

———. 1982 [1954]. *Ideas and Opinions*, ed. Carl Seelig, 2nd ed. New York: Crown.

———. 1987. *Letters to Solovine*, ed. Maurice Solovine, trans. Wade Baskin. New York: Philosophical Library.

Einstein, Albert, Boris Podolsky, and Nathan Rosen. 1935. "Can quantum-mechanical description of physical reality be considered complete?" *Physical Review* 47: 777–80.

Ekert, Artur K. 1991. "Quantum cryptography based on Bell's theorem." *Physical Review Letters* 67: 661–63.

Engelhardt, Tom. 2007 [1995]. *The End of Victory Culture: Cold War America and the Disillusioning of a Generation*, rev. ed. Amherst: University of Massachusetts Press.

Esbenshade, Donald H., Jr. 1982. "Relating mystical concepts to those of physics: Some concerns." *American Journal of Physics* 50: 224–228.

Falkoff, D. L. 1952. "Principles of quantum mechanics." *American Journal of Physics* 20: 460, 461.

Faye, Jan. 1991. *Niels Bohr: His Heritage and Legacy*. Boston: Kluwer.

Federici, William. 1967. "Night mayor hit on hippie foulup." *New York Daily News*, June 2: 5.

Feinberg, Gerald. 1967. "Possibility of faster-than-light particles." *Physical Review* 159: 1059–1105.

———. 1969a. "Mankind's search for long-range goals." *Futurist* 3, no. 3 (June): 60–63.

———. 1969b. *The Prometheus Project: Mankind's Search for Long-Range Goals*. Garden City, NY: Doubleday.

———. 1971. "Long-range goals and the environment." *Futurist* 5, no. 6 (December): 241–46.

———. 1975. "Precognition: A memory of things future." In *Quantum Physics and Parapsychology*, ed. Laura Oteri, 54–73. New York: Parapsychology Foundation.

Fenwick, Sheridan. 1976. *Getting It: The Psychology of est*. Philadelphia: J. B. Lippincott.

Feshbach, Herman. 1962. "Clear and perspicuous." *Science* 136 (May 11): 514.

Feuer, Lewis. 1974. *Einstein and the Generations of Science*. New York: Basic Books.

Feynman, Richard, and A. R. Hibbs. 1965. *Quantum Mechanics and Path Integrals*. New York: McGraw-Hill.

Feynman, Richard, with Ralph Leighton. 1985. *"Surely You're Joking, Mr. Feynman!" Adventures of a Curious Character*. New York: W. W. Norton.

Feynman, Richard, with Robert Leighton and Matthew Sands. 1965. *The Feynman Lectures on Physics*, 3 vols. Reading, MA: Addison-Wesley.

Fine, Arthur. 1986. *The Shaky Game: Einstein, Realism, and the Quantum Theory*. Chicago: University of Chicago Press.

Finkbeiner, Ann. 2006. *The Jasons: The Secret History of Science's Postwar Elite*. New York: Viking.

Finkelstein, David. 1958. "Past-future asymmetry of the gravitational field of a point particle." *Physical Review* 110: 965–67.

——. 1969. "Space-time code." *Physical Review* 184: 1261–71.

——. 1972a. "Space-time code, II." *Physical Review D* 5: 320–28.

——. 1972b. "Space-time code, III." *Physical Review D* 5: 2922–31.

——. 1974. "Space-time code, IV." *Physical Review D* 9: 2219–31.

Finkelstein, David, Graham Frye, and Leonard Susskind. 1974. "Space-time code, V." *Physical Review D* 9: 2231–36.

Finn, Peter. 1995. "Ireland in search of Europe." *New York Times*, August 13: BR15.

Folse, Henry J. 1985. *The Philosophy of Niels Bohr: The Framework of Complementarity*. New York: North-Holland.

Fölsing, Albrecht. 1997. *Albert Einstein: A Biography*, trans. Ewald Osers. New York: Penguin.

Forman, Paul. 1971. "Weimar culture, causality, and quantum theory, 1918–1927: Adaptation by German physicists and mathematicians to a hostile intellectual environment." *Historical Studies in the Physical Sciences* 3: 1–115.

——. 1987. "Behind quantum electronics: National security as basis for physical research in the United States, 1940–1960." *Historical Studies in the Physical and Biological Sciences* 18: 149–229.

Fowler, Glenn. 1989. "George Koopman dies in wreck; Technologist for space was 44." *New York Times*, July 21: B8.

Freedman, Stuart J., and John F. Clauser. 1972. "Experimental test of local hidden-variable theories." *Physical Review Letters* 28: 938–41.

Freire, Olival. 2004. "The historical roots of 'foundations of quantum physics' as a field of research (1950–1970)." *Foundations of Physics* 34: 1741–60.

——. 2005. "Science and exile: David Bohm, the Cold War, and a new interpretation of quantum mechanics." *Historical Studies in the Physical and Biological Sciences* 36: 1–34.

——. 2006. "Philosophy enters the optics laboratory: Bell's theorem and its first experimental tests (1965–1982)." *Studies in History and Philosophy of Modern Physics* 37: 577–616.

——. 2007. "Orthodoxy and heterodoxy in the research on the foundations of quantum physics: E. P. Wigner's case." In *Cognitive Justice in a Global World: Prudent Knowledges for a Decent Life*, ed. Boaventura de Sousa Santos, 203–224. New York: Lexington.

Freistadt, Hans. 1953. "The crisis in physics." *Science and Society* 17: 211–37.

——. 1957. "The causal formulation of quantum mechanics of particles: The theory of de Broglie, Bohm, and Takabayasi." *Nuovo Cimento* 5, Suppl.: 1–70.

French, Anthony P., and Edwin F. Taylor. 1978. *An Introduction to Quantum Physics*. New York: Norton.

Friedrich, Bretislav, and Dudley Herschbach. 2003. "Stern and Gerlach: How a bad cigar helped reorient atomic physics." *Physics Today* 56 (December): 53–59.

Fry, Edward S., and Thomas Walther. 2002. "Atom based tests of the Bell inequalities: The legacy of John Bell continues." In *Quantum [Un]speakables: From Bell to Quantum Information*, ed. R. A. Bertlmann and A. Zeilinger, 103–17. New York: Springer.

Fuller, Robert W. 1961. *Nonadiabatic Changes of Potential and Neutron Production in Fission*. PhD dissertation, Princeton University.

Fuller, Robert W., and Zara Wallace. 1975. *A Look at est in Education: Analysis, Review, and Selected Case Studies of the Impact of the est Experience on Educators and Students in Primary, Secondary, and Post-Secondary Education*. San Francisco: *est*.

Fuller, Robert W., and John A. Wheeler. 1962. "Causality and multiply connected space-time." *Physical Review* 128: 919–29.

Fülling, S., R. Bruch, E. A. Rauscher, P. A. Neill, E. Träbert, P. H. Heckmann, and J. H. McGuire. 1992. "Ionization plus excitation of helium by fast electron and proton impact." *Physical Review Letters* 68: 3152–55.

Galison, Peter. 1987. *How Experiments End*. Chicago: University of Chicago Press.

———. 1995. "Theory bound and unbound: Superstrings and experiment." In *Laws of Nature: Essays on the Philosophical, Scientific, and Historical Dimensions*, ed. Friedel Weinert, 369–408. Berlin: Walter de Gruyter.

———. 1997. *Image and Logic: A Material Culture of Microphysics*. Chicago: University of Chicago Press.

———. 2003. *Einstein's Clocks, Poincaré's Maps: Empires of Time*. New York: Norton.

Galison, Peter, Michael Gordin, and David Kaiser, eds. 2001. *Science and Society: The History of Modern Physical Science in the Twentieth Century*, vol. 4, *Quantum Histories*. New York: Routledge.

Gardner, Martin. 1976. "Magic and paraphysics." *Technology Review* 78 (June): 42–51.

———. 1991. "Perpetual motion." In Gardner, *The New Age: Notes of a Fringe Watcher*, 145–66. Buffalo, NY: Prometheus Books.

Gardner, Martin, and John Archibald Wheeler. 1979. "Quantum theory and quack theory." *New York Review of Books* 26, May 17: 39, 40.

Garfield, Eugene. 1987. "A different sort of great-books list: The 50 twentieth-century works most cited in the *Arts & Humanities Citation Index*, 1976–1983." *Current Comments* 16 (April 20): 101–5.

Garfinkel, Perry. 1982. "*New* new physics." *California Living Magazine* (December 12): 7–9.

Garuccio, A., and F. Selleri. 1976. "Nonlocal interactions and Bell's inequality." *Nuovo Cimento B* 36: 176–85.

Geiger, Roger. 1986. *To Advance Knowledge: The Growth of American Research Universities, 1900–1940*. New York: Oxford University Press.

Gell-Mann, Murray. 1994. *The Quark and the Jaguar: Adventures in the Simple and Complex*. New York: Freeman.

Gelman, David, Pamela Abramson, and Elizabeth Ann Leonard. 1991. "The sorrows of Werner." *Newsweek* (February 18): 72.

Gerjuoy, Edward. 1956. "Quantum mechanics." *American Journal of Physics* 24: 118.

Ghirardi, G. C., A. Rimini, and T. Weber. 1980. "A general argument against superluminal transmission through the quantum mechanical measurement process." *Lettere al Nuovo Cimento* 27: 293–98.

Ghirardi, G. C., and T. Weber. 1979. "On some recent suggestions of superluminal communication through the collapse of the wave function." *Lettere al Nuovo Cimento* 26: 599–603.

———. 1983. "Quantum mechanics and faster-than-light communication: Methodological considerations." *Il Nuovo Cimento* 78B: 9–20.

Gieryn, Thomas. 1999. *Cultural Boundaries of Science: Credibility on the Line*. Chicago: University of Chicago Press.

Gilder, Louisa. 2008. *The Age of Entanglement: When Quantum Physics Was Reborn*. New York: Knopf.

Gilliam, Harold. 1978a. "The power to see the invisible." *San Francisco Chronicle*, January 4: 4.

─────. 1978b. "Skeptics of mind power call it superstition." *San Francisco Chronicle*, January 11: 4.

Gitlin, Todd. 1997 [1993]. *The Sixties: Years of Hope, Days of Rage*, 2nd ed. New York: Bantam.

Glashow, Sheldon. 2008. "Sidney Richard Coleman." *Physics Today* 61 (May): 69.

Glass, Leonard L., Michael A. Kirsch, and Frederick N. Parris. 1977. "Psychiatric disturbances associated with Erhard Seminars Training: I. A report of cases." *American Journal of Psychiatry* 134 (March): 245–47.

Glauber, Roy. 1986. "Amplifiers, attenuators, and Schrödinger's Cat." *Annals of the New York Academy of Sciences* 480: 336–72.

Gleick, James. 1987. *Chaos: The Making of a New Science*. New York: Viking.

Gold, Herb. 1993. *Bohemia: Digging the Roots of Cool*. New York: Touchstone.

Goleman, Daniel. 1985. "Esalen wrestles with a staid present." *New York Times*, December 10: C1, C6.

Gordin, Michael. 2004. *A Well-Ordered Thing: Dmitrii Mendeleev and the Shadow of the Periodic Table*. New York: Basic Books.

Gordon, Suzanne. 1978. "Let them eat *est*." *Mother Jones* 3 (December): 41–49.

Gosse, Van. 2005. *Rethinking the New Left: An Interpretive History*. New York: Palgrave Macmillan.

Gottfried, Kurt, and Tung-Mow Yan. 2003. *Quantum Mechanics: Fundamentals*. New York: Springer.

Goudsmit, S. A. 1973. "Important announcement regarding papers about fundamental theories." *Physical Review D* 8: 357.

Gozzini, Adriano. 1984. "On the possibility of realising a low intensity interference experiment with a determination of the particle trajectory." In *The Wave-Particle Dualism: A Tribute to Louis de Broglie on his 90th Birthday*, ed. S. Diner, D. Fargue, G. Lochak, and F. Selleri, 129–137. Boston: Reidel.

Greenberger, Daniel M. 1986. "Preface." *Annals of the New York Academy of Sciences* 480: xiii, xiv.

Greene, Brian. 2004. *The Fabric of the Cosmos: Space, Time, and the Texture of Reality*. New York: Knopf.

Griffiths, David J. 2005 [1995]. *Introduction to Quantum Mechanics*, 2nd ed. Upper Saddle River, NJ: Prentice Hall.

Gustaitus, Rasa. 1975. "Faster than a speeding photon." *City of San Francisco*, October 7: 22, 23.

Gusterson, Hugh. 1996. *Nuclear Rites: A Weapons Laboratory at the End of the Cold War*. Berkeley: University of California Press.

Guth, Alan, and David Kaiser. 2005. "Inflationary cosmology: Exploring the universe from the smallest to the largest scales." *Science* 307: 884–90.

Haas, Arthur. 1928. *Wave Mechanics and the New Quantum Theory*, trans. L. W. Codd. London: Constable.

Hall, Joseph, Christopher Kim, Brien McElroy, and Abner Shimony. 1977. "Wave-packet reduction as a medium of communication." *Foundations of Physics* 7: 759–67.

Harrison, David. 1978. "Comment on 'Partons in antiquity.'" *American Journal of Physics* 46: 432, 433.

─────. 1979a. "What you see is what you get!" *American Journal of Physics* 47: 576–82.

─────. 1979b. "Teaching *The Tao of Physics*." *American Journal of Physics* 47: 779–83.

———. 1982a. "Bell's inequality and quantum correlations." *American Journal of Physics* 50: 811–16.

———. 1982b. "Comment on 'Relating mystical concepts to those of physics'" (letter to the editor). *American Journal of Physics* 50: 873, 874.

Harrison, David, and James D. Prentice. 1980. "Physics for the amateur" (letter to the editor). *American Journal of Physics* 48: 799.

Harvey, Bill. 1980. "The effects of social context on the process of scientific investigation: Experimental tests of quantum mechanics." In *The Social Process of Scientific Investigation*, ed. Karin D. Knorr, Roger Krohn, and Richard Whitley, 139–63. Dordrecht: Reidel.

———. 1981. "Plausibility and the evaluation of knowledge: A case-study of experimental quantum mechanics." *Social Studies of Science* 11: 95–130.

Hasted, J. B., D. J. Bohm, E. W. Bastin, and B. O'Regan. 1975. Letter to the Editor. *Nature* 254: 470–72.

Hasted, John B. 1976. "An experimental study for the validity of metal-bending phenomena." *Journal of the Society for Psychical Research* 48: 365–83.

———. 1977. "Physical aspects of paranormal metal-bending." *Journal of the Society for Psychical Research* 49: 583–607.

———. 1978/79. "Speculations about the relation between psychic phenomena and physics." *Psychoenergetic Systems* 4: 243–57.

———. 1979. "Paranormal metal-bending." In *The Iceland Papers: Select Papers on Experimental and Theoretical Research on the Physics of Consciousness*, ed. Andrija Puharich, 215–21. Amherst, WI: Essentia Research Associates.

Heck, R. C. Devon, and Jennifer L. Thompson. 1976. "*est*: Salvation or swindle?" *San Francisco Magazine* 18 (January): 20–23, 70, 71.

Heirich, Max. 1976. "Cultural breakthroughs." *American Behavioral Scientist* 19: 685–702.

Heisenberg, Werner. 1930. *The Physical Principles of the Quantum Theory*, trans. Carl Eckart and F. C. Hoyt. Chicago: University of Chicago Press.

———. 1971 [1969]. *Physics and Beyond: Encounters and Conversations*, trans. Arnold J. Pomerans. London: Allen and Unwin.

Henderson, Hazel. 1979. "Used magazines." *The CoEvolution Quarterly* 21 (March 21): 88–90.

Herbert, Nick. 1975. "Cryptographic approach to hidden variables." *American Journal of Physics* 43: 315, 316.

———. 1976. "Mechanical mediums." *Psychic* (July/August): 36–40.

———. 1982. "FLASH—A superluminal communicator based upon a new kind of quantum measurement." *Foundations of Physics* 12: 1171–79.

———. 1985. *Quantum Reality: Beyond the New Physics*. Garden City, NY: Anchor/Doubleday.

———. 1986. "Can single quantum events carry superluminal signals?" *Annals of the New York Academy of Sciences* 480: 578.

———. 1987. *Quantenrealität: jenseits d. neuen Physik*. Trans. Traude Wess. Basel: Birkhäuser.

———. 1988. *Faster Than Light: Superluminal Loopholes in Physics*. New York: Plume.

———. 1989. *Taimu mashin no tsukurikata: kosoku toppa wa mutsukashiku nai!* Trans. Rei Kozumi and Keiko Takabayashi. Tokyo: Kodansha.

———. 1990. *Ryoshi to jitsuzai: fukakuteisei genri kara beru no teiri e*. Trans. Hajime Hayashi. Tokyo: Hakuyosha.

———. 1993. *Elemental Mind: Human Consciousness and the New Physics*. New York: Dutton.

———. 2000. *Physics on All Fours: Selected Verse, 1995–2000*. Boulder Creek, CA: Sea Creature Press.

———. 2007. "Entanglement telegraphed communication avoiding light-speed limitation by Hong Ou Mandel effect." Unpublished preprint arXiv:0712.2530 (December), available at http://www.arXiv.org.

Herbert, Nick, and Jack Karush. 1978. "Generalization of Bell's theorem." *Foundations of Physics* 8: 313–17.

Hess, David. 1993. *Science in the New Age: The Paranormal, Its Defenders and Debunkers, and American Culture*. Madison: University of Wisconsin Press.

Hesseldahl, Arik. 2006. "A quantum leap in data encryption." *BusinessWeek* (November 6). Online edition available at http://www.businessweek.com/technology/content/nov2006/tc20061106_302053.htm (accessed July 12, 2010).

Hillmer, Rachel, and Paul Kwiat. 2007. "A do-it-yourself quantum eraser." *Scientific American* 296 (May): 90–95.

Hiskett, P. A., et al. 2006. "Long-distance quantum key distribution in optical fibre." *New Journal of Physics* 8 (September): 193.

Hofstadter, Douglas. 1979. *Gödel, Escher, Bach: An Eternal Golden Braid*. New York: Basic Books.

Holton, Gerald. 1988 [1973]. *Thematic Origins of Scientific Thought: Kepler to Einstein*, 2nd ed. Cambridge: Harvard University Press.

———. 1998. "Einstein and the cultural roots of modern science." *Daedalus* 127: 1–44.

Houdini, Harry. 1924. *A Magician Among the Spirits*. New York: Harper.

Howard, Don. 1985. "Einstein on locality and separability." *Studies in History and Philosophy of Science* 16: 171–201.

Hubbard, G. Scott, and Edwin C. May. 1982. "Countermeasures: A survey and evaluation." Classified report by the Stanford Research Institute for the Defense Intelligence Agency, January. Declassified in 2000. Now available at http://www.remoteviewed.com/remote_viewing_history_military.htm (accessed November 8, 2007).

Hubner, John. 1990a. "Worlds of Werner." *West* [Sunday magazine of the *San Jose Mercury News*], November 11: 16–27.

———. 1990b. "All in the family." *West* [Sunday magazine of the *San Jose Mercury News*], November 18: 9–18.

Huerta, Loretta Kiklinsky. 1979. "Fifty-four hours: Things to do this weekend." *Los Angeles Times*, July 10: I10.

Hyman, Ray. 1977. "Psychics and scientists: 'Mind-reach' and remote viewing." *The Humanist* 37 (May/June): 16–20.

———. 1995. "Evaluation of program on 'anomalous mental phenomena.'" In *An Evaluation of Remote Viewing: Research and Applications*, ed. Michael D. Mumford, Andrew W. Rose, and David A. Goslin, 41–75. Washington, DC: American Institutes for Research.

Jackiw, Roman, and Abner Shimony. 2002. "The depth and breadth of John Bell's physics." *Physics in Perspective* 4: 78–116.

Jacobs, Kurt, and Howard M. Wiseman. 2005. "An entangled web of crime: Bell's theorem as a short story." *American Journal of Physics* 73: 932–37.

Jaeger, Gregg. 2007. *Quantum Information: An Overview*. New York: Springer.

———. 2010. "Special issue on foundations of quantum information." *Quantum Information Processing* 9, no. 2 (April): 93, 94.

Jammer, Max. 1966. *The Conceptual Development of Quantum Mechanics*. New York: McGraw-Hill.

———. 1974. *The Philosophy of Quantum Mechanics: The Interpretations of Quantum Mechanics in Historical Perspective*. New York: Wiley.

Jauch, J. M., and C. Piron. 1963. "Can hidden variables be excluded in quantum mechanics?" *Helvetica Physica Acta* 36: 827–37.

Jenny, Marcel, et al. 2005. "Apoptosis induced by the Tibetan herbal remedy Padma 28 in the T cell-derived lymphocytic leukaemia cell line CEM-C7H2." *Journal of Carcinogenesis* 4 (September): 15.

Johnson, George. 1988. "Nonfiction: Uncommon Wisdom." *New York Times Book Review*, March 20: 27.

Jordan, Pascual. 1927a. "Über die Polarisation der Lichtquanten." *Zeitschrift für Physik* 44: 292–300.

———. 1927b. *Verdrängung und Komplementarität: Eine philosophische Untersuchung*. Hamburg: Stormverlag.

———. 1955. "Atomic physics and parapsychology." *Newsletter of the Parapsychology Foundation* 2 (July/August 1955): 3–7.

Josephson, Brian D. 1979. "Foreword." In *The Iceland Papers: Select Papers on Experimental and Theoretical Research on the Physics of Consciousness*, ed. Andrija Puharich, 4, 5. Amherst, WI: Essentia Research Associates.

———. 1992 [1973]. "The discovery of tunnelling supercurrents." In *Nobel Lectures, Physics: 1971–1980*, ed. Stig Lundqvist, 157–164. Singapore: World Scientific.

Josephson, Brian D., Richard D. Mattuck, Evan Harris Walker, and Olivier Costa de Beauregard. 1980. "Parapsychology: An exchange." *New York Review of Books* 27 (June 26): 48–51.

Jozsa, Richard. 1998. "Quantum information and its properties." In *Introduction to Quantum Computation and Information*, ed. Hoi-Kwong Lo, Sandu Popescu, and Tim Spiller, 49–75. Singapore: World Scientific.

Kahn, Jennifer. 2002. "Notes from another universe." *Discover* 23, no. 4 (April): 66–71.

Kaiser, David. 1992. "More roots of complementarity: Kantian aspects and influences." *Studies in History and Philosophy of Science* 23: 213–39.

———. 1994. "Bringing the human actors back on stage: The personal context of the Einstein-Bohr debate." *British Journal for the History of Science* 27: 129–52.

———. 2002. "Cold war requisitions, scientific manpower, and the production of American physicists after World War II." *Historical Studies in the Physical and Biological Sciences* 33: 131–59.

———. 2004. "The postwar suburbanization of American physics." *American Quarterly* 56: 851–88.

———. 2005. *Drawing Theories Apart: The Dispersion of Feynman Diagrams in Postwar Physics*. Chicago: University of Chicago Press.

———. 2006a. "The physics of spin: Sputnik politics and American physicists in the 1950s." *Social Research* 73: 1225–52.

———. 2006b. "Whose mass is it anyway? Particle cosmology and the objects of theory." *Social Studies of Science* 36: 533–64.

———. 2007a. "Turning physicists into quantum mechanics." *Physics World* 20 (May): 28–33.

———. 2007b. "When fields collide." *Scientific American* 296 (June): 62–69.

———. 2007c. "Weisskopf, Victor Frederick." In *New Dictionary of Scientific Biography* 7: 262–69. New York: Scribner's.

———. Forthcoming. *American Physics and the Cold War Bubble*. Chicago: University of Chicago Press.

Kaku, Michio. 2008. *Physics of the Impossible: A Scientific Exploration into the World of Phasers, Force Fields, Teleportation, and Time Travel*. New York: Doubleday.

Kanellos, Michael. 2004. "Quantum encryption inches closer to reality." *CNET News*, May 3. Available at http://news.cnet.com.

Kauffman, George B. 1977. "The Tao of Physics." *Isis* 68: 460, 461.

Keller, Evelyn Fox. 1977. "The anomaly of a woman in physics." In *Working It Out*, ed. Sara Ruddick and Pamela Daniels, 78–91. New York: Pantheon.

Kevles, Daniel. 1995 [1978]. *The Physicists: The History of a Scientific Community in Modern America*, 3rd ed. Cambridge: Harvard University Press.

Kirk, Andrew G. 2007. *Counterculture Green: The Whole Earth Catalog and American Environmentalism*. Lawrence: University of Kansas Press.

Kirsch, Michael A., and Leonard L. Glass. 1977. "Psychiatric disturbances associated with Erhard Seminars Training: II. Additional cases and theoretical considerations." *American Journal of Psychiatry* 134 (November): 1254–58.

Kohler, Robert. 1991. *Partners in Science: Foundations and Natural Scientists, 1900–1945*. Chicago: University of Chicago Press.

Kojevnikov, Alexei. 2002. "David Bohm and collective movement." *Historical Studies in the Physical and Biological Sciences* 33: 161–92.

Kornbluth, Jesse, ed. 1968. *Notes from the New Underground*. New York: Viking.

Kornbluth, Jesse. 1976. "The fuhrer over *est*." *New Times: The Feature News Magazine* (March 19): 29–52.

Kragh, Helge. 1999. *Quantum Generations: A History of Physics in the Twentieth Century*. Princeton: Princeton University Press.

Kress, Kenneth A. 1999. "Parapsychology in intelligence: A personal review and conclusions." *Journal of Scientific Exploration* 13: 69–85.

Kripal, Jeffrey. 2007. *Esalen: America and the Religion of No Religion*. Chicago: University of Chicago Press.

Kuczkowski, R. L. 1999. "Lawrence S. Bartell: Biographical notes." *Journal of Molecular Structure* 485/486: xi–xxvii.

Kuhn, Thomas S. 1959. "Energy conservation as an example of simultaneous discovery." In *Critical Problems in the History of Science*, ed. Marshall Clagett, 321–56. Madison: University of Wisconsin Press.

———. 1962. *The Structure of Scientific Revolutions*. Chicago: University of Chicago Press.

———. 1970. "Postscript, 1969." In Kuhn, *The Structure of Scientific Revolutions*, 2nd ed., 174–210. Chicago: University of Chicago Press.

Kuhn, Thomas S., et al., eds. 1967. *Sources for History of Quantum Physics: An Inventory and Report*. Philadelphia: American Philosophical Society.

Kwiat, P. G., and L. Hardy. 2000. "The mystery of the quantum cakes." *American Journal of Physics* 68: 33–36.

Kyle, Richard. 1995. *The New Age Movement in American Culture*. New York: University Press of America.

Lamehi-Rachti, M., and W. Mittig. 1976. "Quantum mechanics and hidden variables: Tests of Bell's inequality by measurement of spin correlation in low-energy proton-proton scattering." *Physical Review D* 14: 2543–55.

LaMothe, John D. 1972. *Controlled Offensive Behavior: USSR*. Washington, DC: Defense Intelligence Agency. Declassified. Now available at http://www.dia.mil/publicaffairs/Foia/cont_ussr.pdf.

Lattin, Don. 1990. "Ex-employees describe abuse in suit against *est*'s Erhard." *San Francisco Chronicle*, April 3: A4.

————. 2010. *The Harvard Psychedelic Club: How Timothy Leary, Ram Dass, Huston Smith, and Andrew Weil Killed the Fifties and Ushered in a New Age for America*. New York: HarperOne.

Laudan, Larry. 1983. "The demise of the demarcation problem." In *Physics, Philosophy, and Psychoanalysis: Essays in Honor of Adolf Grünbaum*, ed. Robert S. Cohen and Larry Laudan, 111–27. Dordrecht: Reidel.

Lavigne, Keith. 1994. "Lederman speech brings light to physics." *The Dartmouth*, April 22.

Leane, Elizabeth. 2007. *Reading Popular Physics: Disciplinary Skirmishes and Textual Strategies*. Burlington, VT: Ashgate.

Leary, Timothy. 1977. "Preface." *Spit in the Ocean* 3 (Fall): 8–11.

Leary, Timothy, with Robert Anton Wilson and George A. Koopman. 1977. *Neuropolitics: The Sociobiology of Human Metamorphosis*. Los Angeles: Starseed/Peace Press.

Lécuyer, Christophe. 2010. "Patrons and a plan." In *Becoming MIT: Moments of Decision*, ed. David Kaiser, 59–80. Cambridge: MIT Press.

Lee, Martin A., and Bruce Shlain. 1992 [1985]. *Acid Dreams: The Complete Social History of LSD: The CIA, the Sixties, and Beyond*, 2nd ed. New York: Grove.

Leggett, A. J. 2009. "Aspect experiment." In *Compendium of Quantum Physics: Concepts, Experiments, History, and Philosophy*, ed. D. Greenberger, K. Hentschel, and F. Weinert, 14–18. New York: Springer.

Lehmann-Haupt, Christopher. 1979. "Books of the Times." *New York Times*, March 28: C23.

————. 1985. "Books of the Times: *Quantum Reality*." *New York Times*, June 24: C13.

Leonard, George. 1978. *The Silent Pulse: A Search for the Perfect Rhythm That Exists in All of Us*. New York: Dutton.

LeShan, Lawrence. 1969. "Physicists and mystics: Similarities in world view." *Journal of Transpersonal Psychology* 1: 1–20.

————. 1974. *The Medium, the Mystic, and the Physicist: Toward a General Theory of the Paranormal*. New York: Viking.

LeShan, Lawrence, and Henry Margenau. 1982. *Einstein's Space and Van Gogh's Sky: Physical Reality and Beyond*. New York: Macmillan.

Leslie, Stuart W. 1993. *The Cold War and American Science: The Military-Industrial-Academic Complex at MIT and Stanford*. New York: Columbia University Press.

Levit, C., and J. Sarfatti. 1997. "Are the Bader Laplacian and the Bohm potential equivalent?" *Chemical Physics Letters* 281: 157–60.

Levy, Steven. 1988. *The Unicorn's Secret: Murder in the Age of Aquarius*. New York: Prentice Hall.

Lewenstein, Bruce. 1992. "The meaning of 'public understanding of science' in the United States after World War II." *Public Understanding of Science* 1: 45–68.

Lewis, James R., and J. Gordon Melton, eds. 1992. *Perspectives on the New Age*. Albany: State University of New York Press.

Lin, Jennifer. 2002a. "Hottest ticket in town: Many are drawn to see the former hippie guru take the stand." *Philadelphia Inquirer*, October 16: A10.

————. 2002b. "This time, Einhorn stands nearly alone: The defendant's brother came to support him in court, but most family and friends stayed away." *Philadelphia Inquirer*, October 1: A10.

Lindsay, Robert, and Henry Margenau. 1936. *Foundations of Physics*. New York: Wiley.

Lisi, A. Garrett. 2007. "An exceptionally simple theory of everything." Preprint arXiv:0711.0770 [hep-ph], available at http://www.arXiv.org.

Litwack, Leo. 1976. "Pay attention, turkeys!" *New York Times Sunday Magazine*, May 2: 44–57.

Lloyd, Seth. 1997. "Capacity of the noisy quantum channel." *Physical Review A* 55: 1613–22.

———. 2006. *Programming the Universe: A Quantum Computer Scientist Takes on the Cosmos*. New York: Knopf.

Lo, Hoi-Kwong, Sandu Popescu, and Tim Spiller, eds. 1998. *Introduction to Quantum Computation and Information*. River Edge, NJ: World Scientific.

Lopez, Steve. 1997. "The search for the unicorn." *Time* 150, no. 13 (September 29): 48–57.

Lowe, J. L. 1927. *The Road to Xanadu: A Study in the Ways of the Imagination*. New York: Houghton-Mifflin.

Lowen, Rebecca S. 1997. *Creating the Cold War University: The Transformation of Stanford*. Berkeley: University of California Press.

Loyd, Linda. 1999a. "Einhorn verdict: $907 million, a city high." *Philadelphia Inquirer*, July 29: A1.

———. 1999. "Abraham asks federal help in getting Einhorn back." *Philadelphia Inquirer*, December 17: B2.

Macintyre, Ben. 1992. "New Age guru goes into hiding." *The Times* [London], July 22.

MacKenzie, Donald. 2001. *Mechanizing Proof: Computing, Risk, and Trust*. Cambridge: MIT Press.

MacNamara, Charles. 1967. "A faint cheer for Gratis U." *Philadelphia Magazine* 58: 147–150.

Mandel, Leonard. 1983. "Is a photon amplifier always polarization dependent?" *Nature* 304: 188.

Mansfield, V. N. 1976. "The Tao of Physics." *Physics Today* 29 (August): 56.

March, Robert. 1979. "The Dancing Wu Li Masters: An Overview of the New Physics." *Physics Today* 32 (August): 54, 55.

Margenau, Henry. 1936. "Quantum-mechanical description." *Physical Review* 49: 240–242.

———. 1950. *The Nature of Physical Reality*. New York: McGraw-Hill.

———. 1953. *Physics: Principles and Applications*. New York: McGraw-Hill.

———. 1956. "Physics and psychic research." *Newsletter of the Parapsychology Foundation* 3 (January/February): 14, 15.

———. 1957. "A principle of resonance." *Newsletter of the Parapsychology Foundation* 4 (May/June): 3–6.

———. 1961. *Open Vistas: Philosophical Perspectives on Modern Science*. New Haven, CT: Yale University Press.

———. 1966. "ESP in the framework of modern science." *Journal of the American Society for Psychical Research* 60: 214–28.

———. 1978. *Physics and Philosophy: Selected Essays*. Boston: Reidel.

Marin, Peter. 1975. "The new narcissism." *Harper's* (October): 45, 46.

Markoff, John. 2005. *What the Dormouse Said: How the Sixties Counterculture Shaped the Personal Computer Industry*. New York: Penguin.

Marks, David. 1981. "Sensory cues invalidate remote viewing experiments." *Nature* 292: 177.

———. 1986. "Investigating the paranormal." *Nature* 320: 119–24.

Marks, David, and Richard Kammann. 1978. "Information transmission in remote viewing experiments." *Nature* 274: 680, 681.

———. 1980. *The Psychology of the Psychic*. Buffalo, NY: Prometheus Books.

Marks, David, and Christopher Scott. 1986. "Remote viewing exposed." *Nature* 319: 444.

Marquard, Bryan. 2008. "Sidney Coleman: Harvard icon taught physics with wit." *Boston Globe*, January 20: A23.

Masterman, Margaret. 1970. "The nature of a paradigm." In *Criticism and the Growth of Knowledge*, ed. Imre Lakatos and Alan Musgrave, 59–89. New York: Cambridge University Press.

Mattuck, Richard D. 1967. *A Guide to Feynman Diagrams in the Many-Body Problem*. New York: McGraw-Hill.

———. 1977. "Probable psychokinetic effects produced in a clinical thermometer." *Psychoenergetic Systems* 2: 31–37.

———. 1978/79. "Thermal noise theory of psychokinesis: Modified Walker model with pulsed information rate." *Psychoenergetic Systems* 3: 301–25.

Mattuck, Richard D., and Evan Harris Walker. 1979. "The action of consciousness on matter: A quantum mechanical theory of psychokinesis." In *The Iceland Papers: Select Papers on Experimental and Theoretical Research on the Physics of Consciousness*, ed. Andrija Puharich, 111–59. Amherst, WI: Essentia Research Associates.

Matusow, Allen. 1984. *The Unraveling of America: A History of Liberalism in the 1960s*. New York: Harper and Row.

Matza, Michael. 1997. "Einhorn trial on trial." *Philadelphia Inquirer*, December 7: E1.

Maudlin, Tim. 1994. *Quantum Non-Locality and Relativity: Metaphysical Intimations of Modern Physics*. Cambridge, MA: Blackwell.

Mauskopf, Seymour H., and Michael R. McVaugh. 1980. *The Elusive Science: Origins of Experimental Psychical Research*. Baltimore, MD: Johns Hopkins University Press.

McConnell, R. A. 1987. *Parapsychology in Retrospect: My Search for the Unicorn*. University of Pittsburgh: Biological Sciences Department.

McCormick, Bernard. 1970. "Don't trust anybody over 30." *Philadelphia Magazine* 61 (May): 57–71.

McCoy, Barry M., Jacques H. H. Perk, and Tai Tsun Wu. 1981. "Ising field theory: Quadratic difference equations for the *n*-point Green's functions on the lattice." *Physical Review Letters* 46: 757–60.

McDowell, Edwin. 1982a. "90 nominees picked for American Book Awards." *New York Times*, February 3: C18.

———. 1982b. "Updike and Kidder win American Book Awards." *New York Times*, April 20: C9.

McKay, Niall. 1999. "Crypto set for a quantum leap." *Wired* 7 (April 5). Online edition available at http://www.wired.com/science/discoveries/news/1999/04/18936 (accessed July 12, 2010).

Mehra, Jagdish. 1975. *The Solvay Conferences on Physics: Aspects of the Development of Physics Since 1911*. Boston: Reidel.

Melton, J. Gordon. 1992. "New Thought and the New Age." In *Perspectives on the New Age*, ed. James R. Lewis and J. Gordon Melton, 15–29. Albany: State University of New York Press.

Merali, Zeeya. 2007. "The pattern that describes the universe: The geometry of a mathematical pattern called E8 may underlie interactions between all the particles and forces in the universe." *New Scientist* (November 17): 8–10.

Mermin, N. David. 1981. "Bringing home the atomic world: Quantum mysteries for anybody." *American Journal of Physics* 49: 940–43.

———. 1985. "Is the moon there when nobody looks? Reality and the quantum theory." *American Journal of Physics* 38: 38–47.

———. 1989. "What's wrong with this pillow?" *Physics Today* 42 (April): 9–11.

———. 1990a. "Quantum mysteries revisited." *American Journal of Physics* 58: 731–34.

———. 1990b. *Boojums All the Way Through: Communicating Science in a Prosaic Age*. New York: Cambridge University Press.

———. 1994. "Quantum mysteries refined." *American Journal of Physics* 62: 880–87.

———. 2004. "Could Feynman have said this?" *Physics Today* 57 (May): 10, 11.

Meystre, Pierre, and Marlan O. Scully, eds. 1983. *Quantum Optics, Experimental Gravity, and Measurement Theory*. New York: Plenum.

Miles, Jack. 1982. "A whole-earth scientific order for the future." *Los Angeles Times*, April 4: N8.

Miller, Arthur I. 2009. *Deciphering the Cosmic Number: The Strange Friendship of Wolfgang Pauli and Carl Jung*. New York: W. W. Norton.

Miller, David A. B. 2008. *Quantum Mechanics for Scientists and Engineers*. New York: Cambridge University Press.

Milonni, P. W., and M. L. Hardies. 1982. "Photons cannot always be replicated." *Physics Letters A* 92: 321, 322.

Mishlove, Jeffrey. 1975. *The Roots of Consciousness: Psychic Liberation through History, Science and Experience*. New York: Random House.

Misner, Charles W., Kip S. Thorne, and Wojciech H. Zurek. 2009. "John Wheeler, relativity, and quantum information." *Physics Today* 62 (April): 40–46.

Mitchell, Edgar D. 1974. "Ex-astronaut on E.S.P." *New York Times*, January 9: 35.

Mitchell, Edgar D., with Dwight Williams. 1996. *The Way of the Explorer: An Apollo Astronaut's Journey through the Material and Mystical Worlds*. New York: G. P. Putnam's Sons.

MIT News Office. 1994. "President emeritus Jerome Wiesner is dead at 79." *MIT Tech Talk* 39 (October 26).

Mody, Cyrus. 2008. "How I learned to stop worrying and love the bomb, the nuclear reactor, the computer, ham radio, and recombinant DNA." *Historical Studies in the Natural Sciences* 38: 451–61.

Molotsky, Irvin. 1998. "Convicted killer arrested again in France." *New York Times*, September 23: A22.

Monroe, Christopher R., and David J. Wineland. 2008. "Quantum computing with ions." *Scientific American* 299, no. 2 (August 11): 64–71.

Moore, Kelly. 2008. *Disrupting Science: Social Movements, American Scientists, and the Politics of the Military, 1945–1975*. Princeton: Princeton University Press.

Moore, R. Laurence. 1977. *In Search of White Crows: Spiritualism, Parapsychology, and American Culture*. New York: Oxford University Press.

Moore, Walter. 1989. *Schrödinger: Life and Thought*. New York: Cambridge University Press.

Moran, Robert. 1998. "Ridge signs Einhorn trial law." *Philadelphia Inquirer*, January 28: B1.

Morrison, Michael A. 1990. *Understanding Quantum Physics: A User's Manual*. Englewood Cliffs, NJ: Prentice Hall.

Morrisson, Mark S. 2007. *Modern Alchemy: Occultism and the Emergence of Atomic Theory*. New York: Oxford University Press.

Moy, Timothy. 2004. "Culture, technology, and the cult of tech in the 1970s." In *America in the Seventies*, ed. Beth Bailey and David Farber, 208–27. Lawrence: University of Kansas Press.

Mullet, Shawn. 1999. *Political Science: The Red Scare as the Hidden Variable in the Bohmian Interpretation of Quantum Theory*. BA thesis, University of Texas at Austin.

————. 2008. *Little Man: Four Junior Physicists and the Red Scare Experience*. PhD dissertation, Harvard University.

Murdoch, Dugald. 1987. *Niels Bohr's Philosophy of Physics*. New York: Cambridge University Press.

Naik, Gautam. 2009. "Science, spirituality, and some mismatched socks." *Wall Street Journal*, May 5: A12.

Nauenberg, Michael. 2007. "Critique of 'Quantum Enigma: Physics Encounters Consciousness.'" *Foundations of Physics* 37: 1612–27.

Newman, E. T. 2002. "A biased and personal description of GR at Syracuse University, 1951–1961." Lecture delivered at the Sixth International Conference on the History of General Relativity,

Amsterdam, June. Text available at http://www.phys.syr.edu/faculty/Goldberg (accessed October 23, 2007).

Nielsen, Michael A., and Isaac L. Chuang. 2000. *Quantum Computation and Quantum Information.* New York: Cambridge University Press.

Nikbin, Darius. 2006. "Quantum encryption sets long-distance record." *Physics World* 19, no. 11 (November): 4.

Noakes, Richard. 2008. "The 'world of the infinitely little': Connecting physical and psychical realities ca. 1900." *Studies in History and Philosophy of Science* 39: 323–34.

Novak, Steven J. 1997. "LSD before Leary: Sidney Cohen's critique of 1950s psychedelic drug research." *Isis* 88: 87–110.

Novick, Julius. 1966. "Theatre: And that's how the rent gets paid." *Village Voice* (September 1): 17.

Olwell, Russell. 1999. "Physical isolation and marginalization in physics: David Bohm's Cold War exile." *Isis* 90: 738–56.

Oppenheim, Janet. 1985. *The Other World: Spiritualism and Psychical Research in England, 1850–1914.* New York: Cambridge University Press.

Ord-Hume, Arthur. 1977. *Perpetual Motion: The History of an Obsession.* New York: St. Martin's Press.

O'Regan, Brendan. 1974. "A comment." *Psychoenergetic Systems* 1: 9, 10.

Orzel, Chad. 2009. *How to Teach Physics to Your Dog.* New York: Scribner.

Oteri, Laura, ed. 1975. *Quantum Physics and Parapsychology.* New York: Parapsychology Foundation.

Oullette, Jennifer. 2004/2005. "Quantum key distribution." *The Industrial Physicist* (December/January): 22–25.

Overbye, Dennis. 1991. *Lonely Hearts of the Cosmos: The Story of the Scientific Quest for the Secret of the Universe.* New York: Harper Collins.

———. 2006. "Mysticism, love, crystals, Atlantis. Far out, man. But is it quantum physics?" *New York Times*, March 14: F3.

———. 2009. "They tried to outsmart Wall Street." *New York Times*, March 10: D1.

Owen, Alex. 1990. *The Darkened Room: Women, Power, and Spiritualism in Late Victorian England.* Philadelphia: University of Pennsylvania Press.

———. 2004. *The Place of Enchantment: British Occultism and the Culture of the Modern.* Chicago: University of Chicago Press.

Owens, Larry. 1990. "MIT and the federal 'angel': Academic R&D and federal-private cooperation before World War II." *Isis* 81: 188–213.

Pagels, Heinz. 1982. *The Cosmic Code: Quantum Physics as the Language of Nature.* New York: Simon and Schuster.

Pais, Abraham. 1982. *"Subtle Is the Lord . . .": The Science and the Life of Albert Einstein.* New York: Oxford University Press.

———. 1991. *Niels Bohr's Times: In Physics, Philosophy, and Polity.* New York: Oxford University Press.

Pamplin, Brian R., and Harry Collins. 1975. "Spoon bending: An experimental approach." *Nature* 257: 8.

Pantell, Richard H., and Harold E. Puthoff. 1969. *Fundamentals of Quantum Electronics.* New York: Wiley.

Pasolli, Robert. 1966. "Theatre: Judson poets' theatre." *Village Voice* (July 7): 19.

Patton, C. M., and J. A. Wheeler. 1975. "Is physics legislated by cosmogony?" In *Quantum Grav-*

ity: An Oxford Symposium, ed. C. J. Isham, R. Penrose, and D. W. Sciama, 538–605. Oxford: Clarendon.

Pauli, Wolfgang. 1979. *Scientific Correspondence with Bohr, Einstein, Heisenberg*, ed. A. Hermann, K. von Meyenn, and V. F. Weisskopf. New York: Springer.

——. 1994. *Writings on Physics and Philosophy*, ed. Charles P. Enz and Karl von Meyenn. New York: Springer.

Pease, Roland. 2008. "'Unbreakable' encryption unveiled." *BBC News*, October 9. Available at http://news.bbc.co.uk.

Peat, F. David. 1997. *Infinite Potential: The Life and Times of David Bohm*. Reading, MA: Addison Wesley.

Peck, Abe. 1985. *Uncovering the Sixties: The Life and Times of the Underground Press*. New York: Pantheon.

Peres, Asher. 2003. "How the no-cloning theorem got its name." *Fortschritte der Physik* 51: 458–61.

——. 2006. "I am the cat who walks by himself." *Foundations of Physics* 36: 1–18.

Petit, Charles. 1973. "Psychic research: The curious results." *San Francisco Chronicle*, May 14: 1, 22.

——. 1981. "Physicists, *est* founder in secret S. F. talks." *San Francisco Chronicle*, January 28: 32.

Physics Survey Committee. 1972/1973. *Physics in Perspective*, 2 vols. Washington, DC: National Academy of Sciences.

Pickering, Andrew. 1984. *Constructing Quarks: A Sociological History of Particle Physics*. Chicago: University of Chicago Press.

——. 1989. "From field theory to phenomenology: The history of dispersion relations." In *Pions to Quarks: Particle Physics in the 1950s*, ed. Laurie Brown, Max Dresden, and Lillian Hoddeson, 579–99. New York: Cambridge University Press.

Pinch, Trevor. 1977. "What does a proof do if it does not prove? A study of the social conditions and metaphysical divisions leading to David Bohm and John von Neumann failing to communicate in quantum mechanics." In *The Social Production of Scientific Knowledge*, ed. E. Mendelsohn, P. Weingart, and R. D. Whitley, 171–215. Dordrecht: Reidel.

——. 1979. "Normal explanations of the paranormal: The demarcation problem and fraud in parapsychology." *Social Studies of Science* 9: 329–48.

Pinch, Trevor, and Harry Collins. 1984. "Private science and public knowledge: The Committee for the Scientific Investigation of Claims of the Paranormal and its uses of the literature." *Social Studies of Science* 14: 521–46.

Pisani, Francis. 2007. "Networks as a unifying pattern of life involving different processes at different levels: An interview with Fritjof Capra." *International Journal of Communication* 1: 5–25.

Popescu, Sandu, and Daniel Rohrlich. 1998. "The joy of entanglement." In *Introduction to Quantum Computation and Information*, ed. Hoi-Kwong Lo, Sandu Popescu, and Tim Spiller, 29–48. Singapore: World Scientific.

Poppe, A., et al. 2004. "Practical quantum key distribution with polarization entangled photons." *Optics Express* 12: 3865–71.

Popper, Karl. 1963. "Science: Conjectures and refutations." In Karl Popper, *Conjectures and Refutations: The Growth of Scientific Knowledge*, 33–65. London: Routledge.

——. 1976. *Unended Quest: An Intellectual Autobiography*. London: Fontana.

Pressman, Steven. 1993. *Outrageous Betrayal: The Dark Journey of Werner Erhard from est to Exile*. New York: St. Martin's.

Price, George R. 1955. "Science and the supernatural." *Science* 122: 359–67.

Przibram, K., ed. 1967. *Letters on Wave Mechanics*, trans. Martin J. Klein. New York: Philosophical Library.

Puharich, Andrija. 1973 [1962]. *Beyond Telepathy*, 2nd ed. Garden City, NY: Anchor.

——. 1974. *Uri: A Journal of the Mystery of Uri Geller*. Garden City, NY: Anchor.

——. 1979a. "Introduction." In *The Iceland Papers: Select Papers on Experimental and Theoretical Research on the Physics of Consciousness*, ed. Andrija Puharich, 6–14. Amherst, WI: Essentia Research Associates.

——, ed. 1979b. *The Iceland Papers: Select Papers on Experimental and Theoretical Research on the Physics of Consciousness*. Amherst, WI: Essentia Research Associates.

Puthoff, Harold. 1996. "CIA-initiated remote viewing program at Stanford Research Institute." *Journal of Scientific Exploration* 10: 63–76.

Puthoff, Harold, and Russell Targ. 1976. "A perceptual channel for information transfer over kilometer distances: Historical perspective and recent research." *Proceedings of the IEEE* 64: 329–54.

——. 1979. "Direct perception of remote geographical locations." In *The Iceland Papers: Select Papers on Experimental and Theoretical Research on the Physics of Consciousness*, ed. Andrija Puharich, 17–47. Amherst, WI: Essentia Research Associates.

——. 1981. "Rebuttal of criticisms of remote viewing experiments." *Nature* 292: 388.

Ramon, Ceon, and Elizabeth A. Rauscher. 1980. "Superluminal transformations in complex Minkowski spaces." *Foundations of Physics* 10: 661–69.

——. 1983. "Remote connectedness in complex geometries." In *Absolute Values and the Creation of the New World* (Proceedings of the 11th International Conference on the Unity of the Sciences), 1423–42. New York: International Cultural Foundation Press.

Randall, Lisa, and Raman Sundrum. 1999. "An alternative to compactification." *Physical Review Letters* 83: 4690–93.

Randi, James. 1975. *The Magic of Uri Geller*. New York: Ballantine.

——. 1982. *Flim-Flam! Psychics, ESP, Unicorns, and Other Delusions*. Buffalo, NY: Prometheus Books.

Rauscher, Elizabeth A. 1960. "Fundamentals of fusion." *The University of California Engineer* (November): 20–49.

——. 1971. "A unifying theory of fundamental processes." Report UCRL-20808 (June), Lawrence Berkeley National Laboratory.

——. 1972. "Closed cosmological solutions to Einstein's field equations." *Lettere al Nuovo Cimento* 3: 661–65.

——. 1973. "The Minkowski metric for a multidimensional geometry." *Lettere al Nuovo Cimento* 7: 361–67.

——. 1979. "Some physical models potentially applicable to remote reception." In *The Iceland Papers: Select Papers on Experimental and Theoretical Research on the Physics of Consciousness*, ed. Andrija Puharich, 49–93. Amherst, WI: Essentia Research Associates.

——. 1983a. "The physics of psi phenomena in space and time, Part I: Major principles of physics, psychic phenomena, and some physical models." *Psi Research* 2: 64–88.

——. 1983b. "The physics of psi phenomena in space and time, Part II: Multidimensional geometric models." *Psi Research* 2: 93–120.

——. 1983c. "Theoretical and experimental exploration of the remote perception phenomenon." In *Absolute Values and the Creation of the New World* (Proceedings of the 11th International Conference on the Unity of the Sciences), 1443–77. New York: International Cultural Foundation Press.

——. 2005. "The unity of consciousness experience and current physical theory." *Subtle Energies & Energy Medicine* 15: 87–116.

———. 2006. "Quantum mechanics and the role for consciousness in the physical world." *Subtle Energies & Energy Medicine* 16: 1–42.

Rauscher, Elizabeth A., and Russell Targ. 2001. "The speed of thought: Investigation of a complex space-time metric to describe psychic phenomena." *Journal of Scientific Exploration* 15: 331–54.

Rauscher, E. A., G. Weissmann, J. Sarfatti, and S.-P. Sirag. 1976. "Remote perception of natural scenes, shielded against ordinary perception." In *Research in Parapsychology 1975*, ed. J. D. Morris, W. G. Roll, and R. L. Morris, 41–45. Metuchen, NJ: Scarecrow Press.

Redner, Sidney. 2005. "Citation statistics from 110 years of *Physical Review*." *Physics Today* 58 (June): 49–54.

Reisz, Matthew. 2010. "He didn't see that coming, or did he? Nobel laureate's interest in paranormal leads to conference rejection." *Times Higher Education* (April 29): 7.

Rensberger, Boyce, 1974. "Physicists test telepathy in a 'cheat-proof' setting." *New York Times*, October 22: 43.

———. 1975. "Magicians term Israeli 'psychic' a fraud." *New York Times*, December 13: 59.

———. 1976. "Paranormal phenomena facing scientific study." *New York Times*, May 1: 19.

Restivo, Sal P. 1978. "Parallels and paradoxes in modern physics and Eastern mysticism, Part I: A critical reconnaissance." *Social Studies of Science* 8: 143–81.

———. 1982. "Parallels and paradoxes in modern physics and Eastern mysticism, Part II: A sociological perspective on parallelism." *Social Studies of Science* 12: 37–71.

Rider, Robin. 1984. "Alarm and opportunity: Emigration of mathematicians and physicists to Britain and the United States, 1933–1945." *Historical Studies in the Physical Sciences* 15: 107–176.

Rigden, John. 1987. *Rabi: Scientist and Citizen*. New York: Basic Books.

Rodgers, Peter. 2002. "A brief history of the double-slit experiment." *Physics World* 15 (September): 15.

Romain, Jacques. 1960. "Introduction to quantum mechanics." *Physics Today* 13 (April): 62.

Romero, Simon. 2006. "From a literary lion in Caracas, advice on must-reads." *New York Times*, September 17: C16.

Romm, Ethel Grodzins. 1977. "When you give a closet occultist a PhD, what kind of research can you expect?" *The Humanist* 37 (May/June): 12–15.

Ronson, Jon. 2004. *The Men Who Stare at Goats*. New York: Simon and Schuster.

Roosevelt, Edith Kermit. 1980. "'New physics' drastic revise." *New Hampshire Sunday News*, 20 April.

Rorabaugh, W. J. 1989. *Berkeley at War: The 1960s*. New York: Oxford University Press.

Rosenberg, Scott. 1978. "Six Institute of Politics fellows trade tales of successes and failures in political life." *Harvard Crimson* 168, no. 8 (September 21): 1.

Rosenfeld, Léon. 1967. "Niels Bohr in the thirties." In *Niels Bohr: His Life and Work as Seen by His Friends and Colleagues*, ed. Stefan Rozental, 114–36. Amsterdam: North-Holland.

Ross, Andrew. 1991. *Strange Weather: Culture, Science, and Technology in the Age of Limits*. New York: Verso.

Rossinow, Doug. 2002. "'The revolution is about our lives': The New Left's counterculture." In *Imagine Nation: The American Counterculture in the 1960s and '70s*, ed. Peter Braunstein and Michael William Doyle, 99–124. New York: Routledge.

Roszak, Theodore. 1969. *The Making of a Counter Culture: Reflections on the Technocratic Society and Its Youthful Opposition*. Garden City, NY: Anchor/Doubleday.

Ruark, Arthur E., and Harold C. Urey. 1927. "The impulse moment of the light quantum." *Proceedings of the National Academy of Sciences* 13: 763–71.

Rubin, Daniel. 1999. "In France, Einhorn has few worries: The convicted killer reads, gardens, and works on the internet." *Philadelphia Inquirer*, December 19: A1.

Rupert, Glenn A. 1992. "Employing the New Age: Training seminars." In *Perspectives on the New Age*, ed. James R. Lewis and J. Gordon Melton, 127–35. Albany: State University of New York Press.

Sakurai, J. J. 1985. *Modern Quantum Mechanics*. Reading, MA: Addison-Wesley.

Sappell, Joel, and Robert W. Welkos. 1990. "The Scientology story" (six-part series). *Los Angeles Times*, June 24: A1, A36–41; June 25: A1, A18; June 26: A1; June 27: A1, A18, A19; June 28: A1; (June 29): A1, A48, 49.

Sarfatt, Jack. 1963. "Quantum-mechanical correlation theory of electromagnetic fields." *Nuovo Cimento* 27: 1119–29.

———. 1965. "Atomic relaxation and fluctuations of laser photons." *Journal of the Optical Society of America* 55: 455, 456.

———. 1967a. "On the nature of the superfluid critical velocity." *Physics Letters A* 24: 287, 288.

———. 1967b. "A new theory of the superfluid vortex phenomenon." *Physics Letters A* 24: 399, 400.

———. 1967c. "Local gauge invariance and broken symmetry in superfluid helium." *Physics Letters A* 25: 642, 643.

———. 1967d. "On the 'type II syperconductor' model of self-trapped laser filaments." *Physics Letters A* 26: 88, 89.

———. 1969. "Destruction of superflow in unsaturated ^{4}He films and the prediction of a new crystalline phase of ^{4}He with Bose-Einstein condensation." *Physics Letters A* 30: 300, 301.

———. 1974a. "Speculations on the effects of gravitation and cosmology in hadron physics." *Collective Phenomena* 1: 163–67.

———. 1974b. "Fear of creativity" (letter to the editor). *Physics Today* 27 (September): 13–15.

Sarfatt, Jack, and A. M. Stoneham. 1967. "The Goldstone theorem in the Jahn-Teller effect." *Proceedings of the Physical Society (London)* 91: 214–21.

Sarfatti, J. 1972. "Gravitation, strong interactions, and the creation of the universe." *Nature Physical Science* 240: 101, 102.

———. 1974a. "Geller performs for physicists." *Science News* 106 (July 20): 46.

———. 1974b. "Implications of meta-physics for psychoenergetic systems." *Psychoenergetic Systems* 1: 3–8.

———. 1975a. "Retraction on Geller." *Science News* 108 (December 6): 355.

———. 1975b. "The physical roots of consciousness." In Jeffrey Mishlove, *The Roots of Consciousness: Psychic Liberation through History, Science, and Experience*, 279–93. New York: Random House.

———. 1975c. "Toward a unified theory of gravitation and strong interactions." *Foundations of Physics* 5: 301–7.

———. 1977a. "Higher intelligence is us in the future." *Spit in the Ocean* 3: 43–50.

———. 1977b. "The case for superluminal transfer" (letter to the editor). *Technology Review* 79 (March/April): 3, 75.

———. 1986. "Can the electroweak unified force be used to neutralize nuclear weapons?" *Defense Analysis* 2: 49–51.

———. 1987. "Superluminal communication" (letter to the editor). *Physics Today* 40 (September): 118–120.

———. 1997. "Sarfatti's Illuminati: In the thick of it!" *Mind/Net Journal* 2 (January 30). Available at http://mindcontrolforums.com/mindnet (accessed May 23, 2008).

———. 2002a. *The Destiny Matrix*. Bloomington, IN: 1stBooks.

———. 2002b. *Space-Time and Beyond: The Series*, episode 2, "Dark Energy." Bloomington, IN: 1st Books Library.

Saunders, Jack, and Clark Norton. 1976. "Coming: A quantum leap in human understanding." *San Francisco Examiner*, July 4: 22.

Scarani, Valerio, Sofyan Iblisdir, and Nicolas Gisin. 2005. "Quantum cloning." *Reviews of Modern Physics* 77: 1225–56.

Scerri, Eric R. 1989. "Eastern mysticism and the alleged parallels with physics." *American Journal of Physics* 57: 687–92.

Schaffer, Simon. 1995. "The show that never ends: Perpetual motion in the early eighteenth century." *British Journal for the History of Science* 28: 157–89.

Scheiding, Tom. 2009. "Paying for knowledge one page at a time: The author fee in physics in twentieth-century America." *Historical Studies in the Natural Sciences* 39: 219–47.

Schmidt, Helmut. 1993. "Observation of a psychokinetic effect under highly controlled conditions." *Journal of Parapsychology* 57: 351–73.

Schmidt, Helmut, and Henry P. Stapp. 1993. "PK [psychokinesis] with prerecorded random events and the effects of preobservation-psychokinesis." *Journal of Parapsychology* 57: 331–48.

Schmitt-Manderbach, Tobias, et al. 2007. "Experimental demonstration of free-space decoy-state quantum key distribution over 144 km." *Physical Review Letters* 98: 010504.

Schnabel, Jim. 1997. *Remote Viewers: The Secret History of America's Psychic Spies*. New York: Dell.

Schrecker, Ellen. 1986. *No Ivory Tower: McCarthyism and the Universities*. New York: Oxford University Press.

Schrödinger, Erwin. 1935a. "Discussion of probability relations between separated systems." *Proceedings of the Cambridge Philosophical Society* 31: 555–63.

———. 1935b. "Die gegenwärtige Situation in der Quantenmechanik." *Die Naturwissenschaften* 23: 807–12.

Schulman, Bruce J. 2001. *The Seventies: The Great Shift in American Culture, Society, and Politics*. New York: Free Press.

Schumacher, E. F. 1973. *Small Is Beautiful: A Study of Economics as if People Mattered*. London: Blond and Briggs.

Schwartz, Jeffrey M., Henry P. Stapp, and Mario Beauregard. 2005. "Quantum physics in neuroscience and psychology: A neurophysical model of mind-brain interaction." *Philosophical Transactions of the Royal Society* B 360: 1309–27.

Schweber, Silvan S. 1988. "The mutual embrace of science and the military: ONR and the growth of physics in the United States after World War II." In *Science, Technology, and the Military*, ed. Everett Mendelsohn, Merritt Roe Smith, and Peter Weingart, 1–45. Boston: Kluwer.

———. 1989. "Some reflections on the history of particle physics in the 1950s." In *Pions to Quarks: Particle Physics in the 1950s*, ed. Laurie Brown, Max Dresden, and Lillian Hoddeson, 668–93. New York: Cambridge University Press.

———. 1994. *QED and the Men Who Made It: Dyson, Feynman, Schwinger, and Tomonaga*. Princeton: Princeton University Press.

Science Citation Index. 1961–. Philadelphia: Institute for Scientific Information.

Scott, Joseph C. 1997. "Sometime-scientist Plutonium says science is 'gobbledygook.'" *The Dartmouth*, September 25.

Scully, M. O., and K. Drühl. 1981. "Quantum eraser: A proposed photon correlation experiment concerning observation and 'delayed choice' in quantum mechanics." *Physical Review A* 25: 2208–13.

Seife, Charles. 2003. "At Canada's Perimeter Institute, 'Waterloo' means 'Shangri-La.'" *Science* 302: 1650–52.

Seitz, Frederick, Erich Vogt, and Alvin M. Weinberg. 1998. "Eugene Paul Wigner." *Biographical Memoirs of Fellows of the National Academy of Sciences* 74: 365–88.

Self, Jane. 1992. *60 Minutes and the Assassination of Werner Erhard*. Houston, TX: Breakthru.

Selleri, Franco. 1978. "Consequences of Einstein locality." *Foundations of Physics* 8: 103–16.

———. 1984. "Gespensterfelder." In *The Wave-Particle Dualism: A Tribute to Louis de Broglie on His 90th Birthday*, ed. S. Diner, D. Fargue, G. Lochak, and F. Selleri, 101–28. Boston: Reidel.

Shapin, Steven. 2008. *The Scientific Life: A Moral History of a Late Modern Vocation*. Chicago: University of Chicago Press.

Shelley, Mary. 1996 [1818]. *Frankenstein*. New York: W. W. Norton.

Shimony, Abner. 1981. "Meeting of physics and metaphysics." *Nature* 291: 435, 436.

———. 1984. "Controllable and uncontrollable nonlocality." In *Foundations of Quantum Mechanics in the Light of New Technology*, ed. S. Kamefuchi et al., 225–30. Tokyo: Physical Society of Japan.

———. 1993. *Search for a Naturalistic World View*, vol. 2, *Natural Science and Metaphysics*. New York: Cambridge University Press.

Shor, Peter W. 2002. "The quantum channel capacity and coherent information." *Lecture Notes MSRI Workshop on Quantum Computation*, available at http://www.msri.org/web/msri.

Siegman, A. E. 1971. *An Introduction to Lasers and Masers*. New York: McGraw-Hill.

Singh, Simon. 1999. *The Code Book: The Evolution of Secrecy from Mary Queen of Scots to Quantum Cryptography*. New York: Doubleday.

Sirag, Saul-Paul. 1977a. "Introduction." *Spit in the Ocean* 3: 12–14.

———. 1977b. "The new physicists." *Spit in the Ocean* 3: 76–82.

———. 1977c. "A combinatorial derivation of the proton-electron mass ratio." *Nature* 268: 294.

———. 1979. "Gravitational magnetism." *Nature* 278: 535–38.

———. 1982. "Why there are three fermion families." *Bulletin of the American Physical Society* 27, no. 1: 31.

———. 1983. "Physical constants as cosmological constraints." *International Journal of Theoretical Physics* 22: 1067–89.

———. 1989. "A finite group algebra unification scheme." *Bulletin of the American Physical Society* 34, no. 1: 82.

———. 1993. "Consciousness: A hyperspace view." In Jeffrey Mishlove, *The Roots of Consciousness*, 2nd ed., 327–65. Tulsa, OK: Council Oak Books.

———. 2002. "Contact." In Jack Sarfatti, *The Destiny Matrix*, 90–119. Bloomington, IN: 1stBooks.

Sirag, Saul-Paul, and Ken Babbs. 1977. "The metaphase typewriter, or the Harry Houdini centennial breakout challenge." *Spit in the Ocean* 3: 51–65.

Sirag, Saul-Paul, Charles Musès, and David Bohm. 1976. "Contending modern physics theories in relation to consciousness—an exchange." *Nous Letter: Studies in Noetics* 2 (Spring): 23–29.

Slobodzian, Joseph. 2003. "Einhorn speaks out about woman's side of the story." *Philadelphia Inquirer*, May 4: B3.

Slobodzian, Joseph, and Suzanne Sataline. 1997. "The long trail that ended in the capture: It all started with his wife's request for a driver's license." *Philadelphia Inquirer*, June 17: A8.

Smith, Adam. 1976. "Notes from the far-out physics underground." *New York Magazine* 10 (December 27): 103, 104.

Smith, Crosbie. 1998. *The Science of Energy: A Cultural History of Energy Physics in Victorian Britain.* Chicago: University of Chicago Press.

Smith, M. Roe, and Leo Marx, eds. 1994. *Does Technology Drive History? The Dilemma of Technological Determinism.* Cambridge: MIT Press.

Smith, Paul H. 2004. *Reading the Enemy's Mind: Inside Star Gate, America's Psychic Espionage Program.* New York: Forge.

Smolin, Lee. 2006. *The Trouble with Physics: The Rise of String Theory, the Fall of a Science, and What Comes Next.* Boston: Houghton Mifflin.

Smyth, Henry DeWolf. 1951. "The stockpiling and rationing of scientific manpower." *Physics Today* 4 (February): 18–24.

Snider, Suzanne. 2003. "*Est,* Werner Erhard, and the corporatization of self-help." *Believer* 1, no. 2 (May): 18–28.

Sokolov, Raymond. 1979. "Nonfiction in brief." *New York Times,* June 17: BR5.

Sommerfeld, Arnold. 1930. *Wave Mechanics,* trans. H. L. Brose. London: Methuen.

Soteropoulos, Jacqueline. 2002a. "Einhorn lawyers attack 2d trial deal." *Philadelphia Inquirer,* August 6: B2.

———. 2002b. "Einhorn murder case goes to jury: His attorney suggested that Holly Maddux's body was planted at Einhorn's home; a prosecutor scoffed at the idea." *Philadelphia Inquirer,* October 17: B1.

———. 2002c. "Ira flops with jury: Longtime fugitive convicted of killing his ex-lover in 1977." *Philadelphia Inquirer,* October 18: A1.

Spiller, Timothy P. 1998. "Basic elements of quantum information technology." In *Introduction to Quantum Computation and Information,* ed. Hoi-Kwong Lo, Sandu Popescu, and Tim Spiller, 1–28. Singapore: World Scientific.

Srinivas, M. D. 1984. "Quantum interference of probabilities and hidden variable theories." In *The Wave-Particle Dualism: A Tribute to Louis de Broglie on His 90th Birthday,* ed. S. Diner, D. Fargue, G. Lochak, and F. Selleri, 367–75. Boston: Reidel.

Stapp, Henry P. 1956. "Relativistic theory of polarization phenomena." *Physical Review* 103: 425–34.

———. 1971. "S-matrix interpretation of quantum theory." *Physical Review D* 3: 1303–20.

———. 1976. "Correlation experiments and the nonvalidity of ordinary ideas about the physical world." Report LBL-5333 (July 9). Available from the Lawrence Berkeley National Laboratory library.

———. 1977a. "Are superluminal connections necessary?" *Il Nuovo Cimento B* 40: 191–205.

———. 1977b. "Theory of reality." *Foundations of Physics* 7: 313–23.

———. 1979. "Whiteheadian approach to quantum theory and the generalized Bell's theorem." *Foundations of Physics* 9: 1–25.

———. 1980. "Locality and reality." *Foundations of Physics* 10: 767–95.

———. 1982. "Mind, matter, and quantum mechanics." *Foundations of Physics* 12: 363–99.

———. 1986. "Gauge fields and integrated quantum-classical theory." *Annals of the New York Academy of Sciences* 480: 326–35.

———. 1994. "Theoretical model of a purported empirical violation of the predictions of quantum theory." *Physical Review A* 50: 18–22.

———. 1995. "Parapsychological Review A?" (letter to the editor). *Physics Today* 48 (July): 78, 79.

———. 2004 [1993]. *Mind, Matter, and Quantum Mechanics,* 2nd ed. New York: Springer.

———. 2007a. "Wigner's friend and consciousness in quantum theory." In *Mindful Universe: Quantum Mechanics and the Participating Observer*, 161–63. New York: Springer.

———. 2007b. *Mindful Universe: Quantum Mechanics and the Participating Observer*. New York: Springer.

Stein, Frederick M. 1987. "Quantum Reality: Beyond the New Physics." *American Journal of Physics* 55: 478, 479.

Stix, Gary. 2005. "Best-kept secrets." *Scientific American* 292, no. 1 (January): 78–83.

Straczynski, J. Michael. 1977. "Super mystic or super fake? Uri Geller: Eternally on trial." *San Diego Magazine* 30 (November): 158–67.

Sullivan, Dan. 1979. "A stand-up philosopher at the Improv." *Los Angeles Times*, August 22: F1.

Sullivan, Gail Bernice. 1973. "Psychic research: A new awareness." *California Living* (July): 18–25.

Susskind, Leonard. 2008. *The Black Hole War: My Battle with Stephen Hawking to Make the World Safe for Quantum Mechanics*. New York: Little, Brown.

Targ, Russell. 1996. "Remote viewing at Stanford Research Institute in the 1970s: A memoir." *Journal of Scientific Exploration* 10: 77–88.

Targ, Russell, and Harold Puthoff. 1974. "Information transmission under conditions of sensory shielding." *Nature* 251: 602–7.

———. 1975. "Remote viewing of natural targets." In *Quantum Physics and Parapsychology*, ed. Laura Oteri, 151–80. New York: Parapsychology Foundation.

———. 1977. *Mind-Reach: Scientists Look at Psychic Ability*. New York: Delacorte.

Tarozzi, G. 1984. "From ghost to real waves: A proposed solution to the wave-particle dilemma." In *The Wave-Particle Dualism: A Tribute to Louis de Broglie on His 90th Birthday*, ed. S. Diner, D. Fargue, G. Lochak, and F. Selleri, 139–48. Boston: Reidel.

Tart, Charles T., Harold E. Puthoff, and Russell Targ. 1980. "Information transmission in remote viewing experiments." *Nature* 284: 191.

Taylor, John G. 1975a. Letter to the Editor. *Nature* 254: 472, 473.

———. 1975b. *Superminds: An Enquiry into the Paranormal*. London: Macmillan.

———. 1980. *Science and the Supernatural: An Investigation of Paranormal Phenomena*. London: T. Smith.

Tegmark, Max. 2000. "Importance of quantum decoherence in brain processes." *Physical Review E* 61: 4194–4206.

Teilhard de Chardin, Pierre. 1959. *The Phenomenon of Man*, trans. Bernard Wall. New York: Harper.

Thacker, Beth-Ann, Harvey S. Leff, and David P. Jackson, eds. 2002. Special issue on teaching quantum mechanics. *American Journal of Physics* 70 (March): 199–367.

Thompson, William Irwin. 1971. *At the Edge of History*. New York: Harper and Row.

Toben, Bob, with Jack Sarfatti and Fred Alan Wolf. 1975. *Space-Time and Beyond: Toward an Explanation of the Unexplainable*. New York: Dutton.

Toben, Bob, with Fred Alan Wolf. 1982. *Space-Time and Beyond: Toward an Explanation of the Unexplainable*, 2nd ed. New York: Dutton.

Tomonaga, Sin-itiro. 1997. *The Story of Spin*, trans. Takeshi Oka. Chicago: University of Chicago Press.

Trombley, William. 1967. "UC Budget cuts seen as financial threat to state: Research decline feared." *Los Angeles Times*, February 2: A6.

Turner, Fred. 2006. *From Counterculture to Cyberculture: Stewart Brand, the Whole Earth Network, and the Rise of Digital Utopianism*. Chicago: University of Chicago Press.

Uhlenbeck, George E. 1963. "Quantum theory." *Science* 140: 886.

U.S. Army Material Systems Analysis Activity. 1979. *Project Grill Flame: AMSAA Phase I Efforts*. Secret classified report, July, Aberdeen Proving Ground, Maryland. Declassified in 2000 and now available at http://www.remoteviewed.com/remote_viewing_history_military.htm (accessed November 8, 2007).

U.S. House, Committee on Un-American Activities. 1953a. *Communist Methods of Infiltration (Education)*, Part 2. Washington, DC: Government Printing Office.

———. 1953b. *Investigation of Communist Activities in the Columbus, Ohio Area*. Washington, DC: Government Printing Office.

Vettel, Eric J. 2006. *Biotech: The Countercultural Origins of an Industry*. Philadelphia: University of Pennsylvania Press.

Vigier, J.-P. 1974. "Three recent experimental verifications of possible measurable consequences of theory of hidden variables." *Comptes Rendues B* 279: 1–4.

von Neumann, John. 1932. *Mathematische Grundlagen der Quantenmechanik*. Berlin: Springer.

———. 1955. *Mathematical Foundations of Quantum Mechanics*, trans. Robert T. Beyer. Princeton: Princeton University Press.

Walker, Evan Harris. 1970. "The nature of consciousness." *Mathematical Biosciences* 7: 131–78.

———. 1971. "Consciousness as a hidden variable" (letter to the editor). *Physics Today* 24 (April): 39.

———. 1972. "Consciousness in the quantum theory of measurement, Part I." *Journal for the Study of Consciousness* 5: 46–63.

———. 1972/73. "Consciousness in the quantum theory of measurement, Part II." *Journal for the Study of Consciousness* 5: 257–77.

———. 1975. "Foundations of paraphysical and parapsychological phenomena." In *Quantum Physics and Parapsychology*, ed. Laura Oteri, 1–53. New York: Parapsychology Foundation.

———. 1977. "Quantum mechanical tunnelling in synaptic and ephaptic transmission." *International Journal of Quantum Chemistry* 11: 103–27.

———. 1978/79. "The quantum theory of psi phenomena." *Psychoenergetic Systems* 3: 259–99.

———. 2000. *The Physics of Consciousness: The Quantum Mind and the Meaning of Life*. Cambridge, MA: Perseus.

Walker, Frank. 1991. "Lawyers chase seminar guru." *Sun Herald* [Sydney, Australia], March 3: 31.

Walker, Morton. 1989. *The Chelation Way: The Complete Book of Chelation Therapy*. New York: Avery.

Wallace, Kevin. 1976. "Nobel Winner's Bay Area tour: A crash course in occult." *San Francisco Chronicle*, December 15: 4.

Wallace-Wells, Benjamin. 2008. "Surfing the universe: An academic dropout and the search for a Theory of Everything." *The New Yorker* (July 21): 32–38.

Walter, Greg. 1979. "Holly." *Philadelphia Magazine* 70 (December): 144–151, 262–273.

Wang, Jessica. 1999. *American Science in an Age of Anxiety: Scientists, Anticommunism, and the Cold War*. Chapel Hill: University of North Carolina Press.

Wasserman, Robin. 2000. *Tuning In: Timothy Leary, Harvard, and the Boundaries of Experimental Psychology*. AB thesis, Harvard University.

Wayner, Peter. 1999. "Quantum codes: Secrets in the light." *New York Times*, June 10: G11.

Weil, Andrew. 1974a. "Andrew Weil's search for the true Uri Geller, Part I." *Psychology Today* 8, no. 1 (June): 45–50.

———. 1974b. "Andrew Weil's search for the true Uri Geller, Part II: The letdown." *Psychology Today* 8, no. 2 (July): 74–78, 82.

Weinberg, Steven. 1989. "Testing quantum mechanics." *Annals of Physics* 194: 336–86.

Weinberger, Sharon. 2006. *Imaginary Weapons: A Journey through the Pentagon's Scientific Underworld*. New York: Nation Books.

Weiner, Charles. 1969. "A new site for the seminar: The refugees and American physics in the thirties." In *The Intellectual Migration: Europe and America, 1930–1960*, ed. Donald Fleming and Bernard Bailyn, 190–234. Cambridge: Harvard University Press.

Weisskopf, Victor F. 1991. *The Joy of Insight: Passions of a Physicist*. New York: Basic Books.

Weyl, Hermann. 1931. *The Theory of Groups and Quantum Mechanics*, trans. H. P. Robertson. New York: Dutton.

Wheaton, Bruce R. 1983. *The Tiger and the Shark: Empirical Roots of Wave-Particle Dualism*. New York: Cambridge University Press.

Wheeler, John A. 1973. "From relativity to mutability." In *The Physicist's Conception of Nature*, ed. Jagdish Mehra, 202–47. Dordrecht: Reidel.

———. 1977. "Genesis and observership." In *Foundational Problems in the Special Sciences*, ed. Robert E. Butts and Jaakko Hintikka, 3–33. Boston: Reidel.

———. 1978. "The 'past' and the 'delayed-choice' double-slit experiment." In *Mathematical Foundations of Quantum Theory*, ed. A. R. Marlow, 9–48. New York: Academic Press.

———. 1980. "Beyond the black hole." In *Some Strangeness in the Proportion: A Centennial Symposium to Celebrate the Achievements of Albert Einstein*, ed. Harry Woolf, 341–75. Reading, MA: Addison-Wesley.

———. 1983. "Law without law." In *Quantum Theory and Measurement*, ed. John A. Wheeler and Wojciech H. Zurek, 182–213. Princeton: Princeton University Press.

Wheeler, John A., with Kenneth Ford. 1998. *Geons, Black Holes, and Quantum Foam: A Life in Physics*. New York: W. W. Norton.

Wheeler, John A., Charles Misner, and Kip Thorne. 1973. *Gravitation*. San Francisco: Freeman.

Wheeler, John A., and Wojciech H. Zurek, eds. 1983. *Quantum Theory and Measurement*. Princeton: Princeton University Press.

White, D. 1979. "The Tao of Physics." *Contemporary Sociology* 8: 586, 587.

Whitney, Craig. 1999. "France grants extradition in 1977 killing." *New York Times*, February 19: A14.

Whittaker, Andrew. 2002. "John Bell in Belfast: Early years and education." In *Quantum [Un]speakables: From Bell to Quantum Information*, ed. R. A. Bertlmann and A. Zeilinger, 7–20. New York: Springer.

Wick, David. 1995. *The Infamous Boundary: Seven Decades of Controversy in Quantum Physics*. Boston: Birkhäuser.

Wiesner, Stephen. 1983. "Conjugate coding." *ACM SIGACT News* 15 (January): 78–88.

Wigner, Eugene. 1952. "Die Messung quantenmechanischer Operatoren." *Zeitschrift für Physik* 133: 101–8.

———. 1962. "Remarks on the mind-body question." In *The Scientist Speculates*, ed. I. J. Good, 284–302. New York: Basic Books.

———. 1967. *Symmetries and Reflections*. Bloomington, IN: Indiana University Press.

———. 1983. "Discussant's remarks." In *Absolute Values and the Creation of the New World* (Proceedings of the 11th International Conference on the Unity of the Sciences), 1479. New York: International Cultural Foundation Press.

Williams, Lambert. Forthcoming. *The New Transcendental Subjects: Chaos and Complexity, 1960–2002*. PhD dissertation, Harvard University.

Williams, Lambert, and William Thomas. 2009. "The epistemologies of non-forecasting simulations, Part II: Climate, chaos, computing style, and the contextual plasticity of error." *Science in Context* 22: 271–310.

Wilson, Colin. 1981. "The world of Uri Geller." *The Unexplained* 2: 621–24, 666–69, 686–89.

Wilson, Robert Anton. 1977. *Cosmic Trigger: The Final Secret of the Illuminati*. New York: Pocket Books.

———. 1979. "The science of the impossible." *Oui* (March): 81–84, 130, 131.

Winter, Alison. 1998. *Mesmerized: The Powers of Mind in Victorian Britain*. Chicago: University of Chicago Press.

Wisnioski, Matt. 2003. "Inside 'the System': Engineers, scientists, and the boundaries of social protest in the long 1960s." *History and Technology* 19: 313–33.

Woit, Peter. 2006. *Not Even Wrong: The Failure of String Theory and the Search for Unity in Physical Law*. New York: Basic Books.

Wolf, Fred Alan. 1981. *Taking the Quantum Leap: The New Physics for Nonscientists*. San Francisco: Harper and Row.

———. 1984. *Star-Wave: Mind, Consciousness, and Quantum Physics*. New York: Macmillan.

———. 1985. *Mind and the New Physics*. London: Heinemann.

———. 1986. *The Body Quantum: The New Physics of Body, Mind, and Health*. New York: Macmillan.

———. 1988. *Parallel Universes: The Search for Other Worlds*. New York: Simon and Schuster.

———. 1991. *The Eagle's Quest: A Physicist Finds Scientific Truth at the Heart of the Shamanic World*. New York: Touchstone.

———. 1994. *The Dreaming Universe: A Mind-Expanding Journey into the Realm Where Psyche and Physics Meet*. New York: Simon and Schuster.

———. 1996. *The Spiritual Universe: How Quantum Physics Proves the Existence of the Soul*. New York: Simon and Schuster.

———. 2001. *Mind into Matter: A New Alchemy of Science and Spirit*. Portsmouth, NH: Moment Point Press.

———. 2004. *The Yoga of Time Travel: How the Mind Can Defeat Time*. Wheaton, IL: Quest Books.

———. 2005. *Dr. Quantum's Little Book of Big Ideas: Where Science Meets Spirit*. Needham, MA: Moment Point Press.

Wolfe, Tom. 1968. *The Electric Kool-Aid Acid Test*. New York: Farrar, Straus, Giroux.

———. 1976a. "The 'me' decade and the third great awakening." *New York Magazine* (August 23): 26–40.

———. 1976b. *Mauve Gloves & Madmen, Clutter & Vine*. New York: Farrar, Straus, and Giroux.

Woodward, Kenneth L. 1976. "Super-salesman of *est*." *Newsweek* (September 6): 58.

Woodward, Kenneth L., and Gerald C. Lubenow. 1979. "Physics and mysticism." *Newsweek* (July 23): 85, 86.

Wootters, William K., and Wojciech H. Zurek. 1979. "Complementarity in the double-slit experiment: Quantum nonseparability and a quantitative statement of Bohr's principle." *Physical Review D* 19: 473–484.

———. 1982. "A single quantum cannot be cloned." *Nature* 299: 802, 803.

———. 2009. "The no-cloning theorem." *Physics Today* 62 (February): 76.

Yanase, Mutsuo. 1961. "Optimal measuring apparatus." *Physical Review* 123: 666–68.

Young, Arthur M. 1972. "An inquiry into the responsibility of current science." *Journal for the Study of Consciousness* 5: 39–45.

———. 1976a. *The Geometry of Meaning*. New York: Delacorte.

———. 1976b. *The Reflexive Universe: Evolution of Consciousness*. New York: Delacorte.

———. 1979. *The Bell Notes: A Journey from Physics to Metaphysics*. New York: Delacorte.

Young, Frank E. 1988. "FDA: The cop on the consumer beat." *FDA Consumer* 22 (April): 6, 7.

Zeh, H. Dieter. 2006. "Roots and fruits of decoherence." In *Quantum Decoherence*, ed. B. Duplantier, J.-M. Raimond, and V. Rivasseau, 151–75. Boston: Birkhäuser.

Zeman, Ned. 1991. "San Francisco is talking." *Newsweek* (May 20): 8.

Zukav, Gary. 1979. *The Dancing Wu Li Masters: An Overview of the New Physics*. New York: Morrow.

———. 2001 [1979]. *The Dancing Wu Li Masters: An Overview of the New Physics*, 2nd ed. New York: HarperCollins.

Zurek, Wojciech H. 1983. "Information transfer in quantum measurements: Irreversibility and amplification." In *Quantum Optics, Experimental Gravity, and Measurement Theory*, ed. Pierre Myestre and Marlan O. Scully, 87–116. New York: Plenum.

———. 1991. "Decoherence and the transition from quantum to classical." *Physics Today* 44 (October): 36–44.

Index

Page numbers in *italics* refer to illustrations.